THE DEFINITIVE GUIDE TO

DAVINCI RESOLVE 14

by Paul Saccone

The Definitive Guide to DaVinci Resolve 14

Paul Saccone

Blackmagic Design

www.blackmagicdesign.com

To report errors, please send a note to training@blackmagicdesign.com.

Series Editor: Patricia Montesion

Contributing Authors: Dion Scoppettuolo, Daria Fissoun, Mary Plummer, Patrick Inhofer

Cover Design: Blackmagic Design

ISBN 13: 978-0-9993913-0-3

Contents

Foreword

Welcome to **The Definitive Guide to DaVinci Resolve 14**

I think one of the most exciting things about DaVinci Resolve 14 is that it combines a true professional video editor with the same incredible color correction tools used in Hollywood on high-end films and television shows, and Fairlight audio, all in one software solution. Since adding professional editing to DaVinci Resolve in version 10 we've been turning it into the world's best professional editor.

Best of all, DaVinci Resolve 14 is absolutely free! We've made sure that the free version of DaVinci Resolve actually has more features than any other paid editing system. That's because at Blackmagic Design we believe everybody should have the tools to create professional, Hollywood caliber content without having to spend thousands of dollars.

I hope you'll enjoy using DaVinci Resolve 14 and we can't wait to see the amazing work you produce!

Grant Petty
Blackmagic Design

Getting Started

Welcome to **The Definitive Guide to DaVinci Resolve 14**, the official Blackmagic Design authorized training book that teaches editors and students how to edit, color correct, and mix audio in DaVinci Resolve. All you need is a Mac or Windows computer, the free download version of DaVinci Resolve 14, and a passion to learn.

In this book, you will edit a project from start to finish in DaVinci Resolve 14. You'll learn how to import and organize media; edit and trim in the timeline; add transitions, effects, and titles; use the fundamental color correction tools; mix and edit audio; and export a finished project. You'll understand the essential editing, color correction, and audio functionality while acquiring skills that you can develop as you grow with DaVinci Resolve.

System Requirements

This book supports DaVinci Resolve 14 for Mac and Windows. If you have an older version of DaVinci Resolve, you must upgrade to the current version to follow along with the lessons.

Download DaVinci Resolve 14

To download the free version of DaVinci Resolve 14 from the Blackmagic Design website:

1. Open a web browser on your Windows or Mac computer.
2. In the address field of your web browser, type: www.blackmagicdesign.com/products/davinciresolve.
3. On the DaVinci Resolve landing page, click the Download button.
4. On the download page, click the Mac or Windows button, depending on your computer's operating system.
5. Follow the installation instructions to complete the installation.

When you have completed the software installation, follow the instructions in the following section, "Copying the Lesson Files," to download the content for this book.

Copying the Lesson Files

The DaVinci Resolve editing lesson files must be downloaded to your Mac or Windows computer to perform the exercises in this book. After you save the files to your hard disk, extract the file and copy the folder to your Documents folder.

To Download and Install the DaVinci Resolve Editing Lessons Files:

When you are ready to download the lesson files, follow these steps:

1. Connect to the Internet and navigate to:
   https://www.blackmagicdesign.com/dvres/intro-to-resolve-14

   The download will begin immediately.

   The IntroToDaVinciResolve14Tutorials.zip file is roughly 5GB in size.

2. After downloading the zip file to your Mac or Windows computer, open your Downloads folder, and double-click IntroToDaVinciResolve14Tutorials.zip to unzip it if it doesn't unzip automatically.
3. Drag the R14 lessons Files folder from the Downloads folder to your Documents folder.

You are now ready to begin Lesson 1, Quickstart: Editing a One Minute Movie.

NOTE The downloaded lesson files are organized into folders that correspond to the chapters in which they are used. Some lesson files are used for multiple chapters, so you will not see individual folders labeled for each chapter in the book.

Acknowledgments

We would like to thank the following individuals for their contributions of media used throughout the book:

- Citizen Chain
- Editstock for “Gnarly in Pink”
- Jitter Bug Riot
- Brian J Terwilliger, Terwilliger Productions for “Living In the Age of Airplanes”

Lesson 1

QuickStart: Editing a One Minute Movie

DaVinci Resolve 14 is the only software that lets you edit, color correct, and mix sound for your projects from start to finish, all within one integrated software application. This book will teach you how to edit video, color correct images, and sweeten and mix audio with DaVinci Resolve.

Whether you use the application on a Mac, Windows, or Linux computer, you'll learn how to take a project from its initial setup to a final output. You'll become familiar with the standard menus and buttons, as well as essential editing, color correction, and audio sweetening workflows.

This first lesson is used to give you an overview of the editing tools in DaVinci Resolve 14 while putting together a one minute program.

Time

This lesson takes approximately 45 minutes to complete.

Goals

Starting a project

At this point you should have installed DaVinci Resolve 14 on your computer, downloaded the content for this book, and located that content in your Documents folder as described in Getting Started. Now, you'll start by opening DaVinci Resolve on your computer and importing the media that you will use in your project.

1 To open DaVinci Resolve, do one of the following:

- In Mac, in the Dock, click the DaVinci Resolve icon.
- In Windows, in the Start menu, click the DaVinci Resolve icon.

The first window that appears is the project manager. Here you'll find all the projects that you have created. Projects represent a single job and they contain the timelines and clips that link to the media on your hard drives. Currently, an empty project with the default title, Untitled Project, is displayed, which you'll use for this exercise.

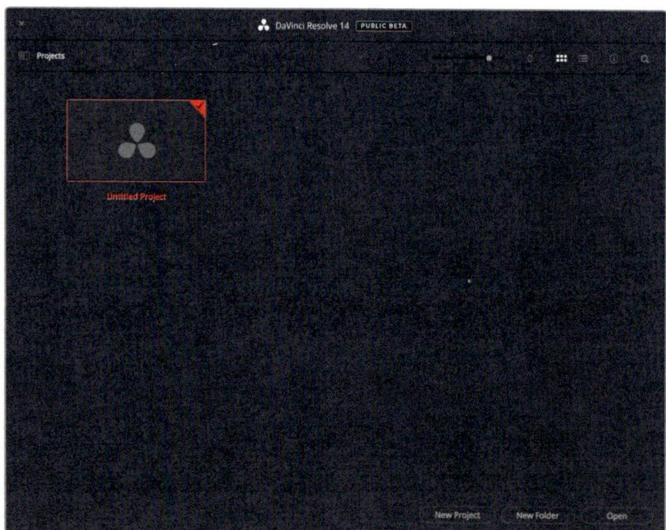

NOTE Depending on your computer's screen resolution, the layout seen in the images in this book may appear differently compared to the layout you see on your display.

2 Double-click the Untitled thumbnail to open a new project.

The Edit page has four primary sections that you'll use in the process of editing a project.

3 To import clips that you want to use in your project, choose File > Import Media.

4 In the dialog that appears, navigate to the Documents folder, where you placed the R14 Lessons folder.

5 Within that folder, open the Lesson 01 folder.

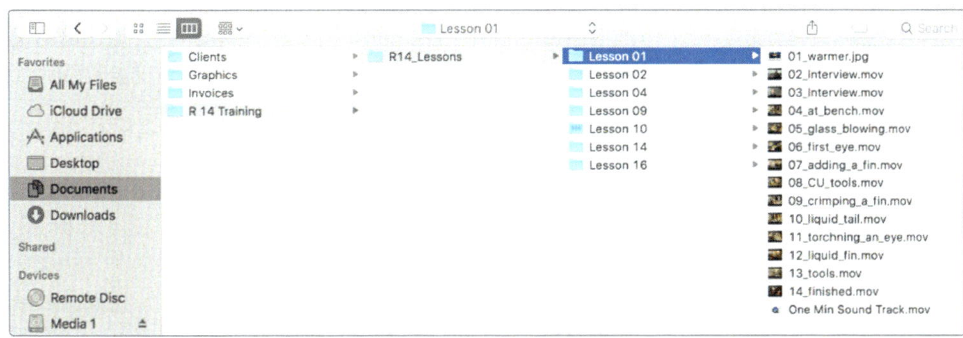

6 Select all the files in the Lesson 01 folder, and click Open.

All the selected media is imported into the Media Pool that runs along the left side of the DaVinci Resolve interface. You are now able to use these clips to edit together a short project.

This lesson's project is designed to give you a quick tour of the interface. You will focus on the big picture and learn how to quickly assemble a project and use the core tools.

You'll get into more detail in the remaining lessons; but for this first lesson, just enjoy the exploratory ride and don't worry about the options, keyboard shortcuts, or even your results.

Assembling clips in a timeline

Once you have media in the Media Pool, you can start assembling the order you want the clips to play. It's easiest to first sort them in the Media Pool, and then just drag them into a timeline.

1 At the top of the Media Pool, in the sort order pop-up menu, choose File Name.

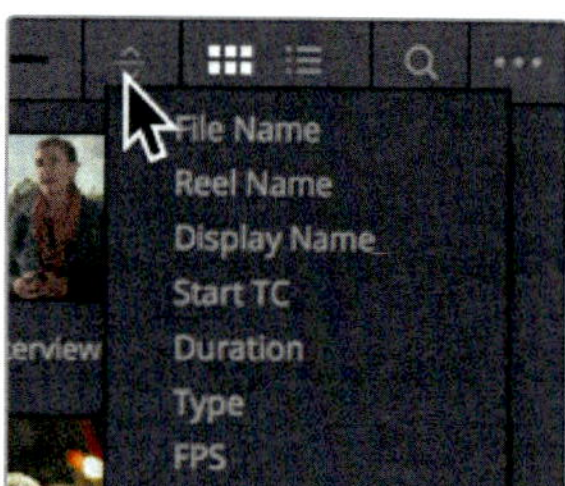

The thumbnails in the Media Pool are now arranged from lowest to the highest alphanumeric filename.

This first edit you will make is an easy one because you'll use a still photograph.

2 Drag 01_warmer from the Media Pool to the top half of the timeline window.

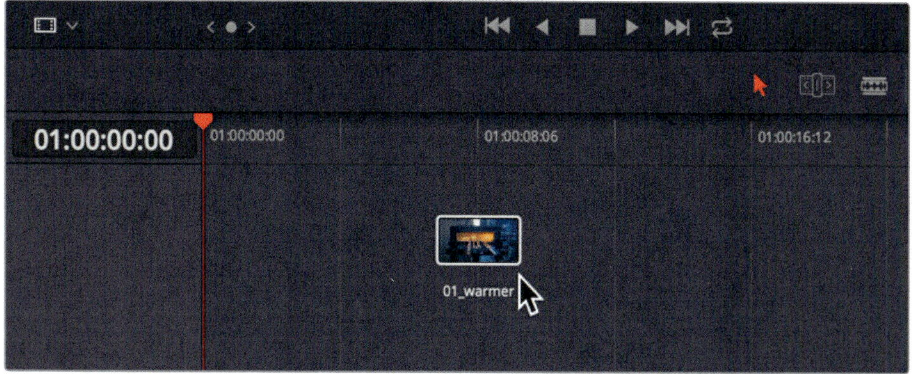

You just made your first edit. When you drag a clip into the timeline, it appears as a clip segment at the beginning of the timeline. In the case of this photograph, it has a five-second duration.

> **TIP** You can change the default duration for a photograph by changing the Standard Still Duration value in the Editing category of the user preferences.

Let's increase the challenge just a bit and add a video clip. Because it is important to know exactly what part of the video clip you are adding, you'll use the source viewer to preview the clip.

3 In the Media Pool, double-click the 02_interview clip to load it into the source viewer.

4 Under the source viewer, click the play button to begin playing the clip.

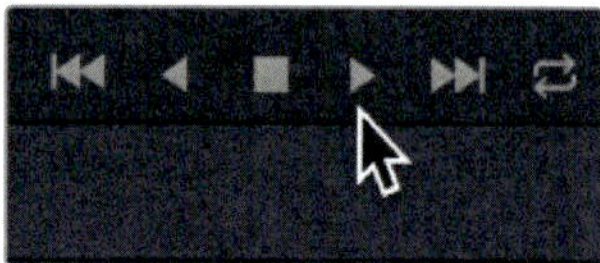

Because of what is said in the interview and the quality of the shot, the entire clip would make a nice second shot in your movie. Instead of dragging it from the Media Pool again, you can drag it directly from the source viewer.

5 Drag the 02_interview clip from the center of the source viewer to the end of the 01_warmer clip in the timeline.

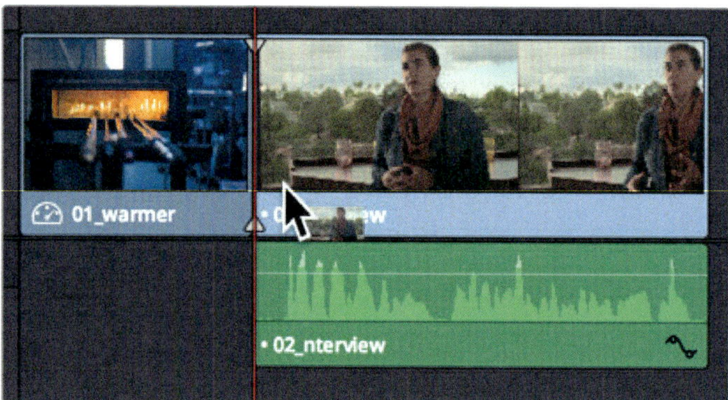

When you drag near the end of a clip in the timeline, the clip you are dragging will snap against the end of the timeline clip. This snapping behavior makes it easy to place clips against one another without leaving any gaps in your timeline.

In the timeline, the orange vertical bar is called the **playhead**. The playhead's location in the timeline corresponds to the current frame displayed in the timeline viewer. After you make an edit, the playhead is placed at the end of the newly added clip, but you can click the buttons under the timeline viewer to position it at the beginning.

6 Under the timeline viewer, click the previous edit button twice to position the playhead at the beginning of the timeline.

7 Under the timeline viewer, click the play button to play the timeline.

The timeline plays the two clips and stops when it reaches the end of the last clip.

You can add multiple clips at once by selecting multiple clips in the Media Pool and dragging them into the timeline. Let's use that technique to add the next 11 clips.

8. Select the fourth thumbnail in the Media Pool (04_at_bench), and then Shift-click the last video thumbnail (14_finished) to select all 11 video clips.

The 11 clips are outlined in red to indicate that they are selected. You can add these clips to the end of the timeline or directly over an existing clip to overwrite it.

9. Drag the playhead to the start of the first interview clip, and press Spacebar to play just that clip.

   You can add these new clips to overwrite the video but leave the audio. Let's position the playhead at the end of the first sentence, "They call us glass blowers, but really we are heat managers."

10. Drag the timeline playhead back to the end of the sentence in which she says, "Really we are heat managers."

11. With the playhead in place as a guide, you can drag the selected clips from the bin up against the playhead.

12. Drag the clips from the Media Pool to the playhead in the timeline, and release the mouse button when the clips snap against the playhead.

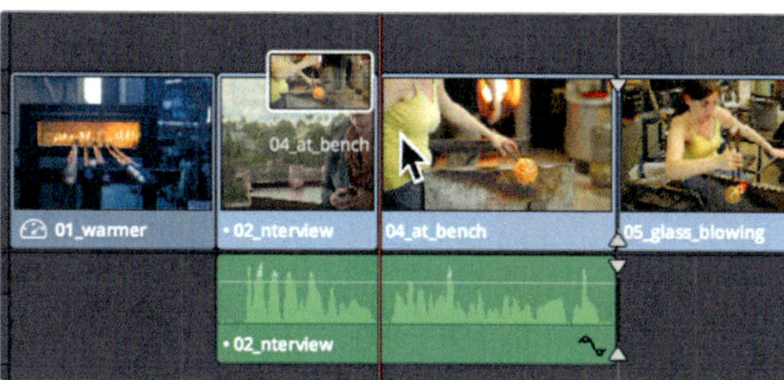

The 11 clips are added to the timeline; but depending on your computer's screen resolution, they may not all fit in the timeline window. Fortunately, DaVinci Resolve includes some handy tools to help you manage the way the timeline is displayed. One of them is an easy menu selection that zooms the timeline so that it displays all the clips in the window.

13 Choose View > Zoom > Zoom To Fit, or press Shift-Z.

Now your entire program fits within the timeline window, making it easier for you to see all the clips and navigate to them.

You can quickly skim over a timeline to see your edits just by dragging the playhead.

14 In the timeline ruler, slowly drag the playhead from the end of the timeline back to the beginning so you can review the clips you just added.

What you currently have is just a collection of pretty clips. You need to develop your story a little. To do so, you can add another interview segment and improve the order of the clips based on the contents of the interview.

Editing narration

Let's add one other piece of the interview. Since you have so many clips that illustrate the topic better than just a talking head, you'll use only the audio from the interview clip.

1 In the Media Pool, double click the 03_interview clip to load it into the source viewer.

2 Under the source viewer, click the "jump to first frame" button to go to the beginning of the clip.

Instead of clicking the button under the source viewer to play the clip, you can use a keyboard shortcut.

3 Press the Spacebar to play the clip in the source viewer.

A one-minute movie inherently has limitations. Because of its length, you can only use a portion of the interview clip. To choose the portion you'll use, you must set a starting location called an In point and an ending location called an out point. To make it easier to set those points, you can use both the waveform overlay in the viewer and the source viewer's jog bar.

4 In the options menu in the upper-right corner of the source viewer, choose Show Zoomed Audio Waveform.

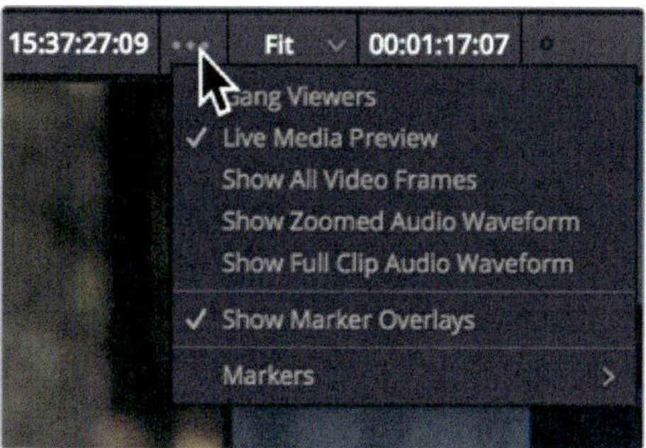

The audio waveform is shown as an overlay at the bottom of the source viewer. The overlay can make it easier to locate precise clip locations based on the soundtrack (or lack thereof).

5 Once again, under the source viewer, click the "jump to first frame" button to go to the beginning of the clip.

You should begin this narration almost halfway into the clip when the subject says, "Also, you can't touch or sculpt." Instead of just playing the clip to locate that spot, you can use the jog bar to slowly skim through the clip while listening to the interview.

6 Under the source viewer, drag the jog bar playhead to the right until you hear the sentence, "Also, you can't touch or sculpt" and then drag to the left until you are at the beginning of the word, "you." You can look at the audio waveform to locate the exact position between the words "also" and "you."

7 Under the source viewer, click the mark in button to set an in point.

You now need to mark an out point to identify the end of the clip. You are looking for the point where she begins to say, “And it’s a huge adrenaline rush.”

You could go on a hunt for this sentence, but DaVinci Resolve can give you a bit of help. Above the source viewer, in the upper-right corner is a timecode numeric display. This display shows you a location within the clip based on hours:minutes:seconds:frames timecode. You can use this timecode to locate the general location of the statement.

8 Drag the jog bar playhead until the timecode display above the source viewer reads 15:38:38:00.

9 Press the Spacebar until you hear the end of the sentence, “And it’s a huge adrenaline rush,” and then press Spacebar again to stop playback.

10 Under the source viewer, click the mark out button.

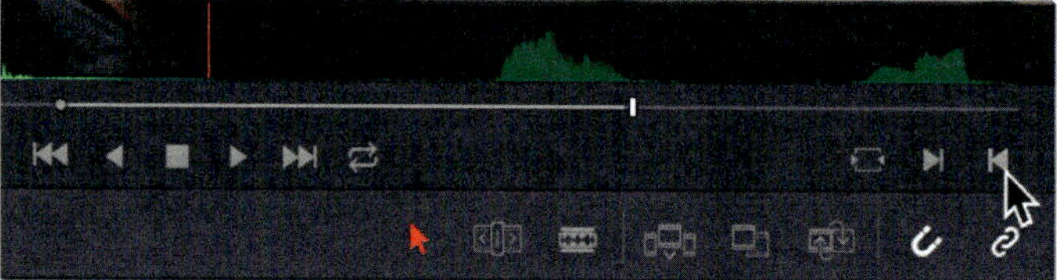

With the in and out points set, it is a good idea to make sure you know where the clip will be placed. You can do so using the timeline playhead.

11 Drag the timeline playhead to the end of the fourth clip in the timeline.

This location provides a bit of breathing room after the previous interview segment in the timeline. It is a good location to place the next piece of narration.

Considering that you already have all these nice pictures in the timeline, you'll need to use only the audio from the interview.

12 Hover your mouse pointer over the source viewer.

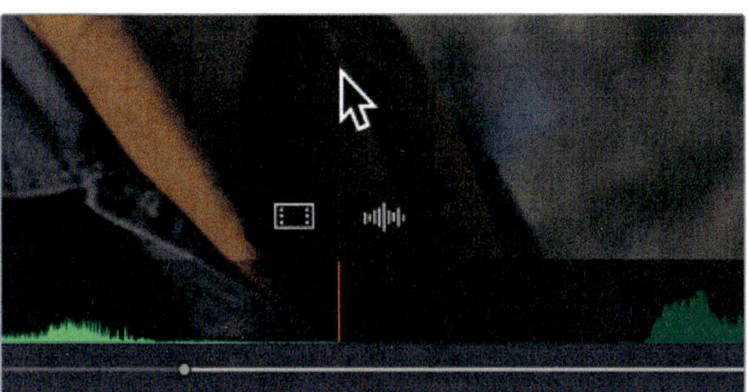

When your pointer is hovered over the source viewer, two overlays appear at the bottom of the viewer. The overlay on the left is for dragging only video into the timeline, and the other is for dragging only audio.

13 Drag the audio-only overlay from the source viewer into the timeline so that it aligns with the playhead position.

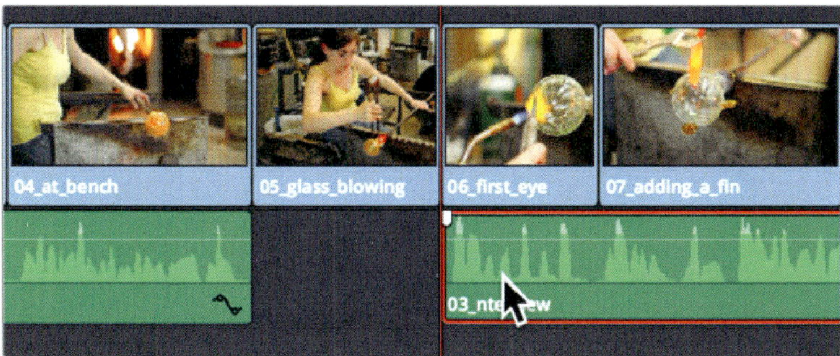

The audio from the clip is now added under all of the photos that you have in your timeline. It's worth playing it back to view your movie so far.

14 Drag the timeline playhead to the start of the timeline, and then press the Spacebar to play the entire timeline.

Considering that all you did was drop some photos into a timeline and add two pieces of narration, this movie is looking good. Let's continue to improve on what you have by adding music.

Adding a soundtrack

Most productions will include multiple audio tracks. In your short film, you only need two: one for the narration and one for music. Let's first listen to the music before you add it to the timeline.

1 In the Media Pool, double click the **One Min Sound Track** clip to load it into the source viewer.

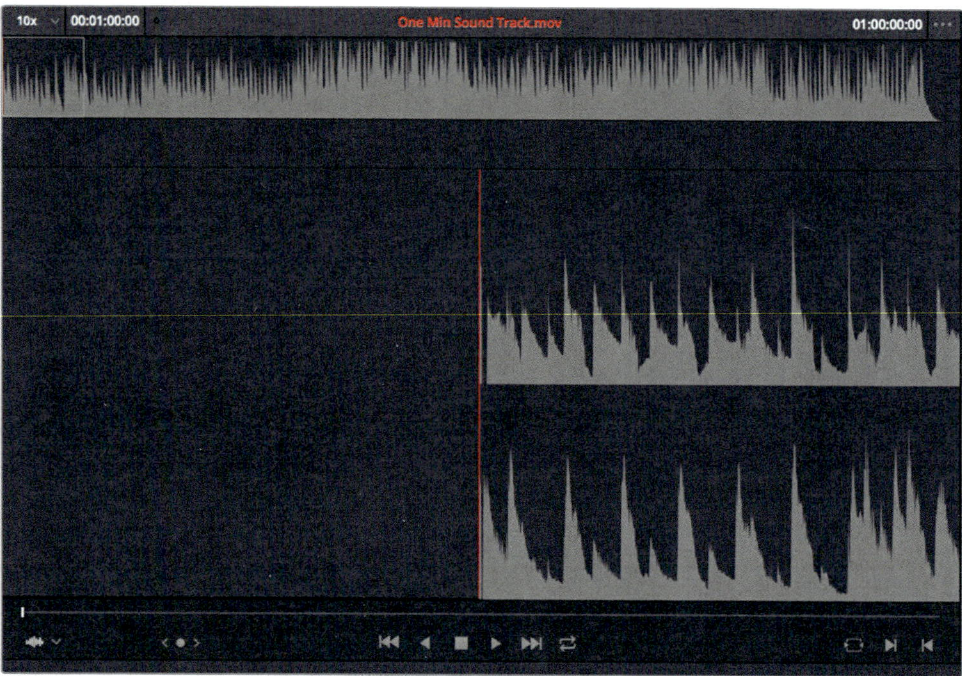

Clips that contain only audio and no video have a unique appearance in the viewer. The entire audio waveform is displayed at the top of the viewer and a zoomed-in portion of the waveform is displayed below it.

2 Press the Spacebar to play a few seconds of the music clip, and press Spacebar again to stop playback.

You will add the entire music track because it is exactly one minute long, which is the target length of your short project. Currently, you have only one audio track that contains the narration, but DaVinci Resolve can automatically add an audio track when you drag an audio clip below an existing audio track.

3 From the center of the source viewer, drag the **One Min Sound Track** clip below the narration track in the timeline. Make sure to position it all the way to the left edge of the timeline so that the music begins at the start of the timeline.

A second audio track is automatically added for the music track. Let's hear how it sounds with your pictures.

4 Drag the timeline playhead to the start of the timeline, and then press Spacebar to watch the entire timeline.

Adding the sound track exposes a few problems you will have to deal with before you finish this project. The first issue to fix is the loudness of the music compared to the narration.

Adjusting audio levels

When you played the timeline, the narration could not clearly be heard over the music. You need to lower the volume of the music so that it enhances the movie without overpowering the narration.

1 In the timeline, click the One Min Sound Track clip to select it.

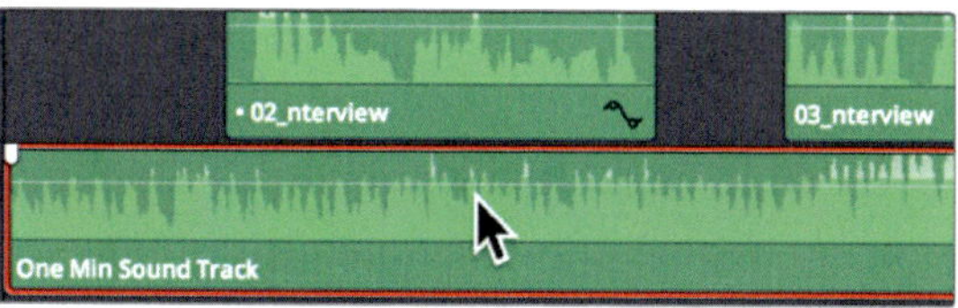

With the music track selected, you can use the Inspector to make volume adjustments.

2 In the upper-right corner of the screen, click the Inspector button to open the Inspector panel.

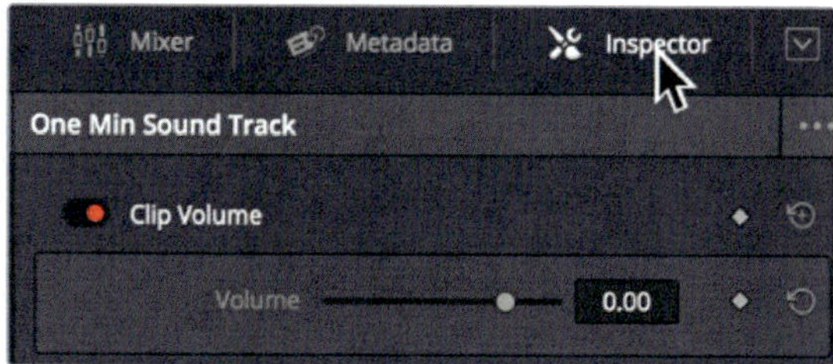

The Inspector panel contains various audio and video controls. Depending on what is selected in the timeline, it displays contextual parameters that you can adjust. Because your music clip is selected, the Inspector displays audio controls, allowing you to play the timeline and adjust the music's volume simultaneously.

3 Drag the playhead to the start of the timeline.

4 Press the Spacebar to begin playback.

5 In the Inspector, drag the Volume slider to the left until it is somewhere between -15 and -20, or wherever it sounds appropriate to you compared to the narration.

6 When you are done setting the Volume level, press the Spacebar to stop playing the timeline.

7 In the upper-right of the screen, click the Inspector button to close the Inspector panel.

You have multiple ways to adjust audio levels in DaVinci Resolve, but using the Inspector is one of the easiest ways. You'll learn more about audio and setting levels in Lesson 7.

Deleting clips from the timeline

Now that you have added music, you can see that the video continued to play even after the music stopped. Because the music has a perfect audio duration for your one-minute movie, you want to figure out how to remove at least one clip to shorten the video duration.

1 Drag the playhead to the start of the third clip from the end of the timeline.

Let's look at the end of your project because on previous viewings one of these shots didn't seem to be necessary.

2 Press the Spacebar to play until the end of the timeline.

The wide shot of the tools doesn't add anything at this point. You can easily take it out without impacting the story.

3 In the timeline, click the 13_tools clip, and press the Delete or Backspace key to remove it.

Pressing the large Delete or Backspace key removed the tools clip, but it left a gap in its place. To close up the gap, you'll need to move the last clip into that space. Moving clips in the timeline is very simple. Just drag it.

4 Drag the 14_finished clip to the left until it snaps up against the end of the clip 12_liquid_ fin.

That places the last clip near the end of the music but it is still not perfect.

Looking at the end of the timeline, you now are so much closer to aligning the last clip to end with the music, but you are still not quite there.

Retiming clips

Another way to create or remove space in a timeline is to retime clips to play slower or faster, thereby making the clip longer or shorter. In this case, you can make the clip faster and shorter until the last clip aligns with the end of the music.

1 Right-click the 12_liquid_fin clip.

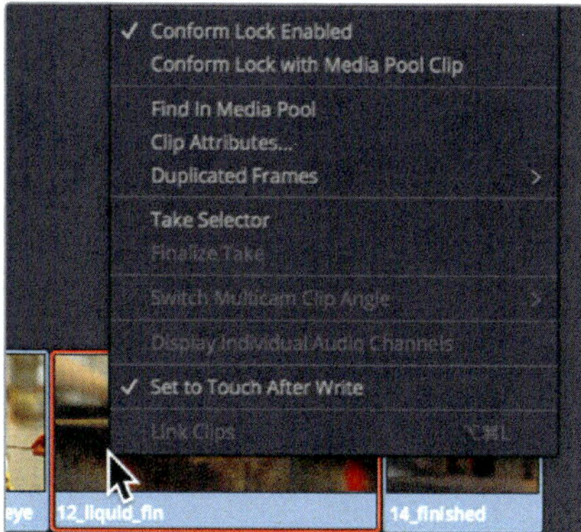

As a general observation, you will find a lot of DaVinci Resolve's functionality is found in right-click, contextual menus.

2 In the contextual menu, choose Retime Controls.

Enabling the retime controls adds a Speed Change bar at the top of the clip. Dragging the Speed Change bar from the end will shorten or lengthen the clip by playing the clip faster or slower.

3 Hover the mouse pointer on the right side of the Speed Change bar until the cursor changes to a resize cursor.

4 Drag the Speed Change bar to the left until the speed percentage at the bottom of the clip reads 125%.

The clip now plays 25% faster, and in the process, opens a gap that you can fill by dragging up the last clip.

5 Drag the last clip in the timeline to the left until it snaps up against the end of the retimed clip.

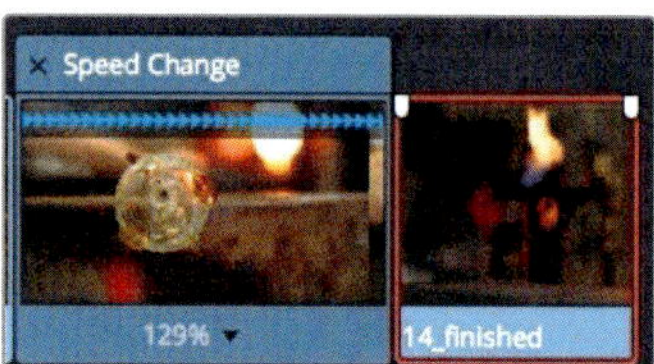

6 Right-click the retimed clip, and in the menu, choose Retime Controls to hide the Speed Change bar.

7 Drag the playhead to the start of the third clip from the end, and press Spacebar to play the end of your movie.

You have done it! The last clip now ends where the music ends. You still have more challenges ahead, but that was a milestone.

Panning and zooming on photos

Now that you have your basic movie in place, it is time to think about improving the visuals in other ways. The first clip in this timeline is a photo, so you might want to add some interest right from the start. Although you could just resize and reposition the photo, DaVinci Resolve has a unique automatic panning effect that automatically pans and zooms photos.

1. In the timeline, click the 01_warmer clip to select it.

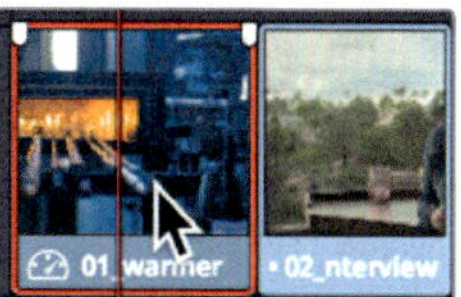

The Dynamic Zoom controls in the Inspector are designed to automatically pan and zoom photos.

2. In the upper-right corner of the screen, click the Inspector button, and scroll the panel down until you see Dynamic Zoom.

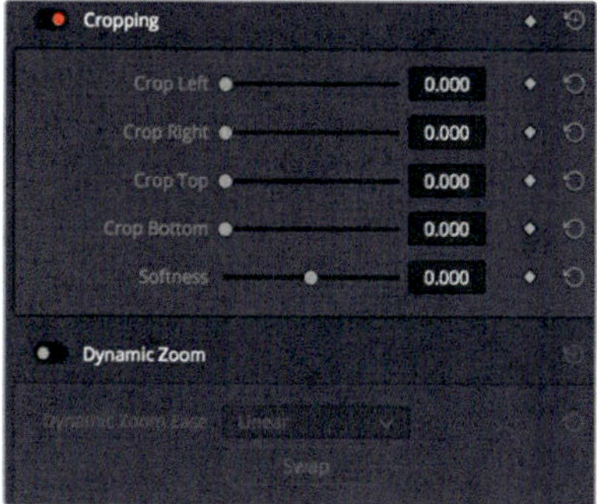

The Dynamic Zoom controls are disabled by default, so you first need to enable them.

3. Click the round gray button to the left of the Dynamic Zoom name to enable the controls.

The Dynamic Zoom button turns red and the effect is applied, so you can now play the clip to see the results.

4. Position the playhead at the start of the timeline.
5. Press the Spacebar to play the clip and see the results.

Dynamic Zoom adds a nice, smooth zoom-out to this clip.

But instead of pulling out of the photo, let's draw the audience in by reversing the direction and zooming into the photo.

6 In the Inspector, in the Dynamic Zoom controls group, click the Swap button.

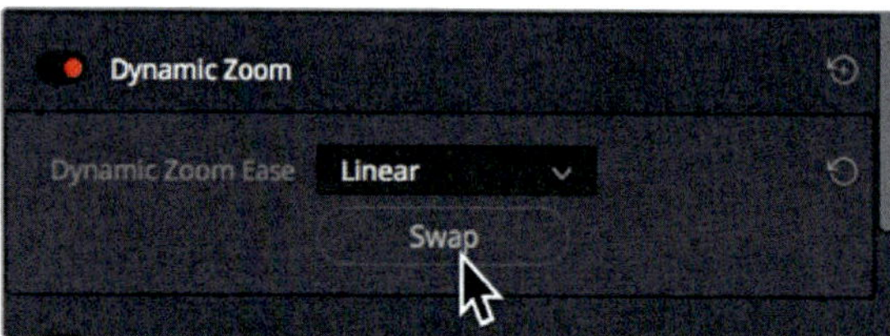

Clicking the Swap button changes the direction of the animation. Instead of animating out to show more of the photo, the selected photo now animates in, zooming into an area of the photo.

7 In the upper-right corner of the screen, click the Inspector button to close the Inspector panel.

8 In the timeline, position the playhead at the start of the timeline.

9 Press the Spacebar to review your animation change.

TIP The dynamic zoom effect works on video clips as well as photos.

You've added some movement to hold attention without detracting from the photo. However, the animations may need some refinement so that it focuses on the most important part of the photo as it zooms in.

Modifying dynamic zoom

The dynamic zoom animation is so simple that you may be tempted to think you can't refine it in any way, but you can! You can completely modify the speed of the effect and precisely choose which portion of the frame is used when zoomed in on the photo.

1 Make sure the 01_warmer photo is still selected, and position the playhead at the start of the timeline.

The animation on this photo can be adjusted so that it zooms in on the furnace rather than the default center of the frame. To modify the animation, you need to show the Dynamic Zoom controls in the viewer.

2 In the lower-left corner of the timeline viewer, click the pop-up menu icon to display a menu of onscreen controls.

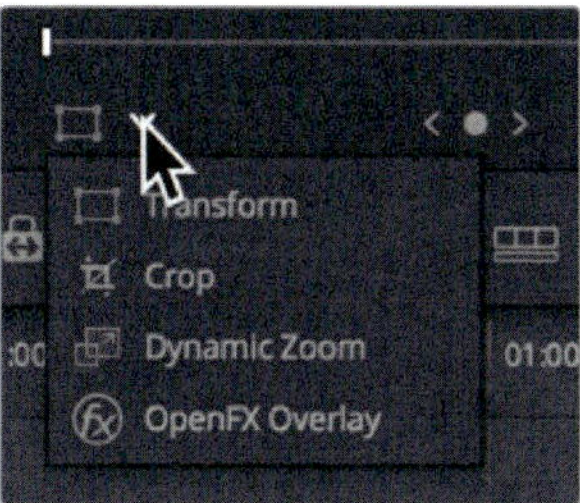

3 In the list of onscreen controls, choose Dynamic Zoom.

The Dynamic Zoom onscreen controls appear over the image in the timeline viewer. The green bounding box represents the starting frame for the animation, whereas the red bounding box represents the frame at the end of the animation. The increased thickness of the green line around the starting bounding box indicates that it is currently selected. To reposition the ending red bounding box, you'll first select it.

4 Click a white control handle in any corner of the red bounding box to select it.

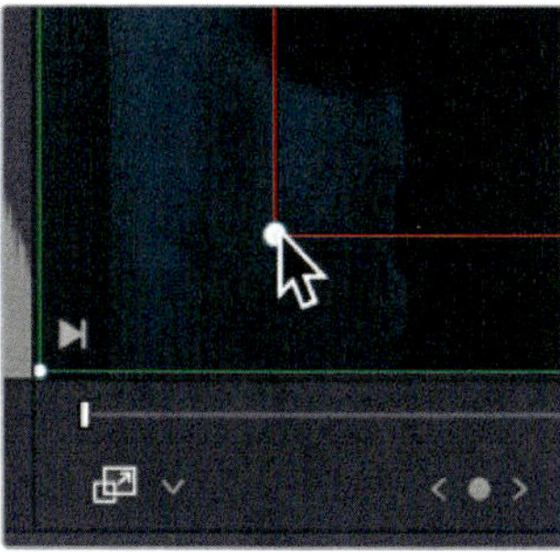

TIP The upper-left corner of the timeline viewer includes a scale pop-up menu in which you can rescale the image in the viewer to better see the onscreen controls. You can also use the scroll control of your mouse, trackpad, or tablet to zoom in and out of the image.

With the bounding box selected, you can drag anywhere within it to reposition it.

5 Drag within the red bounding box and reposition it so it is more fully centered over the warmer.

Now you can play your change to see the results.

6 Click the transform/crop/dynamic zoom button to turn off the onscreen controls.

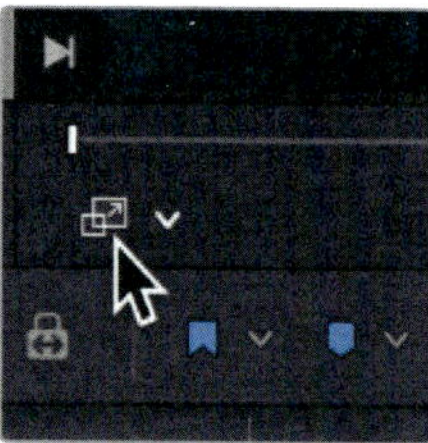

The button turns gray to indicate that it is disabled, and the onscreen controls disappear from the viewer.

7 In the timeline, position the playhead at the start of the timeline.

8 Press the Spacebar to review the animation on this clip.

The subtle animation that you added to this photo draws you into the whole movie more than the static on screen image.

Adding a title

Your program needs a title. Producing a good main title sequence is an art form that uses typography, color, and animation in creative ways, drawing from graphic design principles that date back centuries. Audience expectations for this one-minute movie are not so lofty, so let's just aim for a simple title with correct spelling.

1 In the upper-left area of the screen, click the Effects Library button.

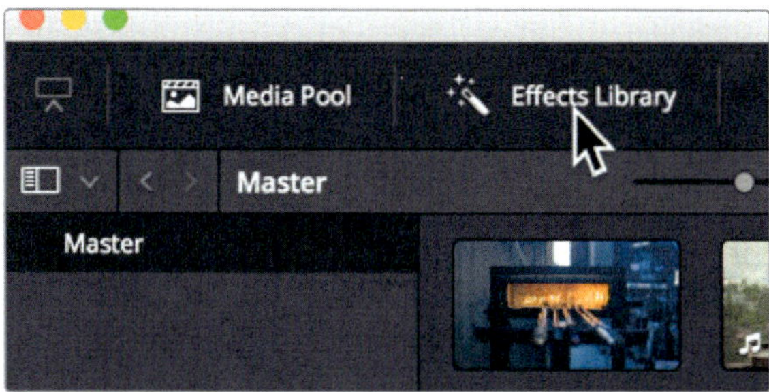

The Effects Library opens below the Media Pool. On the left side of the Effects Library is a list of effect categories. All of the title templates are located in the Titles category.

2 On the left sidebar of the Effects Library, click the Titles category.

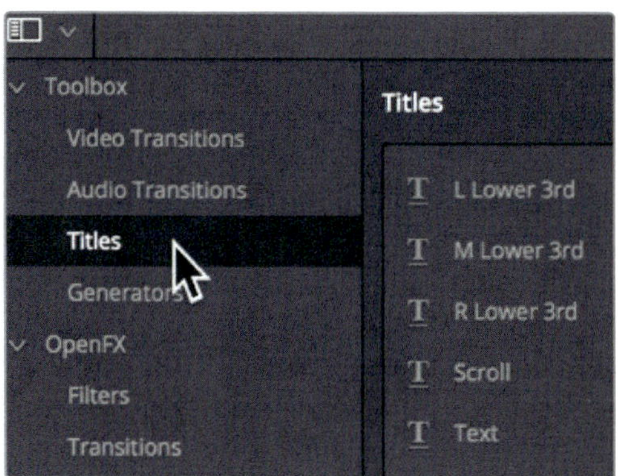

This category includes five title templates that you can add to the timeline and then modify to suit your needs. The basic choice for a main title is the Text template.

3 Choose Viewer > Zoom > Zoom to Fit, or press Shift-Z, to see the entire timeline in the timeline window.

4 Drag the Text template from the Effects Library on top of the third clip in the timeline.

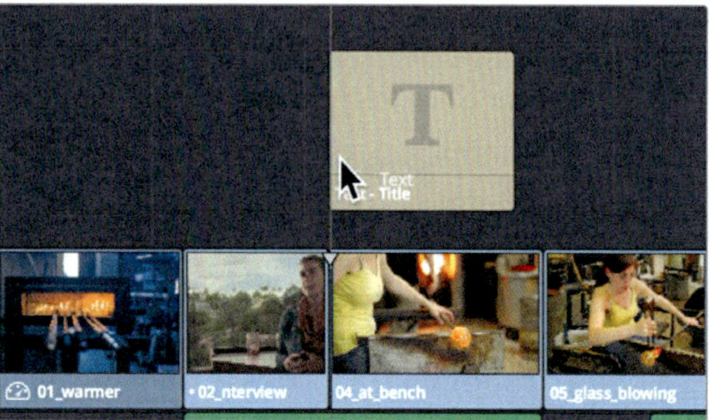

Titles placed over a video clip have a transparent background so both clips are visible in the viewer.

5 In the timeline, move the playhead over the title clip to see the default text in the viewer.

Title templates include default text that you can replace within the viewer.

6 In the timeline, click the title clip to select it, and then double-click the text in the viewer to make it editable.

The text in the viewer is now editable, but the insert cursor is located in the center of the word. You'll need to select the entire word to replace it.

7 Drag from the "T" on the left to the "e" on the right to select the entire word.

8 Type **Taryn Jayne Glass** as the title of your one-minute movie.

As with the Dynamic Zoom controls, the controls for modifying the appearance of the title are located in the Inspector.

9 Click the Inspector button (in the upper-right corner) to display the Inspector panel with the title controls.

Because the title is the selected clip in the timeline, the title controls are displayed in the Inspector. The Inspector contains typical text formatting controls such as font, size, and color.

10 In the Inspector, click the color swatch to open the color picker window.

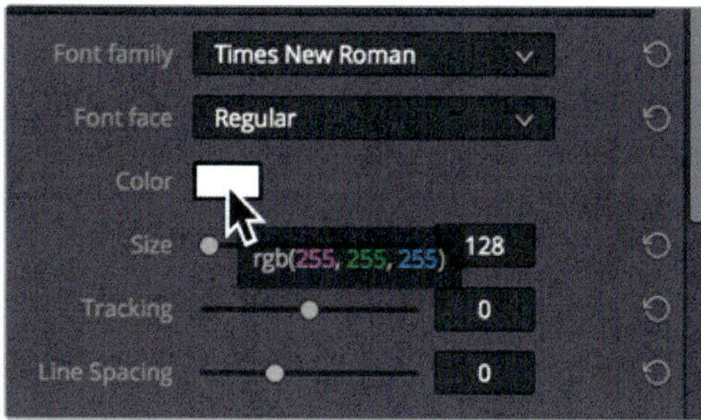

11 Select a light teal color for the text, and click OK to close the color window.

To ensure that the title is a safe distance from the edge of any television screen that it might be displayed on, let's enable safe guides in the viewer before adjusting the text size.

12 Choose View > Safe Area > On.

Guides for safe action and safe title are displayed in the viewer along with a guide for the actual frame size.

13 In the Inspector, drag the Size slider to the right to increase the size of the text until it is almost touching the innermost guide (safe title) in the viewer.

14 Choose View > Safe Area > On to hide the guides.

15 Click the Inspector button to close the Inspector.

With the video and audio editing complete, you can play your movie and review your edits.

Playing full screen

Now it is time for the premiere of your project. If you are working on a single computer display or a laptop and don't have a way to view your program on a dedicated video monitor, you can preview it on the same screen as the DaVinci Resolve interface.

1. Move the playhead to the start of the timeline.
2. Choose Workspace > Viewer Mode > Cinema Viewer, or press Cmd-F (Mac) or Ctrl-F (Windows).

   The viewer now takes up the entirety of your computer screen. Moving the mouse in any way will display an overlay with a play button and a jog bar with which you can fast forward and rewind over the timeline.
3. Press the Spacebar to play the timeline and watch your movie.

   The last detail is to save the work you just did.
4. Move your mouse down over the jog bar overlay, and click the expand button to return to the DaVinci Resolve interface.

5. Choose File > Save Project.
6. Enter **Glass blowing** as the name for the project, and click Save.

Well done! This lesson was designed to give you a basic overview of the DaVinci Resolve editing interface and many of the panels that you will return to throughout this book. If you aren't completely clear on every step that you did, don't worry too much. You'll be repeating many of these techniques in future lessons, so you'll have ample opportunities to master them.

Lesson 2

Organizing a New Project

Producing film and video content is a very creative and exciting process. Lesson 1 gave you a quick overview of that process. Now, you're going to start back at the beginning and get into more detail so you can build the foundation you need to efficiently edit, color correct, and mix audio with DaVinci Resolve 14.

You'll begin by configuring project settings, importing media, grouping clips into folders, and using metadata and Smart Bins to make it easier to search and sort through footage.

Time

This lesson takes approximately 45 minutes to complete.

Goals

Configuring essential settings

When you started the project in the previous lesson, you didn't configure any settings at all. In fact, you even didn't give your project a name until the end of the lesson. That project was designed to skip the sometimes dry project setup tasks and jump right into the fun stuff. Now you'll take a step back and correctly set up a new project. To do so, let's return to the project manager.

1 If DaVinci Resolve is closed, open the application to get to the project manager. If DaVinci Resolve is already open, choose File > Project Manager, or press Shift-1.

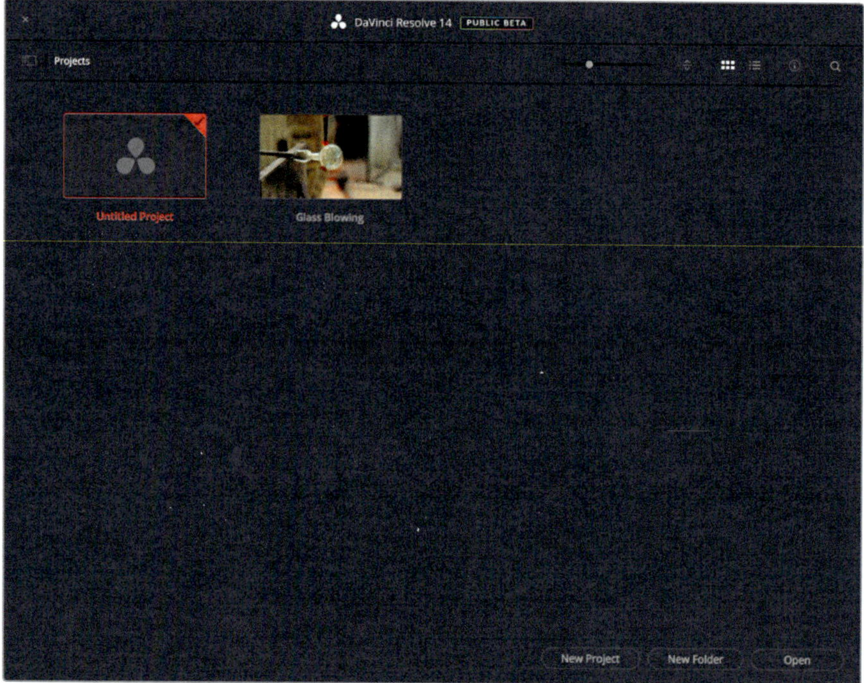

The project manager opens with a thumbnail that represents the project you created in the previous lesson. You'll create a new project in this exercise.

2 At the bottom of the project manager, click the New Project button.

3 In the Create New Project dialog that appears, enter the project name **My New Project**, and click Create.

My New Project is added to the project manager.

4 In the project manager, double-click the My New Project thumbnail to open the project and switch to the Edit page.

5 To close all the extra panels that you opened in the previous lesson, choose Workspace > Reset UI Layout.

DaVinci Resolve uses default values for project settings such as frame rate for playback, and output resolution. You can customize these values in the Project Settings window.

6 Choose File > Project Settings to open the Project Settings window.

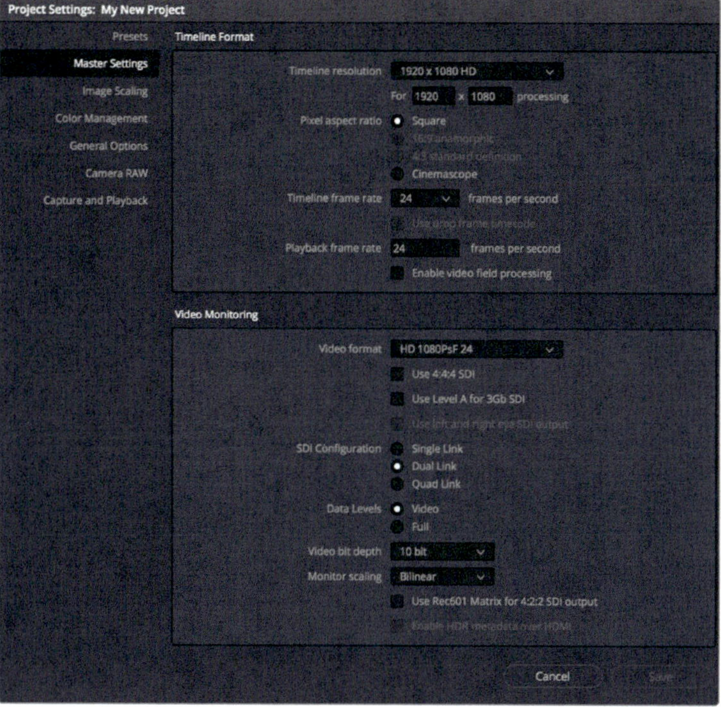

In this lesson, you'll use clips with a frame size of 1280 x 720 pixels running at 23.976 frames-per-second (fps). Let's adjust your project settings accordingly.

7 In the "Timeline resolution" pop-up menu, choose "1280 x 720 HD 720P".

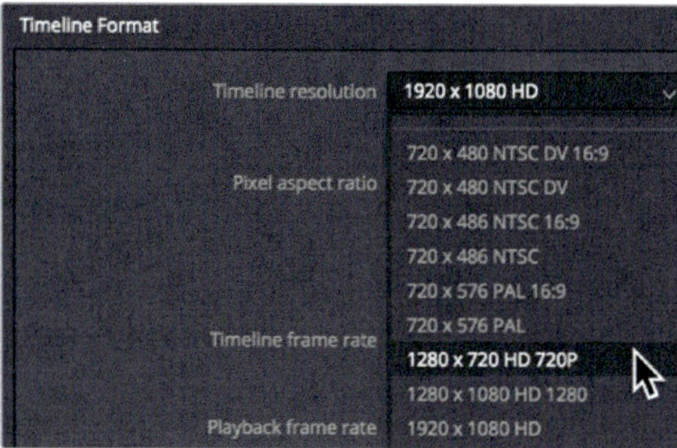

Below the timeline resolution, you can see the timeline frame rate. It is critically important to set this correctly before you import any media because you cannot change the project's frame rate later.

8 In the "Timeline frame rate" menu, choose 23.976.

9 In the Project Settings window, click Save to save your changes and return to the Edit page.

You should set up two other areas before you begin importing clips. Let's start in the Preferences window, which is used for configuring DaVinci Resolve settings that do not typically change from project to project

10 Choose DaVinci Resolve > Preferences, or press Cmd-, (comma) in Mac, or Ctrl-, (comma) in Windows to open the preferences window.

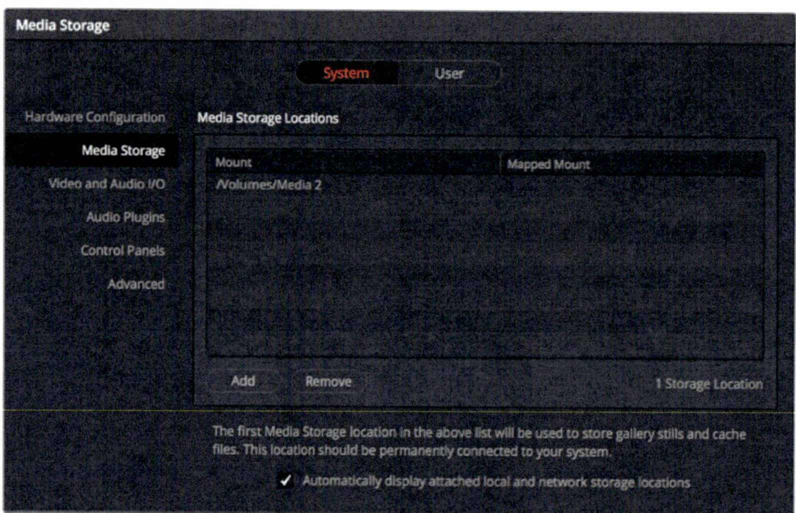

Similar to the Project Settings window, the preferences window has categories along the left side. The preferences window opens to the Media Storage category which is the one you'll typically want to set up.

The Media Storage category allows you to add drives, often called "scratch disks," to your system when using DaVinci Resolve. Throughout the duration of a project, DaVinci Resolve will occasionally create media that must be saved to a hard drive. This may include cached or rendered files, still frames, and various media that is optimized to improve playback performance. Such files are saved to the first hard drive listed in the Media Storage preferences.

11 If you wish to change the drive used for cached content, click the Add button and select your fastest and largest hard drive.

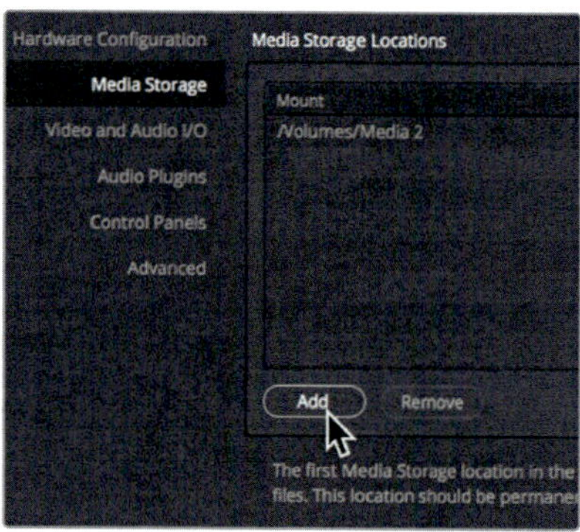

It's almost always advisable to set your scratch disk to the largest, fastest hard drive available to your computer. If you don't specify a scratch disk, your system disk will be used by default.

The preference window has two tabs at the top. The default selected tab is the System tab. Its settings, like the Media Storage settings, are specific to your computer hardware for all projects. The User tab displays preferences that are not saved with a project but saved for use with this workstation.

12 At the top of the preferences window, click the User tab

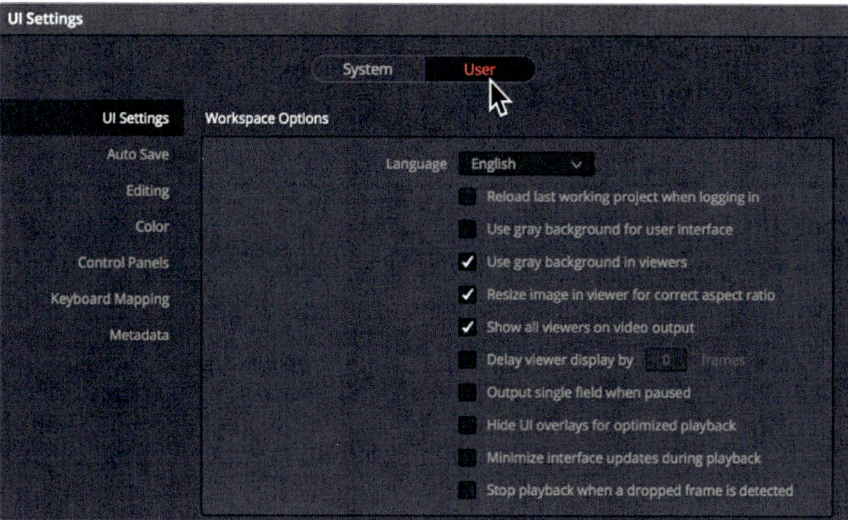

13 Along the left side of the preferences window, click the Auto Save category.

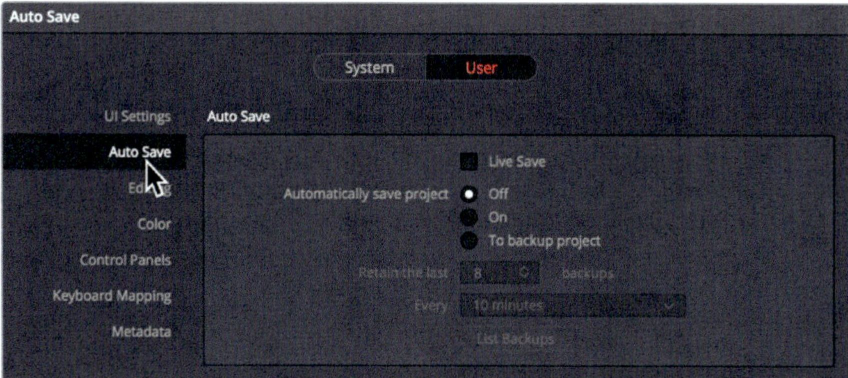

The auto save settings can configure regularly timed project backups as well as implement a continuously live save mode that always saves your work. You'll enable the Live Save mode so you don't have to worry about explicitly saving our project.

14 Click the Live Save button to enable the feature.

> **TIP** The Keyboard Mapping setting offers a way to customize the keyboard shortcuts so that they match other popular editing systems or your own unique layout.

15 Click Save to close the preferences window.

> **TIP** Changing some options in preferences will require you to restart DaVinci Resolve.

Now your set up is complete; but before you learn how to import clips, let's get clear about some of the unique aspects of the DaVinci Resolve interface.

Exploring the DaVinci Resolve interface

DaVinci Resolve is organized as a series of interface pages—Media, Edit, Color, Fairlight, and Deliver—that correspond to the order of most traditional film and video post-production workflows. Each page is accessed by clicking one of five buttons at the bottom of the screen.

1. Click the Media Page button to switch to the Media page.

   The Media page is the most efficient page for importing and organizing media, syncing clips, adding metadata, and cloning camera original media.

2. Click the Edit page button to switch to the Edit page.

   The Edit page contains all of the professional editing tools you'll need to assemble your program and add titles, transitions, and effects.

3. Click the Color page button to switch to the Color page.

   The Color page contains DaVinci Resolve's world-class color correction and creative color grading tools as well as keying, stabilization, and filter effects.

4. Click the Fairlight page button to switch to the Fairlight page.

   The Fairlight page is a complete digital audio workstation that has everything you'll need to deliver incredible cinematic soundtracks .

5. Click the Deliver page button to switch to the Deliver page.

   On the Deliver page, you can master to tape, create files for the web, and even output high-resolution files for theatrical distribution.

You can freely move between these pages to import and manage media, edit in the timeline, color correct, mix audio, and output your program. You'll begin working in the Media page.

The Media page is divided into five areas:

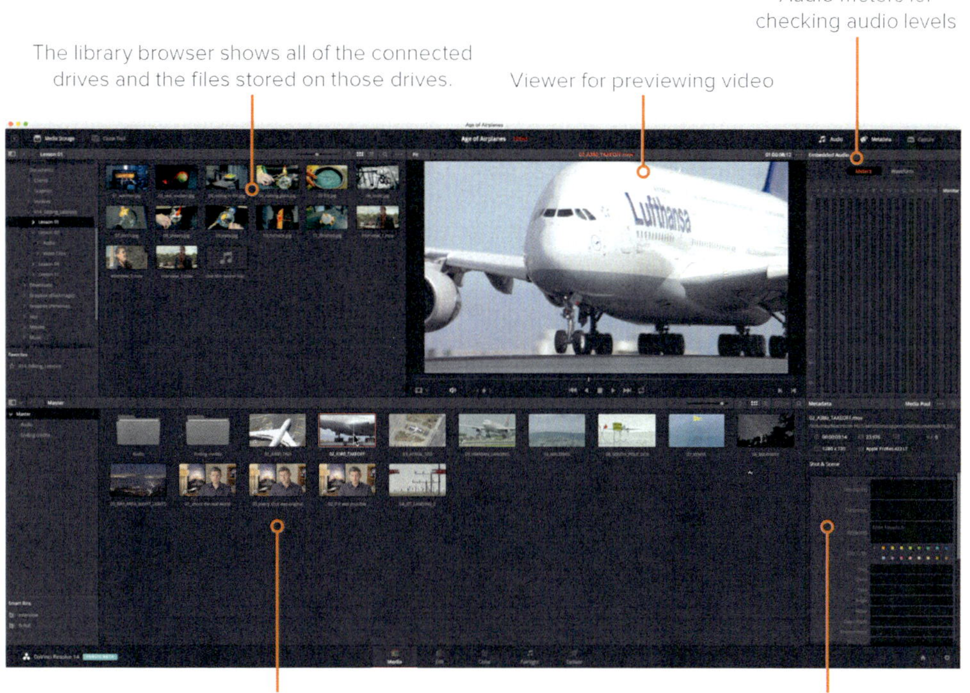

The Media page gives you the most flexibility and functionality when it comes to importing media from your hard drives. You'll also perform other tasks here such as media management and clip organization, syncing audio and video clips, and troubleshooting clips that unexpectedly appear offline.

Importing clips

When you are ready to import clips into a project, the most efficient way is through the library browser in the Media page. In this browser, you can navigate to any folder or hard drive in which you store your media. When you import clips into your project, they are stored in bins in the Media Pool.

NOTE The following steps assume that you have copied the R14 lessons folder to your Documents folder. If you copied the files to a different location, use the library browser to navigate to that location.

1 In the library browser sidebar to the left, click the icon of your computer's internal hard drive.

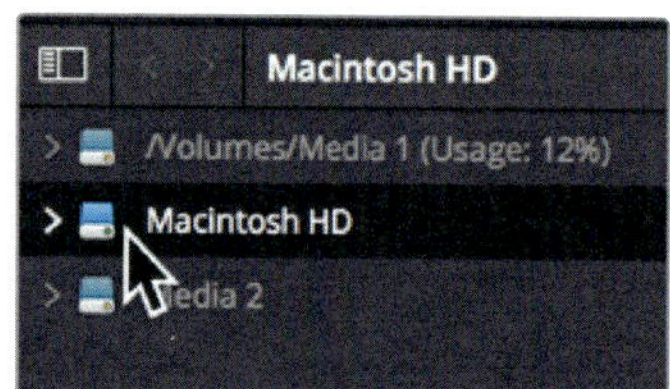

2 In the right panel, navigate to your Documents folder.

3 In the Documents folder, double-click the R14 lessons folder.

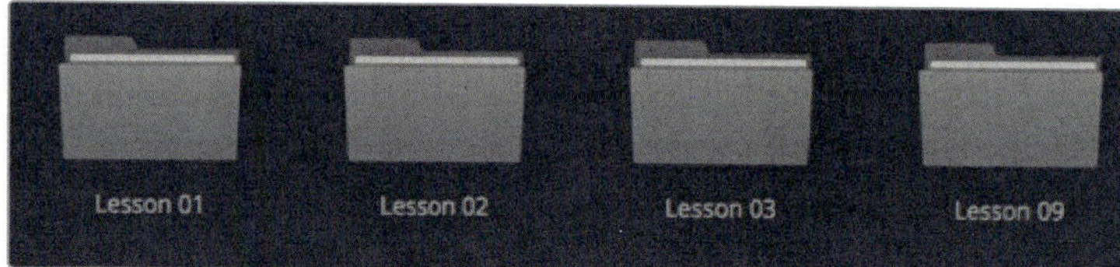

The library displays seven subfolders in the R14 lessons folder. You'll import clips for this project from the Video Clips folder located in the Lesson 02 folder.

4 Double-click the Lesson 02 folder, and then double-click the Video Clips folder to open it.

The right panel of the library displays a thumbnail-sized image icon for each of the clips you want to import. You can preview each clip in the viewer.

5 Click any thumbnail to see the clip in the viewer.

6 Press the Spacebar to begin playing the clip, and then press Spacebar again to stop playback.

After you're finished previewing clips, you can choose to import one or more of them into your project.

7 Choose Edit > Select All to select all the clips displayed in the library, or press Cmd-A (Mac) or Ctrl-A (Windows).

8 Drag the first clip in the library into the Media Pool where it is labeled "No clips in media pool".

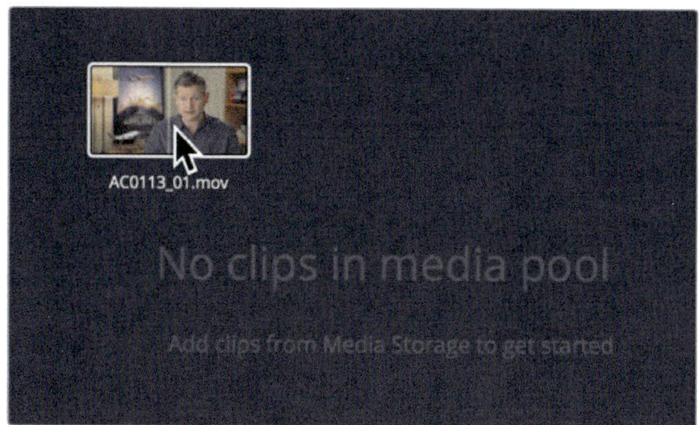

TIP You can also drag clips directly into the Media Pool from the Mac finder or Windows Explorer.

All the selected clips are added to the Master bin in the Media Pool. Every project always includes a Master bin. This Master bin contains every clip you add and every new bin you create. It is also important to understand that the clips are not copied, moved, or transcoded when you import them. DaVinci Resolve is completely non-destructive; it simply links to the unaltered files in their current locations on your hard drive.

Importing Folders

Instead of selecting each of the clips you want to import and adding them all to the Master bin, you can import an entire folder and automatically create a custom bin.

1 In the upper area of the library, click the back arrow to view the contents of the R14 lessons folder.

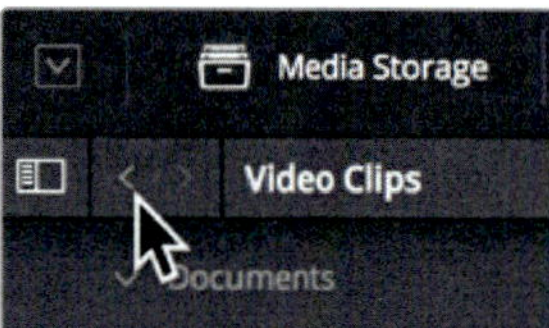

2 Right-click the Audio folder, and choose "Add Folders and SubFolders to Media Pool (Create Bins)".

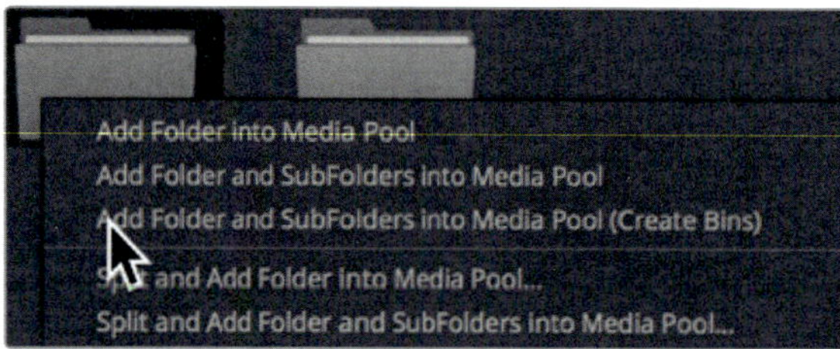

Choosing this menu item creates a bin with the folder's name and adds it to the Master bin. All of the clips contained in the folder are imported as part of the new bin.

Reviewing and scrubbing clips

You might want to review and check your clips after you've imported them, especially if you're editing a project that was shot by someone else and you're as-yet unfamiliar with the footage.

1 In the Media Pool sidebar, select the Master bin.

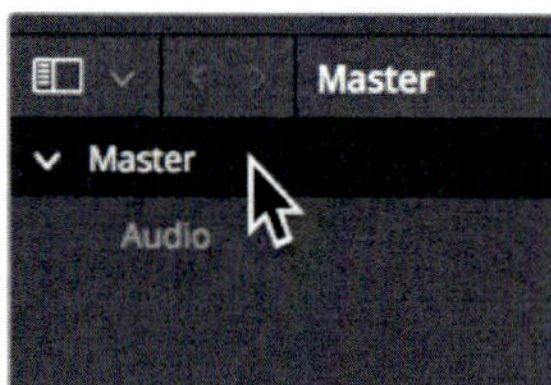

2 In the bin, hover the mouse pointer over any video clip.

When your mouse pointer is located over any clip in a bin, a live preview of the clip is displayed in the viewer. You can move your mouse back and forth over the thumbnail to quickly scrub through it.

NOTE Due to screen and window size differences, the order of clips in your bin may appear slightly different from the figures shown in this lesson.

3 Move the mouse pointer back and forth over the thumbnail to scrub quickly through the clip and see it in the viewer.

TIP Live preview can be disabled in the options menu located in the upper-right corner of the viewer.

The live media preview feature allows you to quickly skim over a clip without having to play it. Still, to edit clips most efficiently, you will need to know their contents intimately. For that purpose, nothing can replace just sitting down and watching a clip play from start to end.

4 While your mouse hovers over the thumbnail, press the Spacebar to play the clip.

The clip plays at its native frame rate.

5 Press the Spacebar again to stop playback.

The viewer transport buttons include playback as well as other buttons to play a clip in reverse, and move the playhead to the beginning or the end of a clip.

Choosing representative thumbnails

The thumbnail used to represent each clip in the bin is the first frame of that clip. At times, that first frame might not be the best representation of the clip, so it is useful to be able to change it.

1 In the upper-right corner of the Media Pool, drag the Scale slider to increase the size of the thumbnails.

2 Position your pointer over the thumbnail labeled 07_ KENYA.

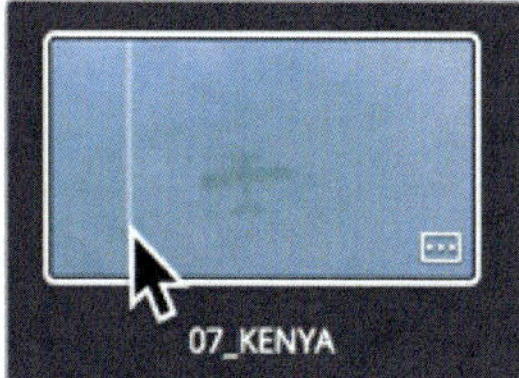

3 Move the mouse pointer left and right over the thumbnail.

As you move the pointer left and right, DaVinci Resolve displays various frames from the clip as if you were fast forwarding and rewinding through it.

4 Move the pointer over the thumbnail until you see the yellow airplane.

This image will make a better representative frame than the blue water.

5 Right-click the thumbnail, and in the menu, choose Set Poster Frame.

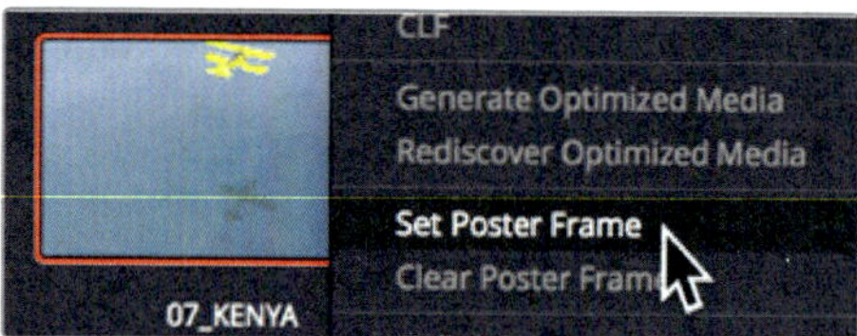

Although setting appropriate poster frames can help a lot when you are searching clips, to truly become familiar with the content you are using, you should always watch each source clip. Once you complete this lesson, take time to review the clips and set poster frames for any other clips that may not currently display the best representative frame.

Viewing clip metadata

In addition to audio and video content, clips can also include information called metadata that describes the content of your clips. Metadata can identify the format, framerate, and resolution, among other data. Some metadata is added automatically to a clip by the camera and audio recorders. You can also add custom metadata, such as keywords, that you can use to help organize your projects more efficiently.

You can view the metadata for a clip and add additional information using the Metadata editor.

1 In the Master bin, click a clip to load it into the viewer.

Detailed information about the selected clip appears in the Metadata editor in the lower-right corner of the Media page.

The upper section of the Metadata editor displays some essential clip information such as name, duration, and frame size. However, because a production can include an enormous amount of metadata for every clip, a pop-up menu in the upper-right corner of the Metadata editor lets you choose other categories of metadata.

2 In the upper-right corner of the Metadata editor, in the pop-up menu, choose Shot & Scene.

Each category in the pop-up menu contains additional fields and checkboxes. Some may display metadata entered automatically by the camera or other devices, whereas other fields allow you to add custom information that may be helpful in organizing your clips. In the next exercise, you'll add some metadata to clips that will make it easier to locate them.

Adding custom metadata

Standard metadata that is captured or created automatically on set during production is certainly helpful when organizing clips, but adding your own metadata is also important. In almost every project, you'll organize content using some form of metadata. Some of that metadata must be added manually in DaVinci Resolve.

The current lesson has only a few shots from one small scene, so it's not very difficult to stay organized. However, when you're working on larger projects with hundreds, or even thousands, of clips, adding and using metadata can help you save countless hours searching and sorting through clips as you build your edit.

1. In the Master bin, select the AA0113_01 clip.

This clip is one of three interview clips that you'll be using. It's a good idea to identify it as an interview clip, so let's add that information as a keyword.

2. In the Keywords field of the Metadata editor, enter **Interview**.

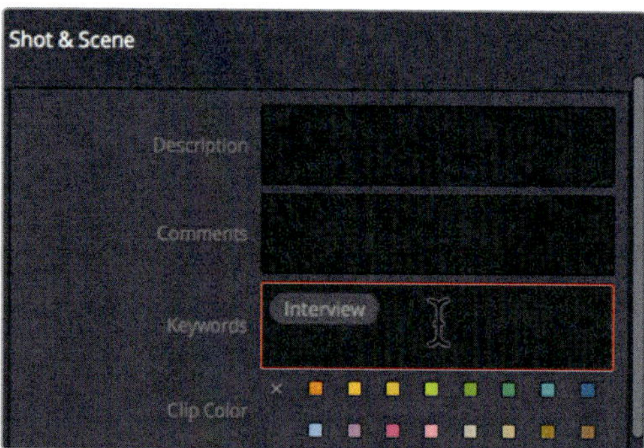

Two other interview clips could also use that "interview" keyword. You can be more efficient by entering the keyword for both clips at the same time.

3 In the Master bin, select AB0102_01, then Cmd-click (Mac) or Ctrl-click (Windows) AC0113_01 to select it, too.

4 With both clips selected, in the Keyword field of the Metadata Editor, enter the letters **INT**.

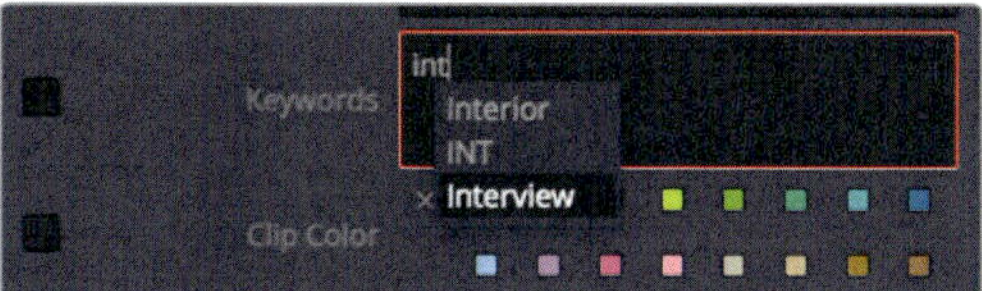

A list of standard keywords and previously entered keywords that start with I-N-T are displayed. It's helpful to select keywords from the list if they are available, rather than running the risk of misspelling.

5 In the pop-up menu, choose "interview" to add that keyword.

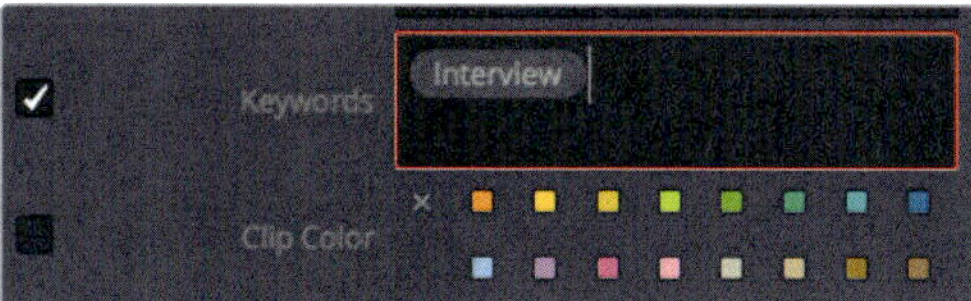

6 At the bottom of the Metadata panel, click Save to apply the keyword to both clips.

You have tagged interview clips with some important information about their contents. Metadata added this way can be used throughout DaVinci Resolve to improve clip organization. Next, you'll learn ways to put that metadata to use as you begin organizing your clips.

Making new bins

Bins are like folders, or containers, for all your clips. Although all the video clips that you import go into the default Master bin, it is not very efficient to rely entirely on a single bin to hold all of them. It's like having a filing cabinet and stuffing everything in one folder. A better strategy is to create custom folders or bins and organize your clips in a way that makes it easier for you to find them later.

1 In the Media Pool, select the Master bin.

   All the video clips currently in the Master bin would be easier to work with (and faster to locate) if they were organized by content. For instance, you could place all of the scenic plane shots in a b-roll bin to distinguish them from the interview clips that you could place in an Interview bin.

2 To create a new bin, choose File > New Bin, or press Cmd-Shift-N (Mac) or Ctrl-Shift-N (Windows).

   A new bin with the default name Bin 1 is created and added to the media pool. Let's change the name so it fits your project. The names of new bins are immediately available for editing.

3 Type **Ending Credits** as the name of the bin.

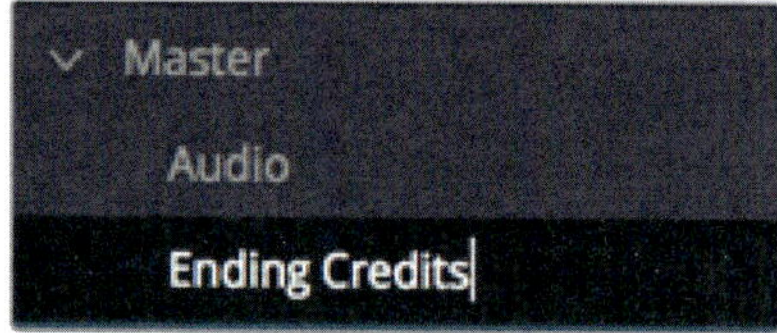

   All new bins that you create appear inside the Master bin. With a new bin created and named to suit your project, you're ready to start organizing your clips into bins.

4 Select the Master bin, and then drag the 11_MOVIE_CREDITS thumbnail onto the Ending credits bin name in the Bin list sidebar. When the Ending Credits bin name highlights, release the mouse button.

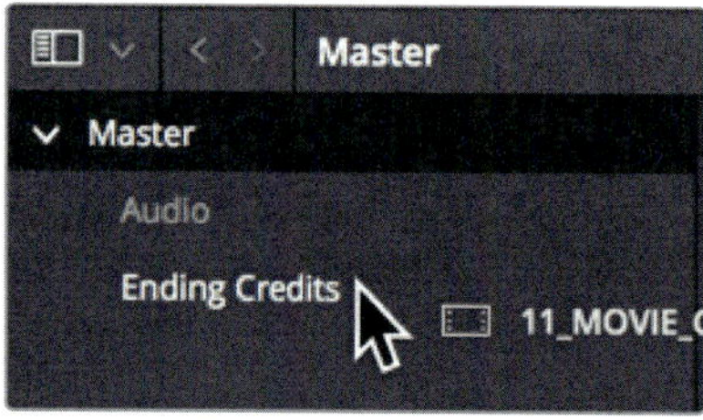

TIP You can open multiple bins at the same time by right-clicking a bin name in the Bin list sidebar, and choosing Open As a New Window.

5 In the Media Pool sidebar, click the Ending Credits bin to view its contents.

You've now successfully moved clips from one bin to another. Although the process of organizing clips is straightforward, it can also be very labor intensive. Let's look at a more efficient way to organize clips.

Creating Smart Bins

Smart Bins search your entire project to group clips based on metadata that you define. For example, you can create a Smart Bin that automatically finds all of the audio clips in your project or all of the clips captured using a specific camera. Best of all, a Smart Bin's contents will continually update as new footage is added to your project. That means you don't have to manually organize footage when using metadata and Smart Bins.

1 Right-click in the sidebar area of the Media Pool under the words "Smart Bins".

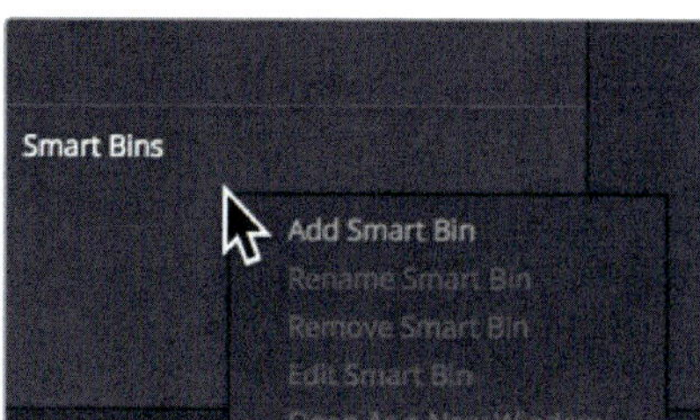

2 In the contextual menu, choose Add Smart Bin.

The Create Smart Bin dialog appears. In this dialog, you set up the rules that determine which clips are automatically added to this Smart Bin. The criteria options are many, enabling you to create Smart Bins that group clips based on a wide range of auto-generated and manually entered metadata.

3 In the Create Smart Bin dialog, in the Name field, enter **Interview**.

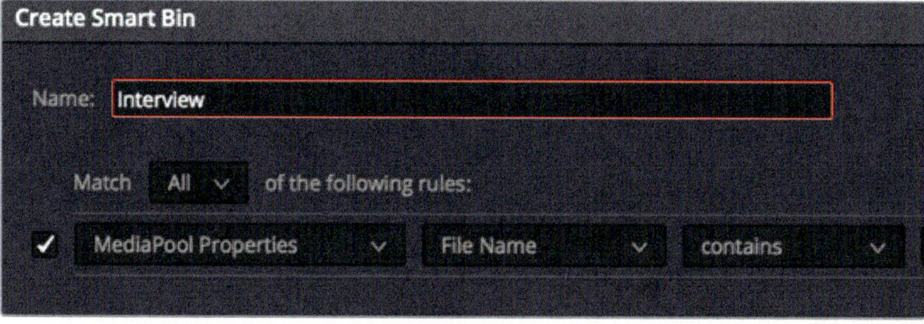

4 Click Media Pool Properties, and in the menu, choose Metadata - Shot & Scene because this was the metadata category chosen when you entered the keyword.

5 Set the metadata type pop-up menu to Keywords, and leave the final metadata criteria pop-up menu set to "contains".

6 In the text entry field, type **interview**, and click Create Smart Bin.

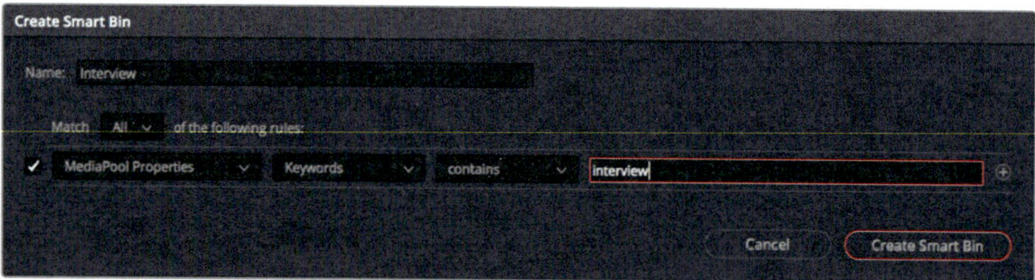

The Smart Bin appears at the bottom of the Media Pool and automatically updates to include all of the clips that contain the keyword.

Using multiple criteria in the Smart Bin

You just created a very simple Smart Bin. But Smart Bins can be much smarter! You can add multiple layers of criteria for more advanced clip selection.

1 Right-click under the Interview Smart Bin, and in the menu, choose Add Smart Bin.

For this Smart Bin, you'll look for all of the clips that are **not** interview clips.

2 In the Create Smart Bin dialog, in the Name field, enter **B-Roll**.

3 Set up the Smart Bin criteria just as you did for the Interview bin, choosing Metadata - Shot & Scene, Keywords, and then typing **interview** in the last field.

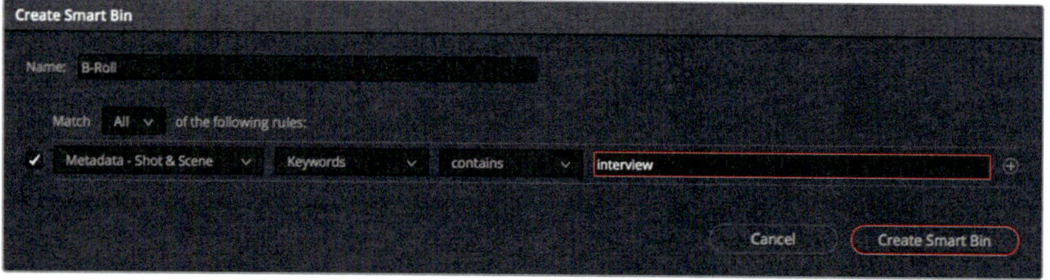

4 Click the “contains” menu, and choose “does not contain”. Then click “Create Smart Bin”.

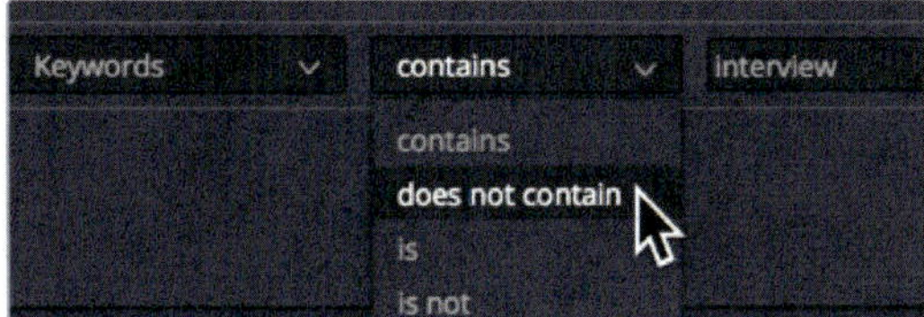

You now have a Smart Bin that includes all the clips that do not have the keyword Interview applied to them. That identifies most of the clips that you want, but the bin also contains the audio clips in your project. You’d actually like to narrow down this Smart Bin even further to exclude those audio clips.

5 To edit the Smart Bin criteria, double-click the B-Roll Smart Bin.

You can add additional layers of criteria for a Smart Bin which will result in a more selective choice of clips.

6 To the far right of the dialog, click the add filter criteria button to add an additional field for criteria.

7 Change “Metadata - Shot & Scene” to “Media Pool Properties”.

8 Change “File Name” to “Clip Type”.

9 Set the last two menus to “is not” and “Audio”.

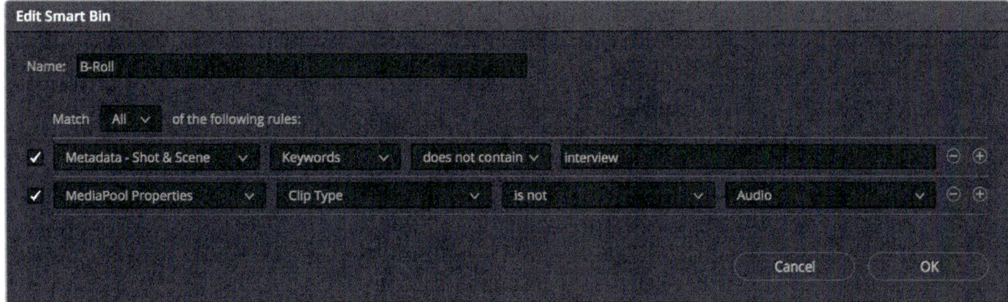

You ‘ve now removed all of the audio clips from this Smart Bin, but you also need to eliminate any timelines.

10 Click the add filter criteria button to add another criteria field.

The new fields use similar criteria as the previous one so you need to change only the last two items.

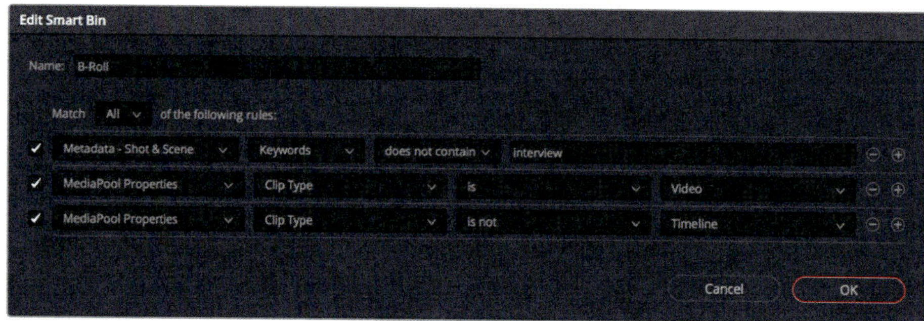

11 Change "is" to "is not" and change the last item from Video to Timeline.

12 Click OK to close the dialog and update the Smart Bin.

In the future, when you add the "interview" keyword to new clips in this project, those clips automatically will be added to the Interview Smart Bin and excluded from the B-Roll Smart Bin. That's the beauty of the Smart Bin. It collects clips based on whatever criteria you identify and continually updates your clip organization.

Saving custom bin views

Beyond organizing clips into bins, you can organize how content appears in those bins. Clips in bins can be shown in list view as well as thumbnail view and can display as much or as little metadata as you'd like.

1 Select the B-Roll bin.

2 At the upper-right of the Media Pool, click the list view button.

The Media Pool switches from showing the clips as thumbnails to showing clips in a text list.

In every project, some of the columns in list view will be more important than others. To ensure that you can see the information you'll need most, you can hide and show columns in a bin and then save that customized view as a Layout.

3 Ctrl-click (Mac) or right-click (Windows) any column heading to display the bin headings contextual menu.

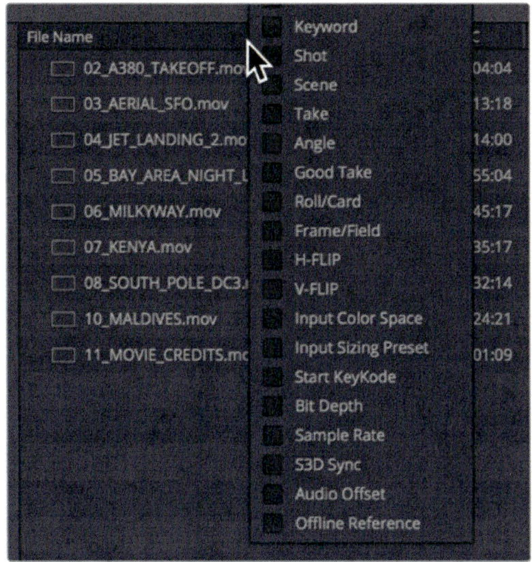

The contextual menu lists all of the columns that you can display in a bin. The headings with a check mark are the currently displayed columns.

4 In the contextual menu, click Reel Name, Frames, Type, File Path, Format, Codec, and Flags to deselect and hide all those columns in the bin.

The columns you deselected are removed from the bin's List view.

You can also sort columns, and therefore the clips, based on the information found in a column.

5 Click the heading for Start TC so that the small arrow next to the name points up.

Clicking the heading for any column sorts that column based on its criterion. Clicking the Start TC heading sorts the bin contents in ascending order based on the starting timecode number for each clip. A small arrow pointing up is shown next to the column heading name to indicate that this column is used for sorting as well as indicating the order of the sort.

After making these changes, you have a nice, slimmed-down number of columns that display only the essential information that you might want to see. You can save as many different bin views as necessary and recall them from the menu later. To save a bin view, you can use the same contextual menu.

6 Ctrl-click (Mac) or right-click (Windows) one of the bin column headings, and in the contextual menu, choose Create Column Layout.

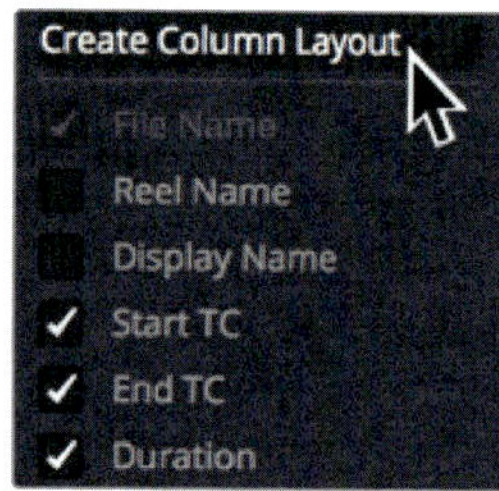

7 Type **Basic View** as the column layout name, and click OK.

You now have a bin that displays all of the desired columns

Changing clip names

Although some of the file names for your clips are clear and descriptive, others are cryptic filenames that make sense only to the camera that created them. DaVinci Resolve lets you create custom names, called Display Names, that are used throughout the software while you're editing and grading. The Display Name column displays the original filename by default, but you can also rename your clips in that column to something more descriptive. (Don't worry, changing the Display Name does not change the filename on your hard disk.)

1 Select the Interview Smart Bin.

2 Ctrl-click (Mac) or right-click (Windows) any column heading to display the bin headings contextual menu.

3 In the contextual menu, choose Display Name to show that column in the bin.

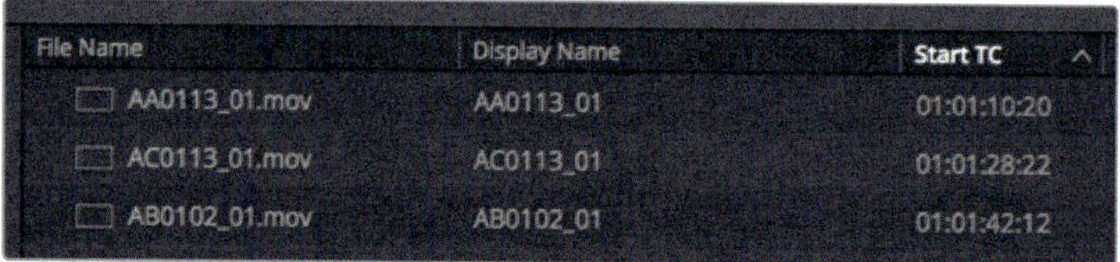

Unlike the File Name column that displays the name of the file on your hard drive, the Display Name column displays an editable name to be used inside of DaVinci Resolve.

Let's replace the clip name with a sentence from the interview.

4 Click the AA0113_01 clip to load it into the viewer, and play it to listen to what the interview subject says.

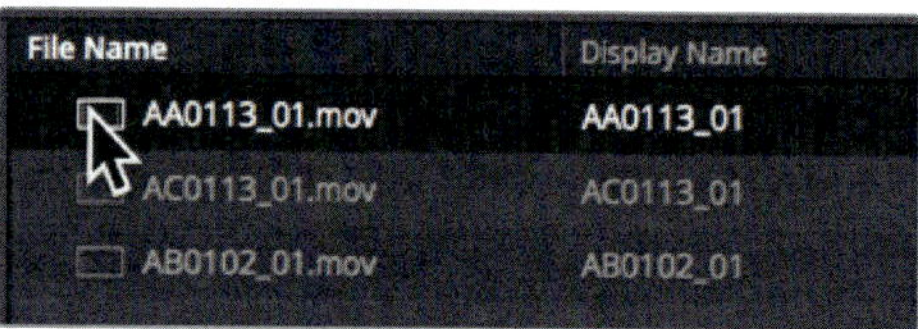

The important part about this clip is where he says, “It was really important to shoot the real world.”

5 Click in the Display Name field, and type **01_shoot the real world**.

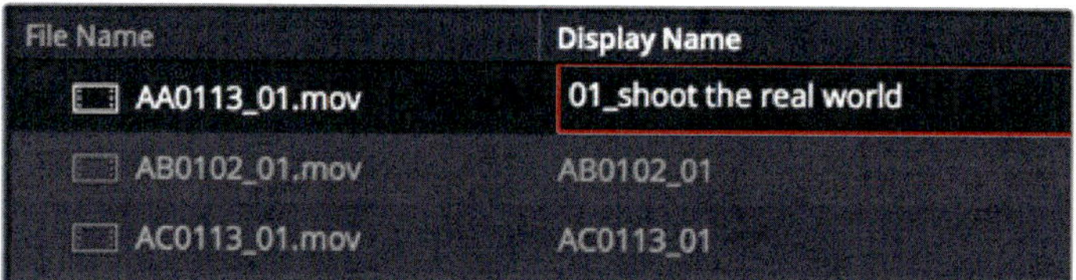

6 Double-click the AB0102_01 clip to load it into the viewer

You can play clips at normal speed but you can also play them in fast forward or reverse using the J and L keyboard short cuts.

7 Press the L key twice to play the clip at 2X normal speed. If you need to rewind, press the J key, and then press L again to play the clip at normal “sound speed.”

The important interview segment here is when he says, “If it was possible to shoot it, you wanted to go shoot it.”

8 Click the second clip’s Display Name, and type **02_if it was possible**.

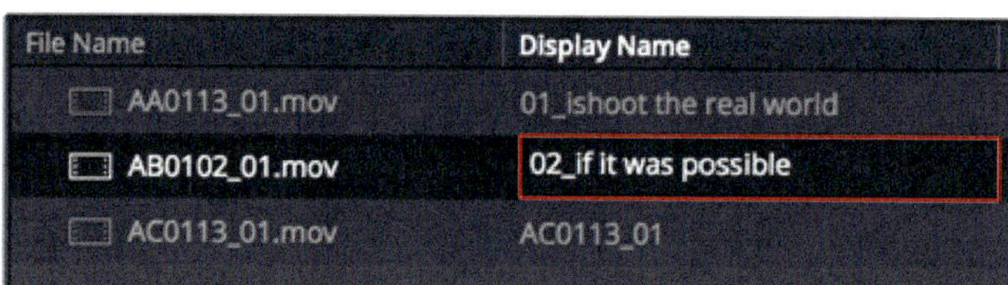

You have one more clip in the Interview bin to rename.

9 Double-click the AC0113_01 clip to load it into the viewer

10 Press the L key twice to quickly play through the clip.

The important interview segment here is when he says, “It was really important that every shot was original and every shot was real.”

11 Click the third clip’s Display Name, and type **03_every shot was original**.

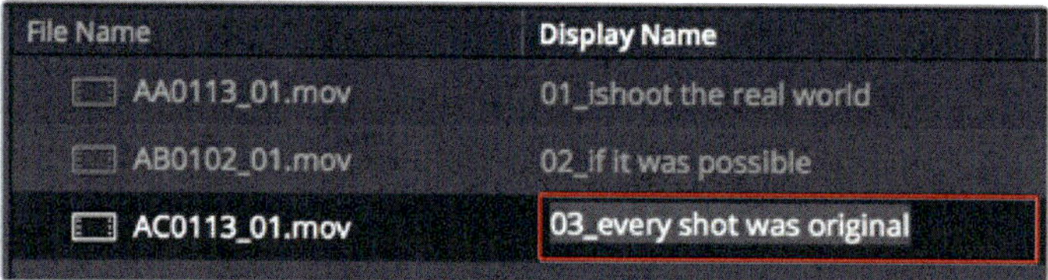

You should be aware of one last step when working with Display Names. Look above the Source Viewer. Although you gave the AC0113_01 clip a new Display Name, the viewer still displays the old filename. The last step you need to do is to make sure that DaVinci Resolve uses Display Names throughout the application.

12 Choose View > Show Display Names.

You have now set up all the bins, sorted the clips, given some of them more descriptive names, and chosen to display those names throughout the application. You are ready to begin editing your movie trailer. In the next lesson, you'll move to the Edit page where you begin assembling clips into a timeline.

Lesson 3

Assembling a Rough Cut

With your content imported and organized, you are ready to start editing. The first pass at creating this timeline is called the **rough cut**. The goal is to place clips in the rough order you'll want them in your final program. It is the equivalent of sketching a picture rather than precisely drawing one. In this lesson, you'll return to the Edit page and look at the various ways to begin a rough cut.

Time

This lesson takes approximately 50 minutes to complete.

Goals

Creating a timeline

Before you can start editing, you must create a timeline into which you place clips in the order you'll want to use them. DaVinci Resolve 14 projects can contain one or more edited timelines that you save into the Master bin. However, as you experiment with multiple versions of your timeline, it can be helpful to keep those timeline iterations together in their own bins. Doing so can make it easier to go back and compare them or locate the exact cut you want.

1 If DaVinci Resolve is closed, launch the application to get to the project manager.

2 Double-click the My New Project thumbnail that you created in Lesson 2.

When you left this project, you were on the Media page, but to begin assembling your rough cut, you'll need to move to the Edit page.

3 At the bottom of the screen, click the Edit page button.

You'll begin by creating a timeline and a bin that will hold your timeline.

4 In the Media Pool sidebar, select the Master bin, and choose File > New Bin to create a new bin within the Master bin.

5 Name the new bin **Rough Cuts**, and press Return or Enter.

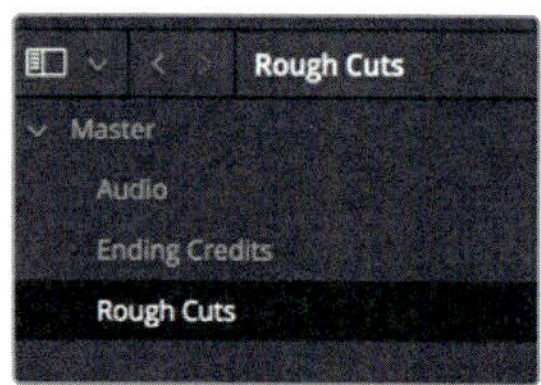

6 With the Rough Cuts bin selected, choose File > New Timeline, or press Cmd-N (Mac) or Ctrl-N (Windows).

When the New Timeline dialog appears, you can give the timeline a new name.

7 Name the timeline **First Rough Cut**, and click Create.

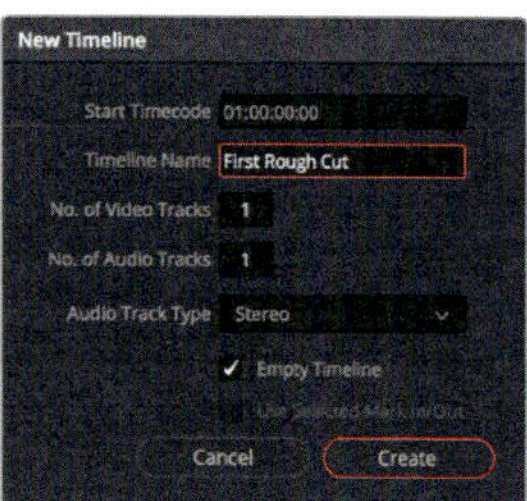

An empty timeline is added to the timeline editor and the name of the timeline is displayed above the timeline viewer, just as the name of the loaded clip is displayed above the source viewer. An icon for that timeline is added to the Rough Cuts bin.

You can make timelines even easier to locate by assigning a color to the Rough Cuts bin so it stands out from the other bins in your list.

8 In the Bin list, right-click the Rough Cuts bin.

9 In the contextual menu, choose Color Tag > Cyan to assign a cyan color to the Rough Cuts bin.

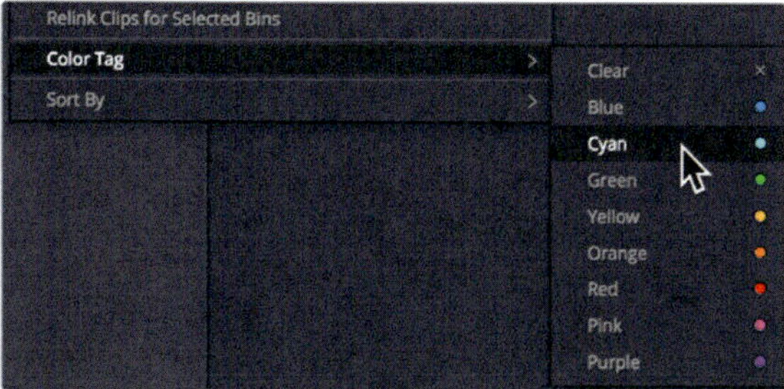

Yes, the bin does appear black, but that is only because it is still selected. When you switch to a different bin in the next exercise, you'll see the correct cyan color. So, let's move on and add clips to your timeline.

Making the first edit

A large part of the editing process is selecting your shots. You are not only selecting which clips to include in the project; you are also selecting the portion of the clip you wish to use. In Lesson 1 you learned how to identify a range using the in and out buttons, but now you'll learn a faster way.

1 Select the Interview Smart Bin, and click the thumbnail view button to display the clips as thumbnails.

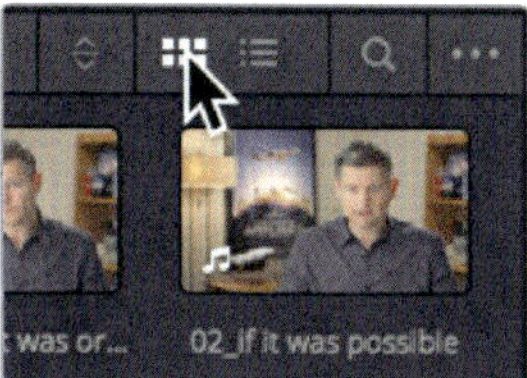

2 In the Smart Bin, double-click the **01_shoot the real world** clip to load it into the source viewer.

Because this clip has dialog, you can use the waveform overlay to quickly locate the portion of the clip you want to include in your timeline. The waveform overlay should still be displayed from Lesson 1.

3 If the waveform overlay is not displayed at the bottom of the source viewer, in the options menu above the source viewer, choose Show Zoomed Audio Waveform.

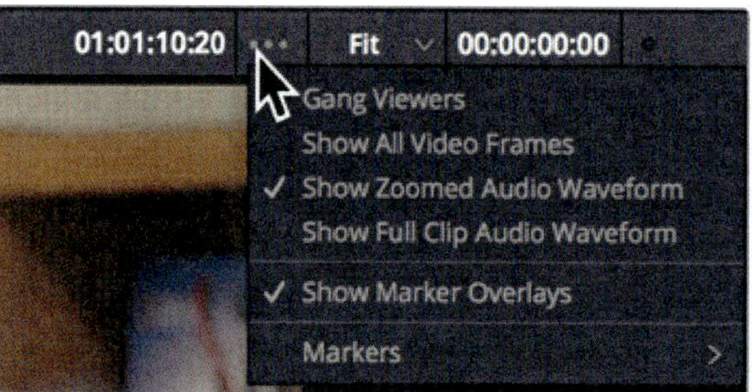

4 Move to the start of the clip and press the Spacebar again to play the clip, then press the Spacebar to stop playback when the interviewee says, "In this film." (Notice that the audio waveform begins at this point.)

5 Press the Left Arrow key to nudge the position of the playhead 10 frames before the word, "In." You can use the audio waveform to guide you.

Instead of clicking the button under the Source viewer to mark an In point, as you did in Lesson 1, you'll use a keyboard shortcut.

6 Press the I key to mark an in point.

7 After marking an in point, play the clip for roughly 10 more seconds until the man says, "Making all the things possible." Then press the Spacebar to stop playback.

8 Again, use the arrow keys to nudge the playhead right after the word "possible", and press the O key to mark an out point.

> **TIP** You can press Option-I (Mac) or Alt-I (Windows) to clear an in point; press Option-O (Mac) or Alt-O (Windows) to clear an out point; and press Option-X (Mac) or Alt-X (Windows) to clear both the in and out points.

It always a good idea to verify your selected range by playing from the In point to the out point.

9 Choose Playback > Play Around/To > Play In to Out, or press Option-/ (slash) in Mac, or Alt-/ (slash) in Windows.

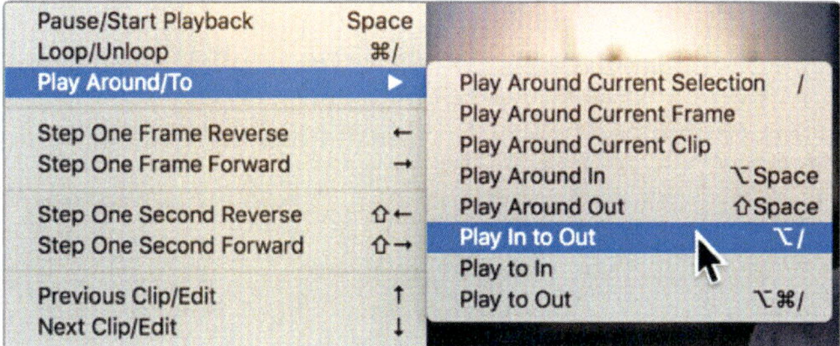

Playing the clip from the In point to the out point is a good way to see if your selected range is correct. If it isn't, feel free to go back and set the marks again.

When you are ready to add your first clip to the timeline, the easiest way to see all of the available editing functions is to use the Edit overlay.

10 Drag from the center of the source viewer into the timeline viewer but do not yet release the mouse button.

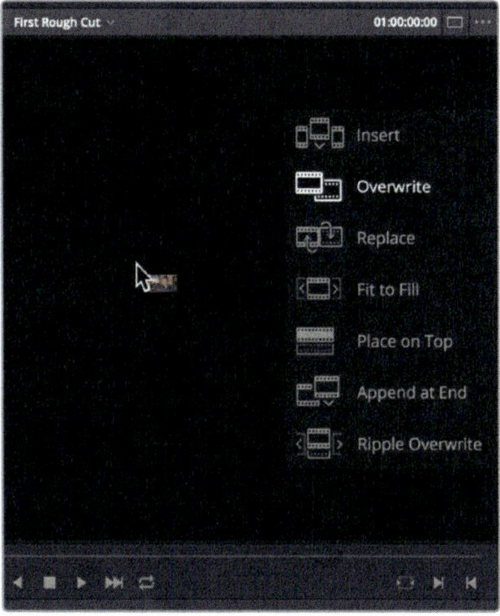

The Edit overlay appears in the Timeline viewer displaying seven edit functions from which to choose:

- **Overwrite** covers up a clip (or part of a clip) in your timeline using the new clip.
- **Insert** adds the clip to the timeline at the location of the playhead. Everything is split and moved down to make room for the new clip.
- **Replace** substitutes a clip in the timeline with a new clip without using any In or Out points.
- **Fit to Fill** changes the speed of a clip to fit a specified length.
- **Place on Top** positions one clip on top of another for blending or compositing.
- **Ripple Overwrite** replaces a clip of one length with a clip of another and ripples the timeline based on the difference in duration.
- **Append at End** adds new clips after the last clip in the timeline, regardless of where the playhead is located.

The overwrite edit is the default highlighted function which means that you can release the mouse button anywhere in the timeline viewer to perform that edit. Because you have nothing else in the timeline, overwrite is a fine choice for this edit.

11 Release the mouse button to perform an overwrite edit.

12 Drag the timeline playhead back to the start of the timeline, and press Spacebar to play a little bit of the first edit.

> TIP If at any time you make a mistake, DaVinci Resolve allows you to undo your most recent steps. You can choose Edit >Undo several times to undo multiple steps, or choose Edit > Undo List and select the last step you want performed from the list of previous steps.

Although you've only edited one clip into your timeline, this process of marking in and out points and choosing an editing function from the Edit overlay is a process that you will perform over and over when creating a rough cut. So, in the next exercise you'll repeat the process but learn to do it a bit more quickly, especially when adding longer clips.

Scrubbing with JKL keys

Using the JKL keys can help you quickly scan through clips. It's like using the fast forward and rewind buttons on your DVR so you can find exactly the right frame. The L key plays forward, the J key plays backward, and the K key stops playback. Tapping the L or J keys multiple times will speed up playback. Holding down K while tapping L or J will playback in slow motion. Let's give it a try.

1. In the Interview Smart Bin, double-click the 02_if it was possible clip to load the clip into the source viewer.

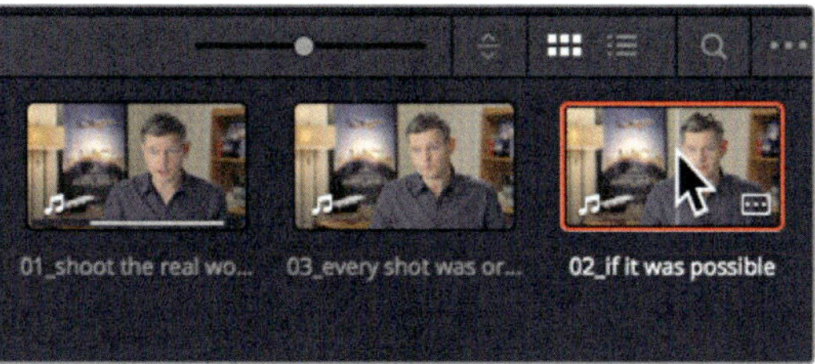

You are going to look for a point near the start of this clip where the man says, "If it was possible to shoot it, we wanted to go shoot it." The J and L keys can scrub over a clip at twice the speed or half the speed which makes it much faster to locate an exact spot rather than playing through the clip at normal speed. Let's start by placing your fingers correctly on the keyboard

2. If you are right handed, leave your right hand on the mouse and place your left ring finger on the J key, middle finger on K, and right index finger on L. If you are left handed place your index finger on J, middle finger on K, and ring finger on L.

TIP Using the user preferences, you can remap the keyboard to better configure the J-K-L key functionality for your comfort.

It can take bit of getting used to these hand positions, but let's start by just playing forward.

3. Tap the L key to play the clip at normal speed, and then press K to pause it when you hear the line you are looking for: "Ah, if it was possible to shoot it."

   More than likely, you overran the beginning of the line. That isn't a problem because you can play backward using the J key.

4. Press the J key to go back near the start of the sentence, and press K to pause playback when you arrive there. Don't forget to use the audio waveform as a guide to locate where the sentence starts.

   When you are near the start of the sentence, you can be more precise by playing at half speed using key combinations. The combination of the J and K keys for playing backward at half speed and the K and L keys to play forward at half speed can make it easier to pinpoint timeline locations.

5. Press J and K together to play backward at half speed, and then press K and L together to play forward at half speed until you are closer to the start of the sentence.

   You can even use these keyboard shortcuts to nudge one frame forward and one frame back rather than moving your hand down to press the arrow keys.

6 Hold down the K key and tap the J key to nudge one frame back, and then hold down the K key and tap the L key to move one frame forward. Keep repeating those keyboard shortcuts until you are exactly at the start of the sentence, being careful not to include the "ah" at the start of the sentence.

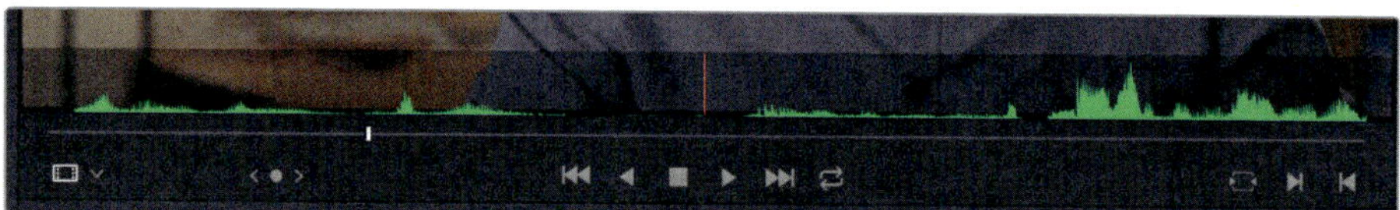

7 Press the I key to mark an In point.

Now let's use the same J-K-L scrubbing technique to locate an Out point. You'll look for the sentence about 10 seconds farther in, where he says, "Every shot was original and every shot was real."

8 Tap the L key twice to play the clip at double speed, then press K to pause when you hear the line you are looking for: "Every shot was original and every shot was real."

When you get close to the correct location, you'll want to switch to half speed playback and one-frame nudging.

9 Press J and K together and K and L together until you are closer to the end of the sentence.

10 Hold down the K key and tap the J or L key to move one frame at a time until you locate the exact end of the sentence, "Every shot was original and every shot was real."

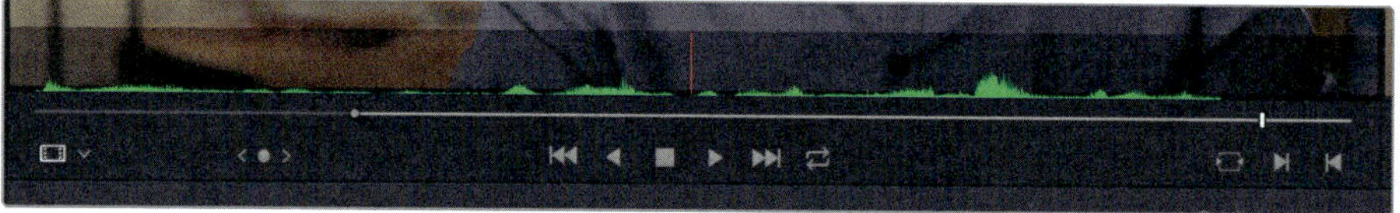

11 Press the O key to mark an out point.

12 Choose Playback > Play Around/To > Play In to Out, or press Option-/ (slash) in Mac, or Alt-/ (slash) in Windows, to review your range.

You can now edit this range into your timeline. Just as you more efficiently placed the in and out point using the JKL keys, you'll now speed up the actual edit using the toolbar.

13 Position the playhead at the end of the last clip in the timeline.

Although you can mark in and out points in the timeline, the timeline playhead will act as an implicit In point if neither mark exists.

The toolbar below the source and timeline viewers contains many of the most common functions that you'll use during editing. It even includes the three most commonly used editing functions: insert, overwrite, and replace.

14 In the toolbar, under the source and timeline viewers, click the overwrite button, or press F10 on the keyboard.

The second clip is added directly after the first clip in the timeline. Using the edit overlay or clicking one of the three edit buttons in the toolbar are both acceptable methods of making edits into your timeline. The method you choose is up to you and your style of editing. This lesson will continue to use these two methods, as well as introducing you to others to help you decide which is right for you.

Inserting clips into a timeline

Editing clips sequentially into the timeline can only take you so far. You'll eventually need to place new clips between two existing timeline clips. If you don't have an In point marked in the timeline, the playhead will be used instead. This means that any clips you insert will be inserted at the location of the playhead.

1 Position the playhead at the start of the timeline and play the two clips you have added so far.

   The two interview clips create a jarring cut because they are very similar looking. This jarring cut is called a **jump cut** because it appears as if the person abruptly jumped from one position to another. You can eliminate an unwanted jump cut by inserting a different clip between the two interview clips.

2 Position the playhead between the two interview clips in the timeline. As you drag, the playhead should snap to the cut point, ensuring that you are in the right spot.

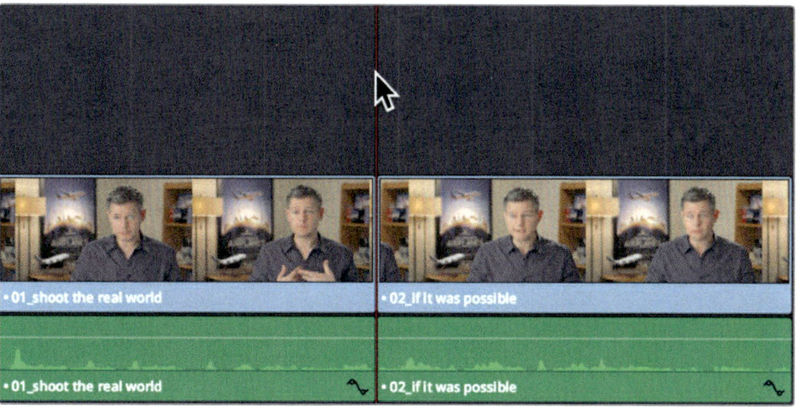

TIP The snapping behavior can be disabled by clicking the snapping button in the toolbar, or pressing N.

3 In the bin list, click the B-Roll Smart Bin, and double-click the 08_SOUTH_POLE_DC3 clip to load it into the source viewer.

4 Move to the start of the clip and press Spacebar to play this clip.

The clip of a plane taking off from the South Pole is a good one to add between these two interview clips because it goes with his comments about getting impossible shots.

5 From the beginning of the clip, drag the source viewer's jog bar until you see the tail of the plane emerge from behind the sign.

This spot in the clip gives you a little buffer room from the very first frame so it will work as your in point.

6 Press the I key to mark an in point in the source viewer.

This shot doesn't have a lot of action and it isn't very long, so you will use all of it.

As noted by the highlighted section of the jog bar, if no out point is set, DaVinci Resolve will use the remaining part of the clip to the end.

7 To make the edit, drag from the center of the source viewer into the timeline viewer. When the edit overlay appears, drag over the word "Insert" and release the mouse button.

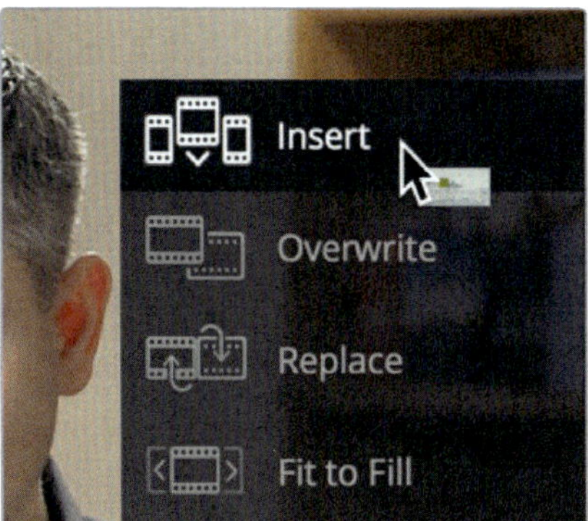

When you choose Insert in the edit overlay, the source clip is added to the timeline by pushing the second clip in the timeline to the right to make room. The overall duration of the timeline is extended by the duration of the added clip.

8 Drag the timeline playhead to the start of the timeline, and press Spacebar to view the new edit.

TIP Pressing the Home and End keys will move you to the start or end of the timeline, respectively. Some Mac keyboards do not have Home and End keys. When using one of those keyboards, press Fn-Left Arrow to move to the start of the timeline and Fn-Right Arrow to move to the end.

Splitting a clip with an insert

Not every insert you perform must be placed between two existing clips in the timeline. In some instances, you may want a new clip to split an existing clip in two. Let's learn how to deal with that situation.

1 Position the timeline playhead at the start of the second interview, if necessary.

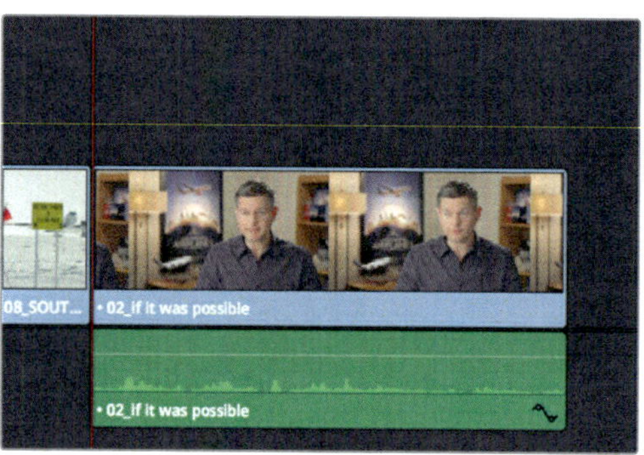

2 Press Spacebar to watch and listen to this clip.

This long interview clip would be better if you split it up into two smaller clips. Also, that last line, "It was really important, every shot was original and that everything was real," would be a lot more impactful if it stood alone at the end of this trailer.

3 Press the J+K and K+L keys to play slowly over the last two sentences and locate the space between the sentences "It didn't make the movie" and "It was really important."

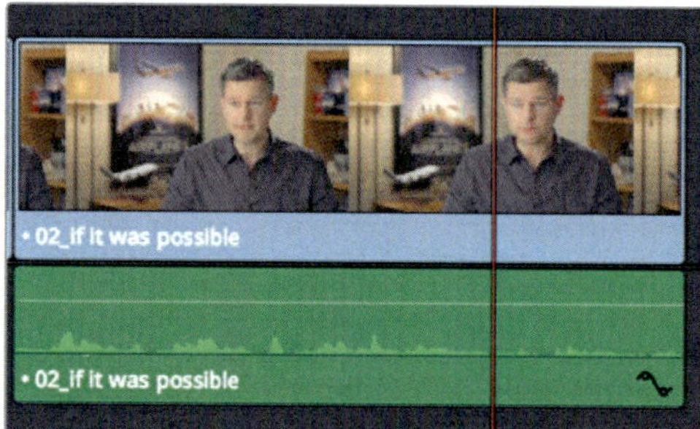

This brief gap between the two sentences is where you will insert a new clip. It can be easier to locate that small gap when you zoom into the timeline waveform.

4 In the toolbar, drag the zoom slider to the right to zoom in to the interview clip and see more detail in the audio waveform.

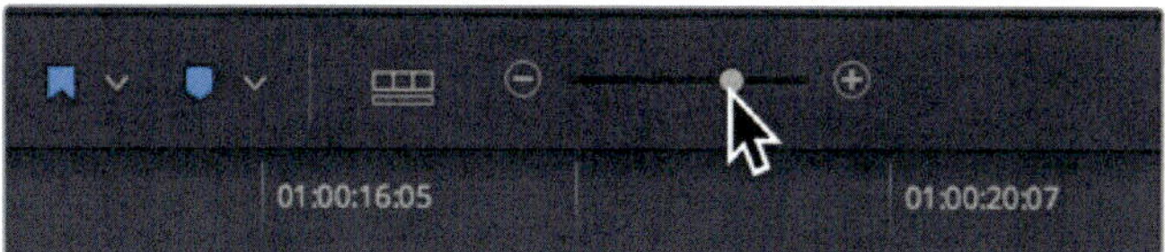

5 Press the J+K and K+L keys to refine your location between the sentences "It didn't make the movie" and "It was really important."

You can leave your playhead in this location and set in and out points on the source clip.

6 In the B-Roll Smart Bin double-click the 06_MILKYWAY clip to load it into the source viewer.

7 Move to the start of the clip and press Spacebar to watch this clip and stop playback somewhere in its middle.

For this edit, you will use the entire clip. If you do not enter any marks on a source clip, DaVinci Resolve uses the entire clip.

Instead of using the edit overlay to choose an edit, this time you'll perform the Insert edit using the toolbar edit button. The insert edit button provides the same editing function as choosing "Insert" in the overlay. The method you choose depends only on your preferred style of editing.

8 In the toolbar, click the insert button, or press the F9 key.

TIP To use the function keyboard shortcuts in Mac, go to System Preferences > Keyboard and enable "Use F1, F2 etc. keys as standard functions keys".

The insert edit splits the timeline clip at the playhead position and places the new clip at that point in the timeline. It will be easier to see if you zoom out the timeline.

9 Choose View > Zoom > Zoom to Fit, or press Shift-Z, to see the entire timeline.

10 Position the timeline playhead at the start of the second interview clip, and press Spacebar to review the last half of the timeline.

An insert edit is one of the most basic edits you will perform. However, it is important to understand that once you insert a new clip, all of the other clips in the timeline are pushed farther down the timeline to make room for the new clip. Your overall timeline duration is therefore increased.

Using timecode

Up until now, you've placed clips in your timeline using sound and picture as a rough guides. An alternative guide is to use timecode numbers. Using timecode lets you specify the exact duration of a clip before you edit it into the timeline.

1 In the B-Roll bin, double-click the 05_BAY_AREA_LIGHTS clip to load it into the source viewer.

You can type an offset number to move the source viewer's jog bar or the timeline playhead forward or backward by a specific number of seconds and frames. You first identify the direction in which you want to move the playhead. To move forward, you press the + (plus) key on the keypad.

2 Move to the start of the clip, and type **+100**. Press Return or Enter to move the playhead forward one second.

TIP When using a keyboard without a number pad, press Shift-= (equals sign) to enter a + (plus sign).

You'll use this frame for your in point.

3 Press I to mark an in point.

You can also identify a playhead location by exact seconds and frames if you have precise values you want to use. Let's move ahead four seconds and 10 frames in this clip.

4 Type **+410**, and then press Return or Enter.

The playhead jumps forward four seconds and 10 frames. You'll use this frame for your out point.

5 Press O to mark an out point.

Let's position the timeline playhead where you want to insert the clip.

6 In the timeline, position the playhead at end of the 06_MILKYWAY clip.

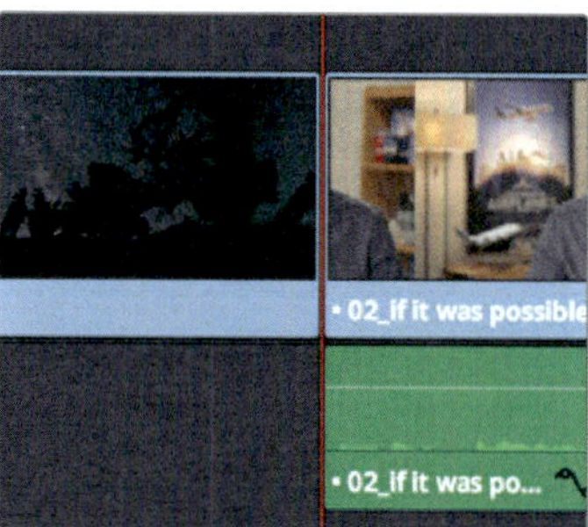

7 In the toolbar, click the insert button, or press the F9 key to Insert edit the clip.

8 Press Shift-Z to see the entire timeline. Then position the timeline playhead at the start of the second interview clip, and press Spacebar to review the last half of the timeline.

When you start to rely on keyboard shortcuts for a lot of your editing, it is important to be aware of which viewer is active. You want to make sure you are moving the playhead and marking in and out points in the correct viewer before you tap a keyboard key or you will be setting those points in the wrong places. The best way to know which viewer is active is by looking at the clip or timeline name above the viewer. The active viewer appears in red.

9 Press the Q key to activate the source viewer.

The Q key is a toggle, so it will switch between activating the timeline or the source viewer.

The insert and overwrite edit functions you have used so far are the two most commonly used edits throughout a program. After performing the edit a few times, you should more fully understand the usefulness of an insert edit. Let's make the case for the overwrite edit.

Overwriting video only

When creating a rough cut, you may want to replace the footage of your interview subject with B-roll, which is secondary footage that illustrates what he or she is discussing. You may choose to do this to refocus the audience's attention, hide problems with the existing video, or enhance the subject's commentary. This is called a cut-away because the visual temporarily cuts away from the main action. You can perform a cut-away using an overwrite edit.

1 In the timeline, position the playhead between the 08_SOUTH_POLE_DC3 clip and the second Interview clip, 02_if it was possible.

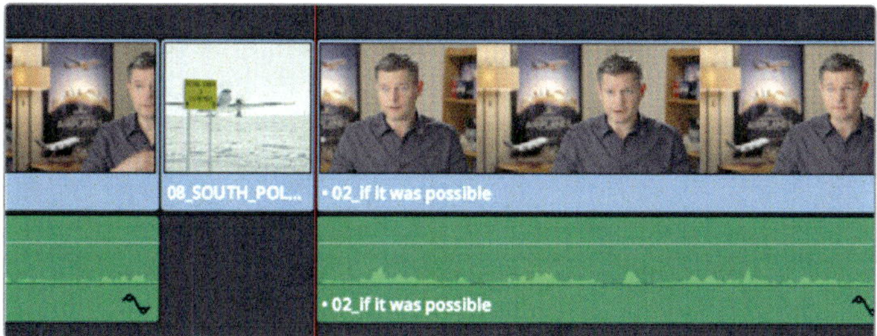

2 Press Spacebar to play the interview.

The first line is, "If it was possible to shoot it you wanted to go shoot it." Instead of having him onscreen during that time, it would look nicer to have one of those impressive, almost impossible shots. This time, you'll make an in and out point in the timeline because you know you would like the duration of the new shot to be the length of that first sentence.

3 In the timeline, position the playhead between the 08_SOUTH_POLE_DC3 clip and the second Interview clip, 02_if it was possible.

4 Press I to mark an in point.

5 Press Spacebar to play the interview and stop playback when the interviewee finishes the line, "If it was possible to shoot it, you wanted to go shoot it."

6 Press O to mark an out point

Now you'll look for a clip to edit into that location.

7 In the B-Roll bin, double-click 04_JET_LANDING_2 to load it into the source viewer.

8 Move to the start of the clip and press Spacebar to play the clip.

This is a well-framed shot of a large jet landing. Let's mark an in point just as you see the landing wheels at the top of the screen.

9 Drag the source viewer jog bar from the beginning of the clip until the wheels of the plane are visible in the frame.

You need to mark only an in point on this source clip because you already have the duration set in the timeline.

10 Press I to mark an in point.

You'll use an overwrite edit to replace the area marked in the timeline. But you don't want to overwrite the audio track of this clip. You want to continue to hear the voice of the interview subject while viewing the new 04_JET_LANDING_2 shot. So, you need a way to replace only the video content of the timeline while retaining the audio of the original clip.

The timeline has destination controls that display which video and audio tracks from the source clip are edited into the timeline. By default, the destination controls are enabled in both the V1 (Video 1) and A1 (Audio 1) source tracks, as indicated by the orange outline around the track number.

11 In the timeline track header, click the A1 destination control to disable it.

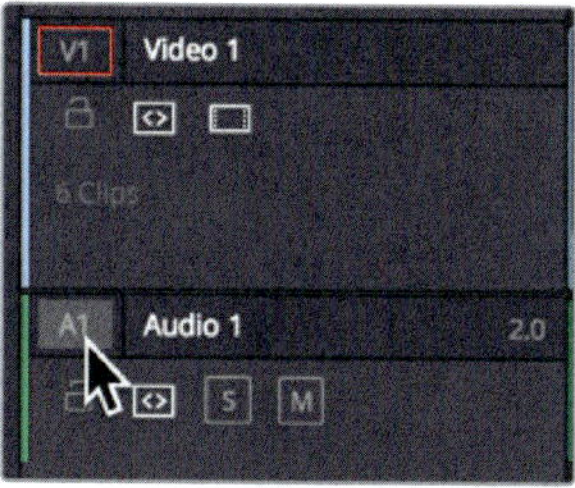

The destination control displays no outline when it is disabled. As a result, audio source material on A1 will not be edited into the timeline.

12 In the toolbar, click the overwrite button, or press F10.

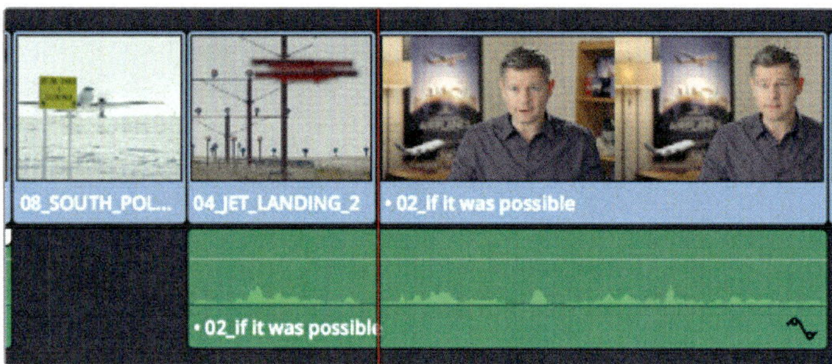

The overwrite edit caused a few seconds of the second interview clip to be covered by the new 04_JET_LANDING_2 clip. The audio from the interview, however, remained, and can be heard even when the cut-away is visible.

13 In the timeline, position the playhead between the 08_SOUTH_POLE_DC3 clip and the 04_JET_LANDING video clip.

14 Press Spacebar to play over the interview.

Making a V2 Cut-away

Using an overwrite edit is the most common method for making video-only cut-aways, but it's not the only method. DaVinci Resolve can layer video tracks on top of one another. This feature comes in handy when creating effects, but it can also be used to create a simple cut-away.

1 In the timeline, position the playhead between the 04_JET_LANDING_2 clip and the 02_if it was possible interview clip.

   You'll add the new clip right after the previous cut-away of the jet landing. Then you'll play the timeline to locate your out point.

2 Press I to mark an in point.

3 Press Spacebar to play the interview, and stop playback when he finishes the line, "If you couldn't get the shot or the weather wasn't right."

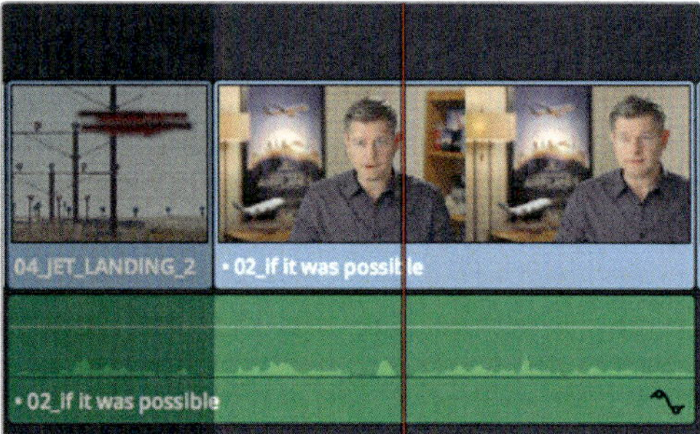

4 Press O to mark an out point after the word, "right."

Now you'll look for a clip to add as your second cut away.

5 In the B-Roll bin, double-click the 03_AERIAL_SFO clip to load it into the source viewer.

6 Move to the start of the clip and press Spacebar to play the clip.

This is a long zoom out of a plane on the tarmac. You could almost pick any In point, but let's find one where the plane is still fairly near in the shot.

7 From the start of the clip type **+2.** (plus sign, 2, period) to move the playhead forward two seconds.

At this point, you're far enough away from the beginning of the clip to have some flexibility if you later want to adjust the starting point, but it also still places the plane fairly near in the shot.

8 Press I to mark an in point.

For this cut-away, you'll perform an edit similar to an overwrite edit by using the Place on Top edit function. The source clip will cover up (overwrite) the middle of the interview clip using a second video track.

9 Drag the clip from the source viewer over the timeline viewer and move the mouse pointer over "Place on Top" to highlight it.

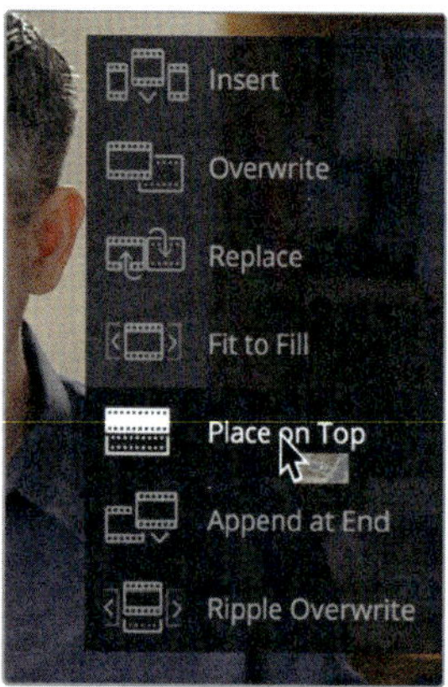

10 Release the mouse button to perform the place on top edit.

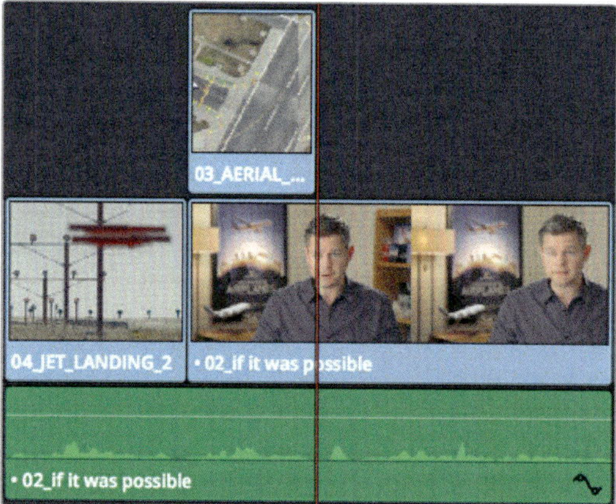

When layering in DaVinci Resolve, any video clip that appears above another clip in the timeline obscures the lower clips in the viewer. This makes it easy to use the place on top edit to create this cut-away.

11 In the timeline, drag the playhead to the start of the 04 JET LANDING 2 shot, and play the timeline to review the new cut-away you just added.

The place on top edit makes it easy to superimpose titles and other clips that you may want to composite over another clip. In this case, you used it to create a simple cut-away while retaining the flexibility to move it and expose any of the interview still remaining under it on video track 1.

Appending a clip to the end

Sometimes, you just want a simple solution to a simple problem. The Append at End editing function is just such a feature. It always places the selected clip at the end of the last clip in the timeline, regardless of any In or Out points or where your playhead is located.

1 Select the Ending Credits bin, and double-click 11_MOVIE_CREDITS to load it into the viewer.

2 Press Spacebar to play the short credits.

You don't need to mark in or out points because you will use the entire clip. Also, you do not have to position the timeline playhead because the append at end function always add the clip to the end of the timeline.

3 Drag from the center of the source viewer into the timeline viewer. In the edit overlay, place the mouse pointer over Append at End, and release the mouse button.

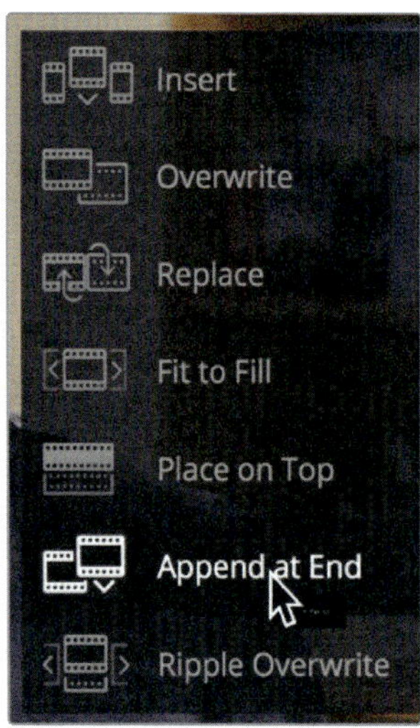

The movie credits are added to the end of the timeline.

4 In the timeline, position the playhead between the 05_BAY_AREA_LIGHTS shot and the last interview clip, and press Spacebar to view the added credits clip.

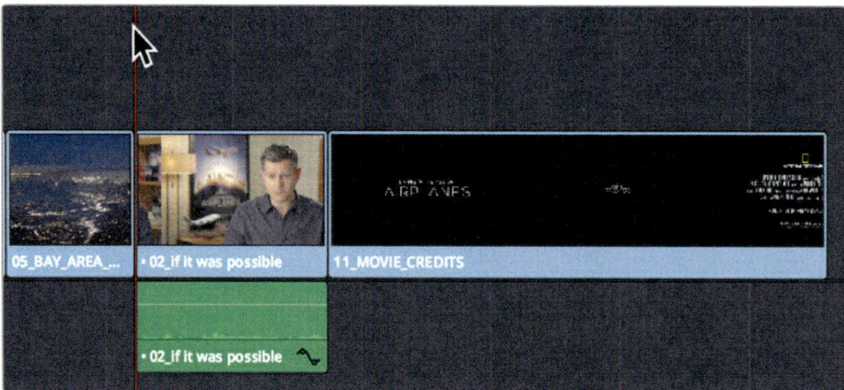

That is probably the easiest edit you are going to make. It seems trivial, but in a longer timeline append at end saves you a lot of scrolling back and forth.

Editing from a Bin

If the clips you are working with do not contain audio of any importance and you just want to add a few clips to the timeline as a montage, you can set marks using the thumbnails in the Media Pool instead of loading each clip into the source viewer. You'll do this for the final three clips you will add to your program.

1 At the top of the Media Pool, drag the Scale slider to the right to increase the size of the thumbnails.

When working with thumbnails in the bin, it can be helpful to give you a larger target.

2 Select the B-Roll Smart Bin, and hover your mouse pointer over 01 A380 TAXI.

3 Move your mouse back and forth over the thumbnail to preview the frames in the viewer.

This is a quick way to preview clips and set in and out points without explicitly loading the clip in the viewer.

4 Skim the mouse pointer all the way to the left of the thumbnail to the start of the clip, and press I to set an in point

A thin white line appears at the bottom of the thumbnail to indicate the presence of an in point.

5 Skim to the right on the thumbnail until the plane is pointing straight up to the top of the frame, and press O to set an out point.

The white line at the bottom of the thumbnail stops three-quarters of the way across the thumbnail to display a relative duration compared to the entire length of the clip. Let's set in and out points on two more clips using the same technique.

6 Hover your mouse pointer over 02_A380_TAKE OFF.

7 Skim the mouse pointer all the way to the left of the thumbnail to the start of the clip, and press I to set the in point

8 Skim to the right on the thumbnail until the you can no longer see the Lufthansa name on the plane, and press O to mark an out point.

9 Hover your mouse pointer over 09_HAWAIIAN_LANDING.

10 Mark an in point at the start of the clip, and an out point in the middle of the clip.

To place these clips into the timeline, you learned in Lesson 1 that you can drag from a bin directly to the timeline, but this method limits you to an overwrite edit. A more flexible method is to drag clips from the bin to the edit overlay in the timeline viewer.

11 Position the timeline playhead at the start of the 11_MOVIE_ CREDITS clip.

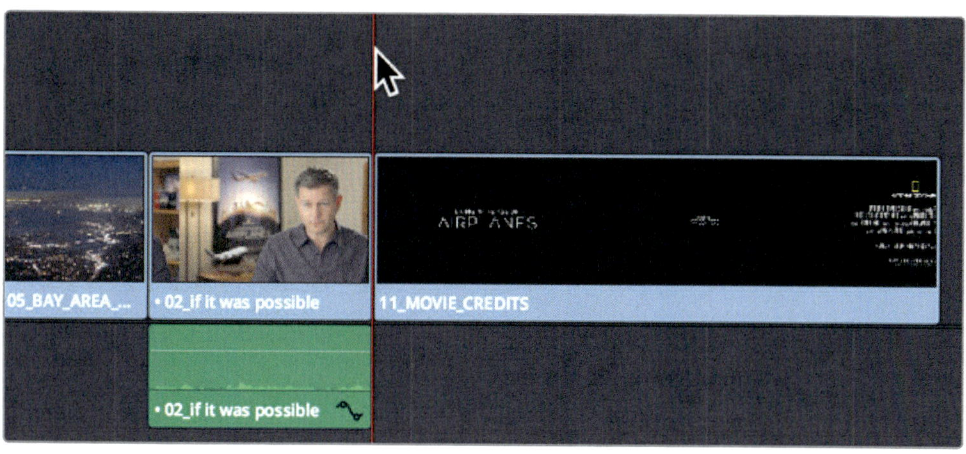

12 In the Media Pool, click the 01_A380_TAXI clip, and then Cmd-click (Mac) or Ctrl-click (Windows) the 02_A380_TAKE OFF and the 09_HAWAIIAN_LANDING clips.

The three clips are outlined in red to indicate that they are selected.

13 Drag the clips directly into the timeline viewer, and when the edit overlay appears, move the pointer over Insert and release the mouse button.

All three clips are inserted into the timeline based on the order they are displayed in the bin. Let's zoom the timeline to fit the window and play this program from the beginning.

14 Choose View > Zoom > Zoom to Fit, or press Shift-Z, to see the entire timeline in the window.

15 Go to the start of the timeline, and choose Workspace > Viewer Mode > Cinema Viewer, or press Cmd-F (Mac) or Ctrl-F (Windows), to play the entire program in full screen view.

Combining the technique of marking points in the bin with the Edit overlay options makes drag-and-drop style editing much more flexible and viable as an advanced editing technique.

Replacing a shot

When reviewing your rough cut, you'll often realize that your first choice of shots is not always the best. Editing comes down to a lot of trial and error. You'll try out an arrangement, pacing, and juxtaposition of clips, and then think of a better way. Doing so leads to replacing shots in your timeline with those that you think will work better. The Replace edit function is designed to make that process easy.

1 In the timeline, drag the playhead over the 04_JET_LANDING_2 clip.

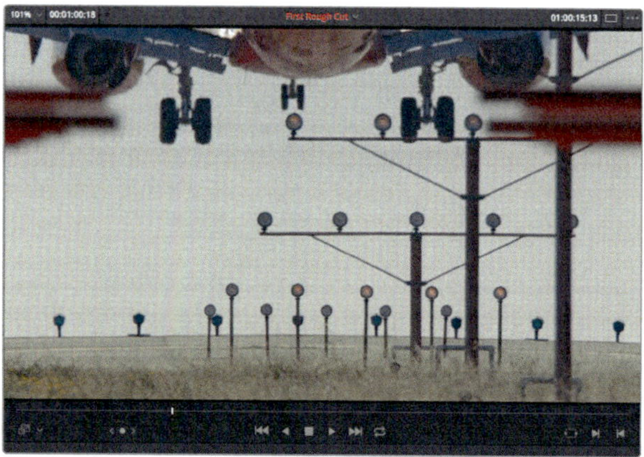

Although this is a nicely framed shot, you need a more spectacular shot at this point in the trailer. Luckily, you have a shot that like that in your B-Roll Smart Bin.

2 In the B-Roll Smart Bin, double-click the 10_MALDIVES clip, and press Spacebar to view it in the source viewer.

You would like to have the plane fly in from overhead about a third of the way into the clip. Without setting any marks, the Replace edit will swap one shot for another by aligning the current frame in the source viewer with the current frame in the timeline.

3 In the timeline, position the playhead over the center of 04_JET_LANDING_2 clip.

Zooming into the clip in the timeline will make it easier to position the playhead more precisely.

4 In the toolbar, drag the zoom slider to the right to zoom in on the clip, or press Cmd-= (equals sign) in Mac, or Ctrl-= (equals sign) in Windows.

TIP Zooming is always centered on the current position of the playhead, even if the playhead is offscreen.

5 Move the playhead roughly one-third of the way into the 04_JET_LANDING_2 clip. Around the time when you first see the jet engines enter the frame.

This is the time location when you want the plane to come into the frame on the 10_MALDIVES shot.

6 In the source viewer, drag the jog bar until you first see the plane coming into frame.

With the playheads aligned, you can use the replace edit button in the toolbar.

7 In the toolbar, click the replace edit button, or press F11.

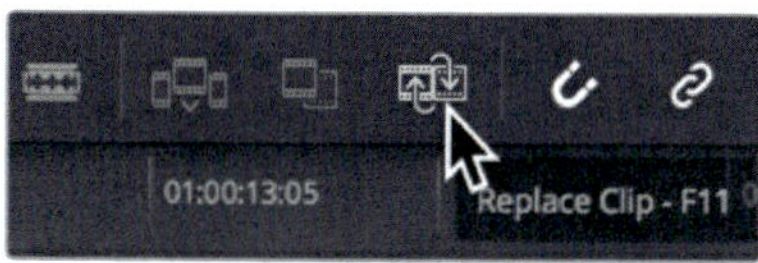

The replace edit figures out the in and out points you need based on the timeline clip you are replacing.

8 Choose View > Zoom > Zoom to Fit, or press Shift-Z, to see the entire timeline.

9 Drag the playhead to the start of the timeline and play the program you have created so far.

Whenever you have a spare moment while you are editing, it is always a smart idea to take a step back and watch the entire program. Sometimes you forget what the big picture is like because you spend so much time fiddling with specific clips.

Lesson 4

Moving Clips in the Timeline

The timeline is much more than just a view of the edits you make. Once you start putting a project together, the timeline quickly becomes the hub of all activity. It is the place where you will move segments around, split clips in half, and delete segments altogether. Knowing how to operate in the timeline will improve your editing skills.

Time

This lesson takes approximately 30 minutes to complete.

Goals

Importing projects and relinking media

When someone sends you a project from another computer or you want to move a project from one computer to another, you need to import the project file and relink the media.

For instance, if you were moving from a desktop computer to a portable laptop to edit on-the-go, you would export the project from the desktop computer and import it on the laptop.

In this lesson, you'll import a project that contains several pre-made timelines that are slight variations on the timeline you created in the previous lesson. The timeline for this lesson is set up so you can learn how to move, delete, and split clips.

1 Open DaVinci Resolve 14, if necessary; and choose File > Project Manager.

2 In the project manager, right-click in an empty area, and from the contextual menu, choose Import.

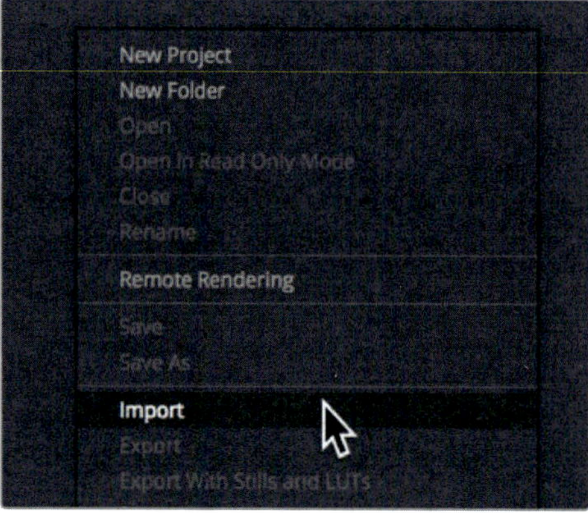

NOTE To export a project, you must right-click the project in the project manager, and choose Export.

3 In the open dialog, select R14 lessons folder > Lesson 04 > Age of Airplanes.drp.
The project is imported but the media may not yet be linked to the clips. The project contains only the metadata for clips and timelines. It has no media associated with it. Links between the clip and timeline metadata and the media can break when media is copied or moved from one computer to another, or when folder names are changed. When necessary, you can easily relink the media to all of the clips and timelines in a project.

4 Double-click the Age of Airplanes project to open it.

5 In the Media Pool sidebar, select the Master bin, and double-click any clip to open it in the source viewer.

These clips are offline. You need to guide DaVinci Resolve to the locations of the media so it can relink the media and clips to the project. You can do so just by selecting the bins that contain the offline clips, which in this case, is every bin in the project.

Because the Master bin contains every bin and all the clips in the project, you can just relink from the Master bin.

6 Ctrl-click (Mac) or right-click (Windows) the Master bin, and in the contextual menu, choose “Relink Clips for Selected Bins”.

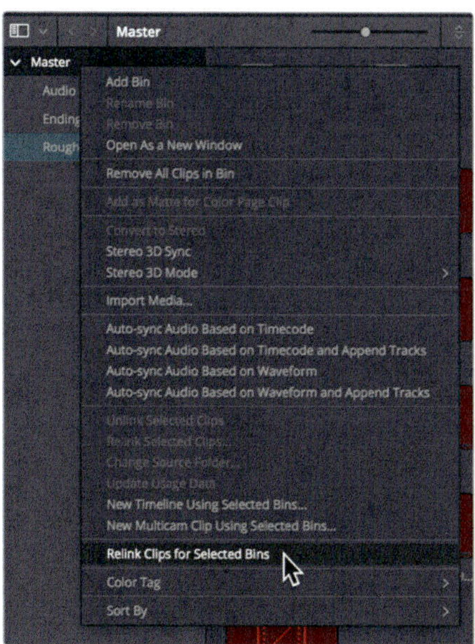

7 In the select source folder dialog, select the Documents > R14 lessons folder. Click OK.

Relinking the Master bin automatically relinks all of the clips in all of the bins, as well as the timelines in those bins.

8 In the Rough Cuts bin, double-click the Cut Copy Paste timeline to load it into the timeline viewer.

9 Play the timeline to review the cuts that you will work with in this lesson.

This timeline is a slimmed down version of the one you created in Lesson 3. It does, however, include a new music track that, as you will see, can cause some hurdles when you start cut copying and pasting clips in the timeline.

Color coding clips

Additional clips within a timeline can slow you down when trying to locate a single shot. You can make it easier to find that shot by assigning a color to different clip groups. The assigned color will appear in the timeline whenever a clip from that group is used.

1 In the Bin list, select the Interview Smart Bin.

2 Click any clip in the Media Pool, and choose Edit > Select All, or press Cmd-A (Mac) or Ctrl-A (Windows), to select all of the clips in the bin.

3 Right-click any one of the selected clips, and in the menu, choose Clip Color > Navy.

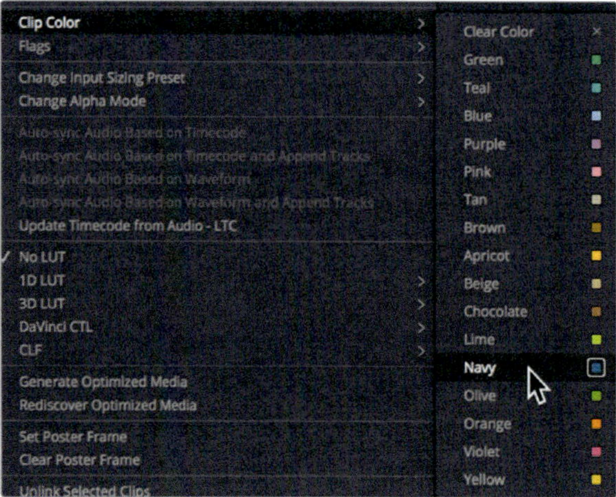

Now let's color code all the B-Roll clips.

4 In the Bin list, select the B-Roll Smart Bin.

5 Click any clip in the Media Pool, and choose Edit > Select All, or press Cmd-A (Mac) or Ctrl-A (Windows), to select all the clips in that bin.

6 Right-click any one of the selected clips, and in the contextual menu, choose Clip Color > Brown.

All the timeline clips from the Interview and the B-Roll bins are now color coded to navy blue or brown. As you make changes in the timeline, the differentiation in color will make it easier for you to locate these categories of clips.

Deleting clips without leaving a gap

Knowing when and how clips should be deleted is almost as important as knowing where to place clips in the timeline. You can delete clips and leave a gap, as you learned in Lesson 1, or you can delete clips and automatically close the gap. In your timeline, if you were trying to line up your picture to music, you might seem to have too many clips. You'll need to delete at least one and possibly more.

1 In the timeline, position the playhead at the end of the middle interview clip.

2 Press Spacebar to play until the end of the movie.

Because the last three clips are so similar, one of these three can be deleted. The most likely culprit is the slow-moving Hawaiian Landing clip. In Lesson 1, you learned how to remove a selected clip from the timeline: press the Delete key and leave a gap in the timeline. But what if you don't want to leave a gap? In this situation, you want to remove the clip but have all the clips to the right move left to close up the gap. This type of deletion is often called a **ripple delete** because the change ripples through the rest of the timeline.

3 In the timeline, select the 09_HAWAIIAN_LANDING clip.

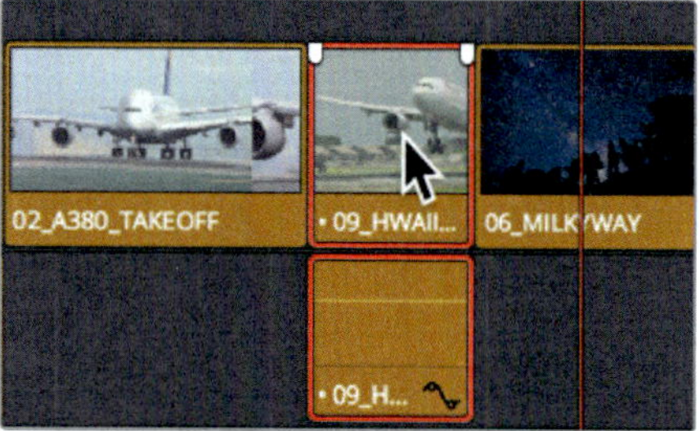

Because the audio and video tracks are part of the same clip, they are both selected in the timeline.

4 Choose Edit > Ripple Delete, or press Shift-Delete or Shift-Backspace.

Both the audio and video tracks of the clip are removed, or extracted, from the timeline, but not deleted from the bin or your hard drive. In addition, the gap is closed up because all of the clips positioned after the 09_HAWAIIAN_LANDING clip shift to the left.

Deleting video or audio content separately

As you played over those three jetliner clips, you heard some ambient sound of the crew talking. What if you wanted to delete only the audio contents from clips in the timeline and leave their video contents in place?

1 In the timeline, position the playhead over the middle interview shot.

2 Play the timeline and listen carefully to the audio during the 01_A380_TAXI shot.

   You can hear directions being called out, along with some undesirable helicopter noise. You'll want to retain the video of this clip in the timeline, but remove its audio. In the previous exercise, you found that when you select a clip, DaVinci Resolve automatically selects both its video and audio tracks. To remove the crew noise but leave the clip's video track in the timeline, you'll need to unlink the two tracks so you can select them separately.

3 In the toolbar, click the linked selection button, or press Cmd-Shift-L (Mac) or Ctrl-Shift-L (Windows).

   The linked selection is no longer highlighted, indicating that the function is disabled. With the association between video and sync audio track temporarily disabled, you can move and delete them independently.

4 In the timeline, select the audio track of the 01_A380_TAXI clip.

   Only the audio track is outlined in red.

5 Press Delete or Backspace to remove the audio track.

   The audio is now removed from the timeline, but the video track remains.

6 In the toolbar, click the linked selection button, or press Cmd-Shift-L (Mac) or Ctrl-Shift-L (Windows), to re-enable the link behavior.

   It is usually a good idea to keep the linked selection button enabled so that clips that have audio and video in sync are not accidentally separated.

Deleting a range within a clip

Often with interview clips, or "talking heads," you won't want to delete an entire clip as often as you'll want to remove repetitive sentences or fumbled words. For this, you need to delete just a portion of audio somewhere within the clip.

1 In the timeline, position the playhead at the start of the 10_MALDIVES clip.

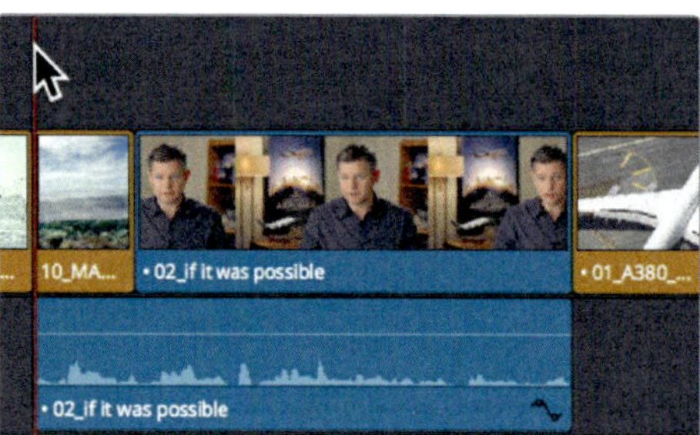

2 Play the timeline and listen carefully to the audio during the interview shot.

You can tighten up this interview by removing the redundant statement, "Stay at that location longer." When deleting small ranges within a clip, it is best to zoom into the timeline area you will be working on.

3 In the toolbar, drag the zoom slider to the right until the 02_if it was possible interview clip almost fills the entire timeline window.

Zooming in will help you identify the sentence by giving you a more detailed view of the audio waveform. Using the waveform as a guide, you'll mark the range you want to delete using in and out points.

4 Use your J-K-L keys to slowly play back and forth over the interview clip until you locate the start of the line, "Stay at that location longer."

5 When the playhead is located before the word "stay," press I to mark an in point.

This is the start of the range you will delete. Now let's look for the out point.

6 Press your J-K-L keys to slowly play back and forth over the interview clip until you locate the end of the line, "Stay at that location longer."

7 When the playhead is located after the word "longer," press O to mark an out point.

With the range identified, you can delete it, or more precisely, ripple delete it so you don't leave a gap in the timeline. However, if you deleted the range right now, you'd run into a problem. To see the problem, let's delete the range, observe the problem, and then figure out how to solve it.

8 Press Shift-Delete/ Backspace to remove the range.

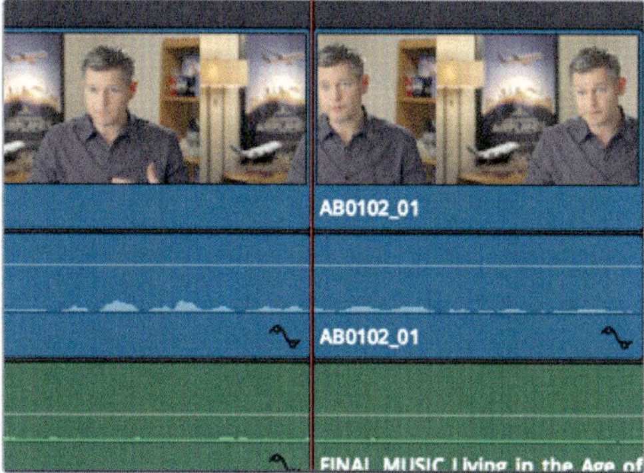

The range between the in and out points is deleted on the interview, but is also deleted on the music track. You need to not only identify the range, but also identify the individual tracks in which you want the deletion to occur. DaVinci Resolve includes auto select buttons on each track to help with this situation.

9 Choose Edit > Undo, or press Cmd-Z (Mac) or Ctrl-Z (Windows), to undo the previous operation.

10 In the timeline track header, on the Audio 2 track, click the auto select button to disable that track.

The auto select buttons are incredibly important buttons to be aware of because they perform so many different functions when editing in the timeline. For your current purposes, you only need to be aware that these buttons will identify the tracks to operate on when they are enabled and will exclude the tracks on which they're disabled.

11 Press Shift-Delete/ Backspace to remove the range on tracks V1 and A1.

Let's review the deletion.

12 Move the playhead to the start of the 10_MALDIVES shot, and press the Spacebar to play over the deleted region.

Don't be too concerned with the jump cut you just created, you'll fix that later. Often when editing, you'll create audio that sounds exactly as you want it to sound, and then later fix jump cuts when all of the audio is in place.

13 In the timeline track header, on the A2 track, click the auto select button to enable it so you don't forget to do so later.

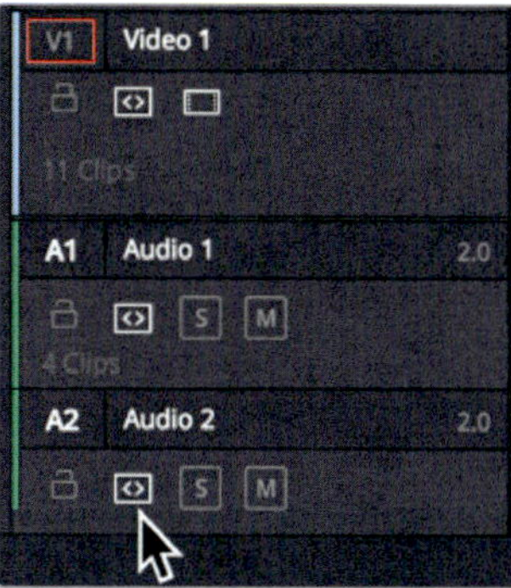

Although creating a backup isn't all that critical in this training project, when you start making dramatic changes to any rough cut, such as deleting lots of clips, it's a good idea to duplicate your timeline so you can return to a previous version.

Splitting clips

In some cases, you might want to create space between sentences to improve pacing. In this example, instead of identifying the range you want to delete, you'll identify a frame where you want to split a sentence. The easiest way to do so is using the razor edit mode

1 At the bottom of the timeline, drag the scroll bar to center the cut between the 10_MALDIVES and the 02_if it was possible interview clips in the window.

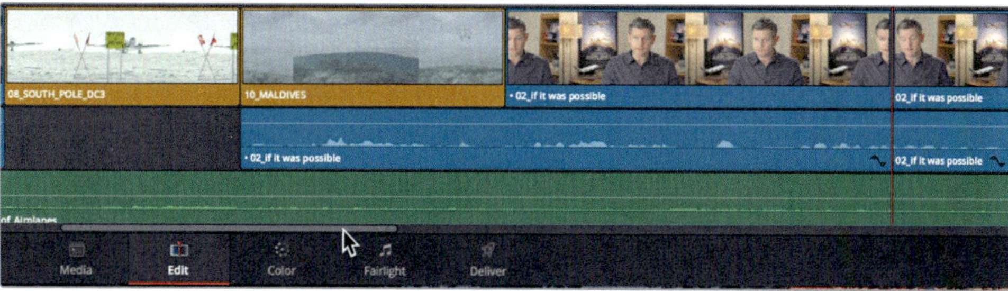

2 Drag the timeline playhead to the start of the 10_Maldives clip.

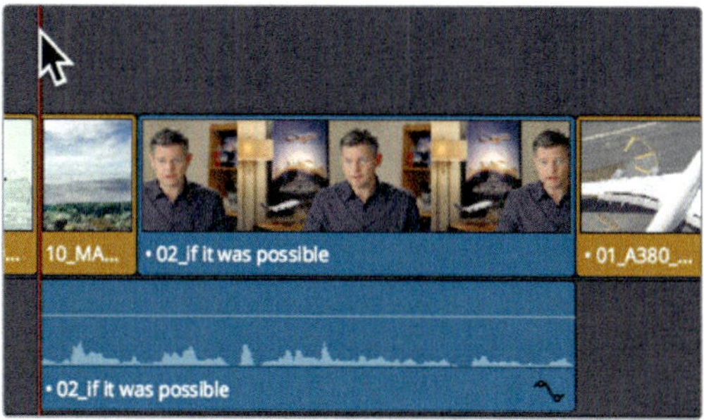

3 Press the Spacebar to play the timeline, and stop playback when the audio from the interview says, "If the weather wasn't right."

Creating some separation from the first sentence—"If it was possible to shoot it, you wanted to go shoot it"—seems like the right thing to do because that is a complete thought that you want the audience to soak in. You'll need to split that sentence from the rest of the clip.

4 Position the playhead on the cut between the 10_MALDIVES and the 02_if it was possible interview clips because that is where the sentence ends.

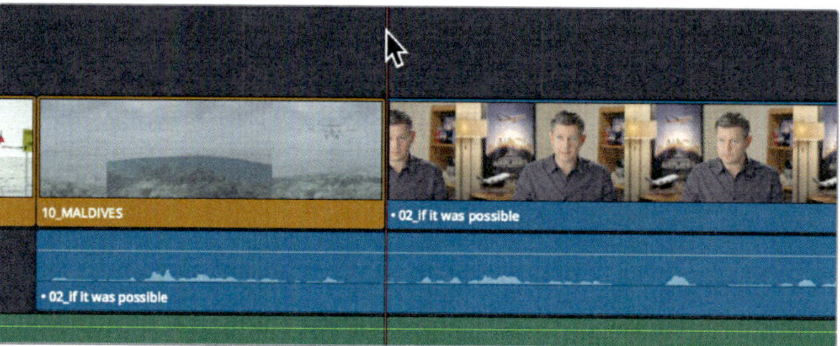

The razor edit mode is a tool that can split clips into sections. You can use it here to split the audio track.

5 In the toolbar, click the razor edit mode tool, or press B.

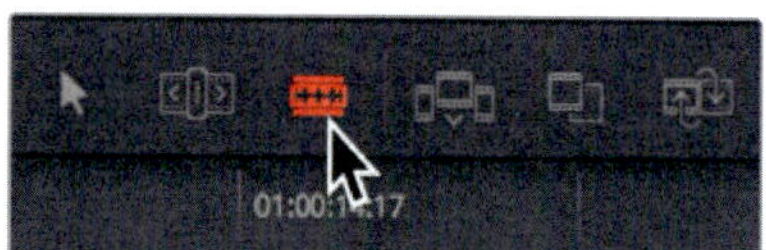

With the razor edit mode tool selected, anywhere you click in the timeline will divide that clip.

6 Position the left edge of the razor blade pointer directly over the playhead on the audio track, and click to split the clip.

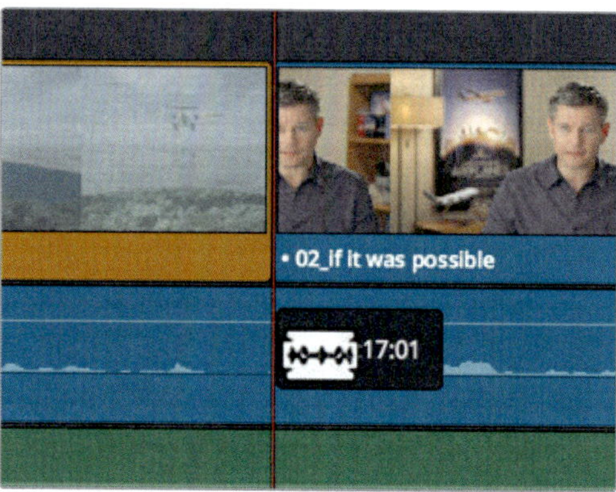

The clip is divided based on where the left edge of the razor blade pointer was located when you clicked. You now have a separated clip that you can reposition.

TIP To split every clip under the playhead that has a track with auto select enabled, choose Timeline > Split Clip, or press Cmd-\ (backslash) in Mac, or Ctrl-\ (backslash) in Windows.

7 Click the selection mode button, or press the A key.

With the clip successfully split, you can now separate it from the interview that comes directly after it.

8 In the timeline, select the split audio track.

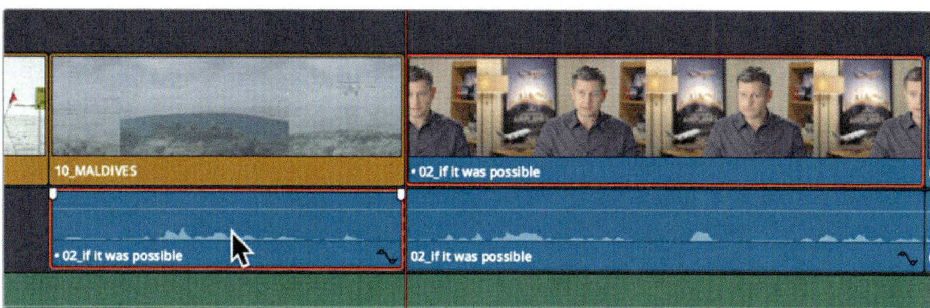

Selecting the audio track still selects the video clip even though it is split. The relationship between the two clips remains intact. To move only the audio track, you need to disable the linking between the two tracks just as you did when you deleted the audio on the A380_TAXI clip.

9 In the toolbar, click the linked selection button, or press Cmd-Shift-L (Mac) or Ctrl-Shift-L (Windows).

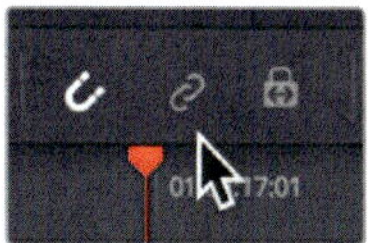

The linked selection button is no longer highlighted, indicating that the function is disabled. With the association between video and sync audio tracks temporarily disabled, you can move and delete them independently.

TIP You can temporarily disable the linked selection button without going to the toolbar by Option-clicking (Mac) or Alt-clicking (Windows) a clip.

10 Click anywhere in the gray background of the timeline to deselect any selected clips, and click the audio you want to move.

To create one second of silence, you can use the same technique that you used when you entered timecode values to move the source viewer's playhead.

11 To move the clip one second earlier in the timeline, type **-1.** (minus, 1, period), and press Enter or Return.

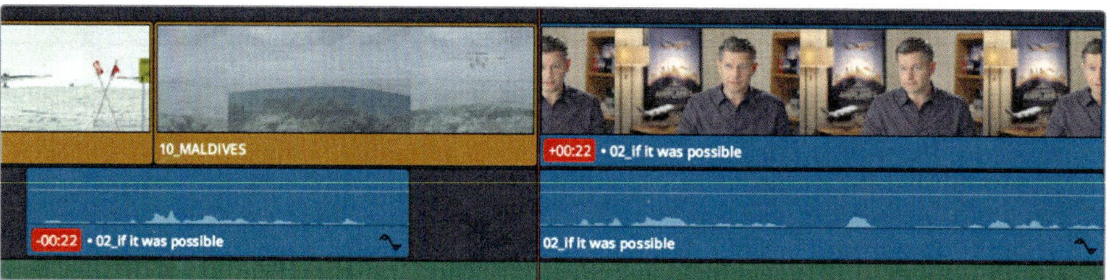

TIP You can nudge the selected clip one frame at a time by pressing , (comma) to move one frame left or . (period) to move one frame right. Pressing Shift-, (comma) or Shift-. (period) nudges the playhead five frames to the left or right, respectively.

The audio moves one second toward the beginning of the timeline. The red out-of-sync badges that appear on the video and audio tracks indicate that these two segments no longer have the same sync relationship that they have in the master source clip. Although those out-of-sync badges can cause anxiety in even the savviest of editors, you needn't worry in this situation because you purposefully moved the audio and broke that sync relationship.

TIP To remove or create a sync relationship between two or more clips, you can right-click them in the timeline, and choose Link Clips.

12 In the toolbar, click the linked selection button, or press Cmd-Shift-L (Mac) or Ctrl-Shift-L (Windows), to enable it again.

13 Choose View > Zoom > Zoom to Fit, or press Shift-Z, to see the entire timeline.

TIP Shift-Z acts as a zoom toggle. Press the key combination once to zoom all the way out to see the entire timeline. Press it again to return to the previous zoom level.

14 Position the playhead at the end of the first interview clip and press the Spacebar to play the timeline until you see the 01_A380_TAXI clip in the viewer.

The result of splitting the clip may or may not be perfect. It all depends on how precisely you selected the cut points. It will become easier to make precise adjustments when you learn more trimming options later in Lesson 5.

Cutting and pasting clips

Dragging or entering time values are fine ways to move clips when no other clips are in the way. However, when other clips are in the timeline, you'll not only need to decide what the clip you're moving will do, but also what the other clips in the timeline will do because of the move. Unless you want to overwrite the other clips as you drag or move a clip, you'll need to ripple the timeline.

1 Position the playhead at the start of the 01_A380_TAXI clip in the timeline.

2 Press the Spacebar to play to the end of the timeline

DaVinci Resolve has some useful cut/copy/paste operations that work on clips you've selected in the timeline. They can be very helpful when moving clips from one end of the timeline to the other, such as the two A380 jets clips you have at the end of this timeline.

3 In the timeline, click the 01_A380_TAXi clip to select it, and then Cmd-click (Mac) or Ctrl-click (Windows) the 02_A380_TAKE_OFF clip to select both clips.

To remove the clips from their current locations with the intent to relocate them, you can cut them from the timeline.

4 Choose Edit > Ripple Cut, or press Cmd-Shift-X (Mac) or Ctrl-Shift-X (Windows).

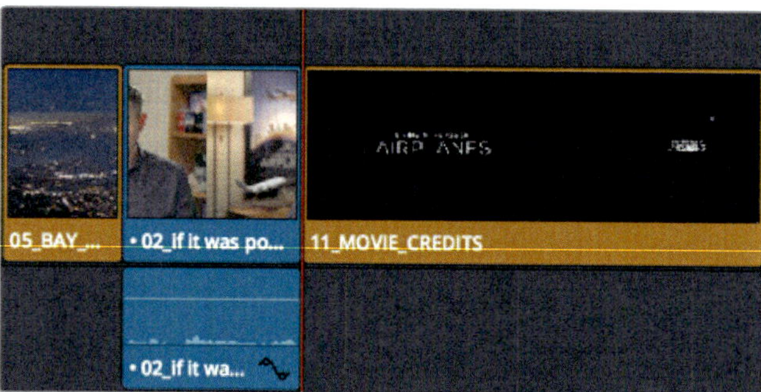

The ripple cut extracts the clips from the timeline, and closes up the gap. Unlike a ripple delete, the clips are kept in memory ready to be pasted into another location.

5 In the timeline, position the playhead at the end of the first interview clip.

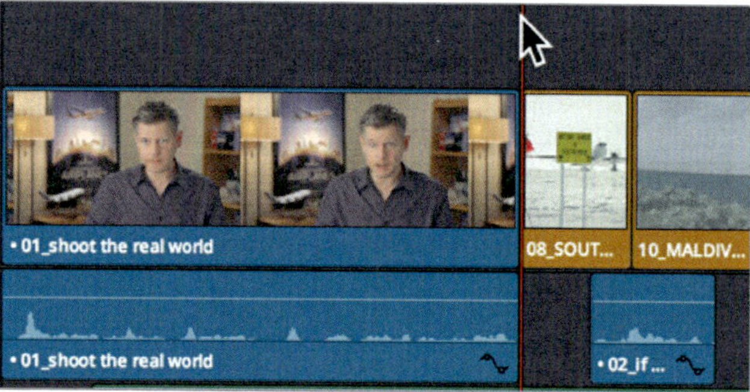

6 In the timeline header, click the Audio 2 auto select button because you want to paste onto only the Video and Audio 1 tracks.

7 Choose Edit > Paste Insert, or press Cmd-Shift-V (Mac) or Ctrl-Shift-V (Windows).

The clips are inserted at the position of the playhead. The paste insert function is identical to inserting clips using the Insert button in the toolbar or the Edit overlays.

Swapping clip positions

If you don't have to move a clip over more than two or three clips, you can just swap its position.

1 In the timeline, click the 05_BAY_AREA_LIGHTS clip to select it.

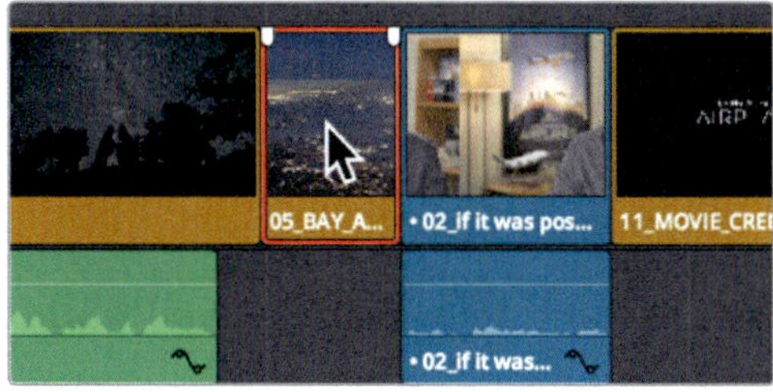

2 Drag the clip to the right. Once you begin dragging, hold down Cmd-Shift (Mac) or Ctrl-Shift (Windows) until you are at the start of the 11_MOVIE_CREDITS clip.

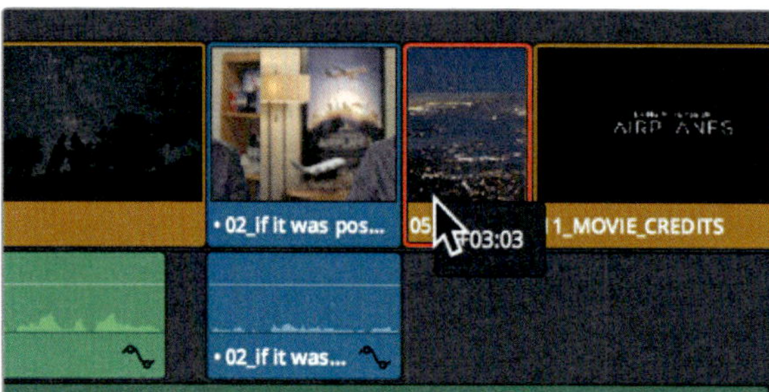

The 05_BAY_AREA_LIGHTS clip and the last interview clip swap positions.

Because you've made a lot of large changes, this seems like a good time to sit back and watch your entire movie.

3 Drag the playhead to the start of the timeline, and press Cmd-F (Mac) or Ctrl-F (Windows) to play the timeline and review your changes.

This cut looks very nice, but it still is not fully aligned to the music and some of the shots are too long. These timing problems cannot be addressed by merely moving clips around. You need to extend and shorten clips using trim techniques you will learn in the next lesson.

Lesson 5

Refining a Timeline

The real artistry of editing is achieved in the pacing of the clips you assembled in the timeline. Pacing is refined by shortening or extending clips by a few seconds or even just a few frames to get the perfect timing between the two.

These adjustments are achieved through **trimming**. DaVinci Resolve 14 includes precision trimming tools that allow you to try out multiple edit choices and review them quickly to address your creative concepts. In this lesson, you'll learn how multiple trimming methods can help you pace your program perfectly.

Time

This lesson takes approximately 50 minutes to complete.

Goals

Customizing the layout for trimming

DaVinci Resolve lets you customize and save your user interface layouts to make various workflows easier. For example, the current interface layout has somewhat small dual viewers that are unnecessary for trimming. Before you dive into trimming, let's optimize your layout for that process.

1 Open DaVinci Resolve, if necessary, and then open the Age of Airplanes project.

2 In the Rough Cuts bin, double-click the Trim Rough Cut timeline to load it into the timeline viewer.

   This timeline is similar to the one you worked on in the previous lesson with slight adjustments made for the following exercises.

   While trimming, you will work with only your timeline clips, so you can optimize your workspace by hiding the Media Pool and the source viewer. This will give you a larger area for the timeline and the timeline viewer.

3 In the upper-left user interface toolbar, click the Media Pool button to hide the Media Pool.

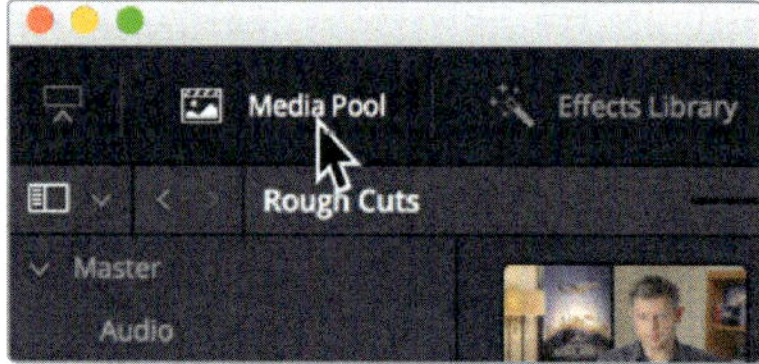

   Hiding the Media Pool already gives you a lot more room for the viewers and timeline.

4 In the upper-right corner of the interface, click the single viewer mode button to hide the source viewer.

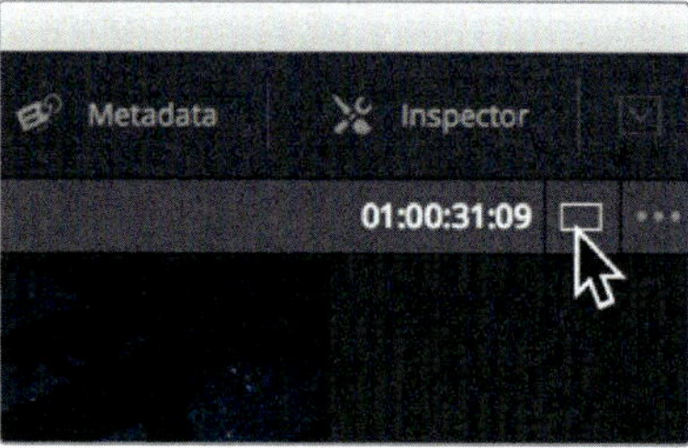

   The timeline viewer now moves to the center of the screen. With the extra horizontal space, you can increase the size of the viewer by taking away some space from the timeline.

   The timeline has a horizontal divider that separates the toolbar from the viewers. You can drag down that divider to allow more room for the viewers and less to the timeline tracks. Because you don't need too much vertical space for your timeline, this will work out well while trimming.

5 Position the mouse pointer between the toolbar and the transport controls.

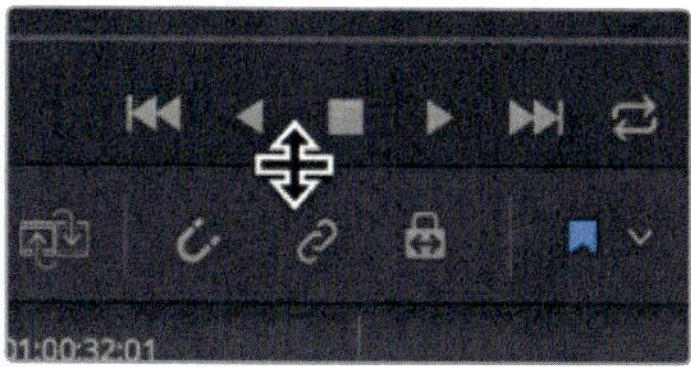

6 When the pointer changes to a resize cursor, drag down the horizontal divider while leaving a little bit of room above the video track in the timeline.

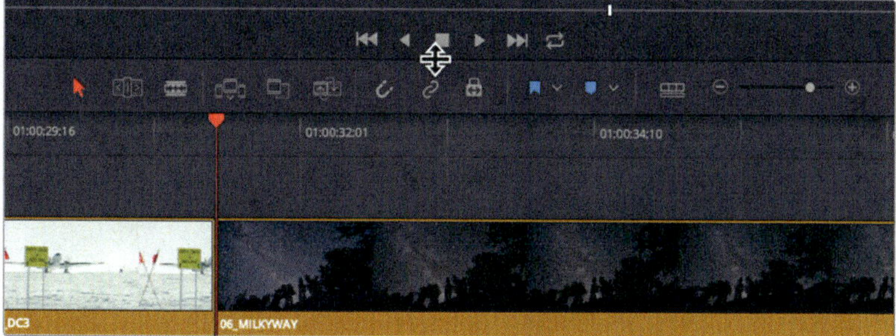

You can save this layout as your Big Trim layout and use it in all of your projects.

7 Choose Workspace > Layout Presets > Save Layout as Preset.

8 In the dialog, enter **Big Trim** as the layout name, and click OK.

Now that you have an optimized Big Trim layout, you can get to work trimming and refining your program.

Trimming to the playhead

One of the fastest ways to remove frames from the start or end of a clip is sometimes called top and tail trimming. This trimming style is used heavily in broadcast news-type programs, but it is useful for all types of programs. The concept is to loosely edit-in clips by adding more of the clip than you really want. Then, you'll remove frames from the start (top) of the clip or the end (tail) of the clip by positioning the playhead where you would like to locate the new start or end. Let's begin this trimming lesson at the start of the timeline. You don't have to play the entire timeline, but it is worth looking at the first clip to see how it can be improved.

1 Choose View > Zoom > Zoom to Fit, or press Shift-Z, to see the entire timeline in the window.

2 Position the playhead at the start of the timeline, and play it to review the first two clips.

The first clip isn't wrong, but it could be improved by shortening both its start and end. Let's start by trimming some of the silence from the start of the clip.

3 Position the playhead at the start of the timeline.

4 In the toolbar, drag the zoom slider to the right until the first clip fills half of the timeline window.

5 Position the playhead between the words "so" and "in this film." Use the audio waveform as a guide to position the playhead between the words.

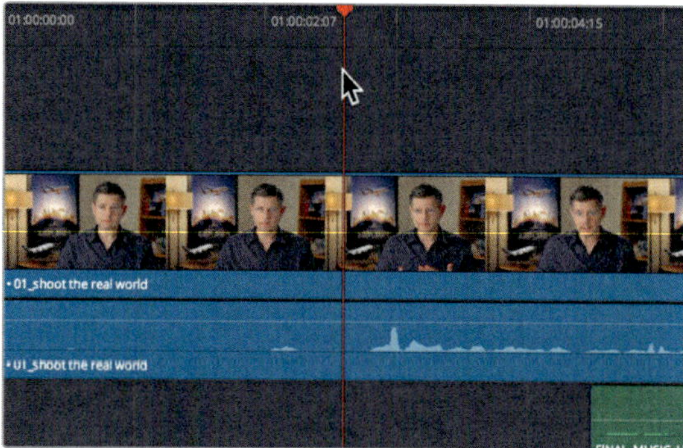

This playhead position is where you want the clip to start. You can perform the trim-to-playhead function using two different tools. Let's first apply the selection mode tool that you are already using.

6 Choose Trim > Trim Start, or press Shift-[(left bracket).

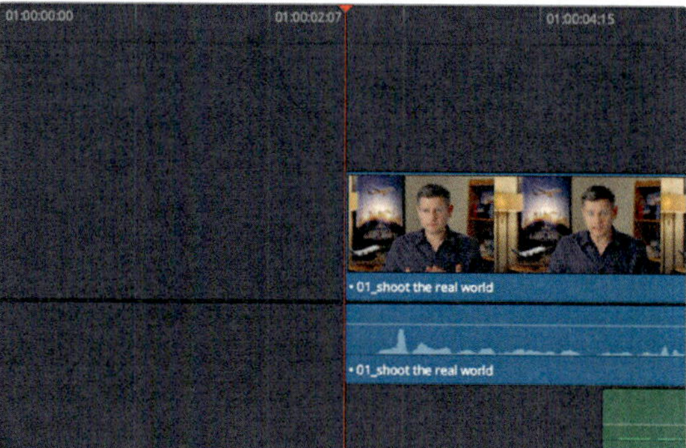

The start of the clip is trimmed to the playhead position. Let's review the change.

7 Position the playhead at the start of timeline, and press the Spacebar to listen to the first clip.

Although the clip starts on the correct word, it no longer starts at the beginning of the timeline. A gap now exists between the start of the timeline and the first clip. You'll have to undo that trim, and try another tool.

8 Choose Edit > Undo, or press Cmd-Z (Mac) or Ctrl-Z (Windows), to undo the previous trim.

9 In the toolbar, select the trim edit mode tool, or press T.

The trim edit mode tool is the most flexible tool to use when you want to shorten and lengthen clips in the timeline. The primary difference between the trim edit mode tool and the selection mode tool, is that trim edit mode ripples the timeline instead of leaving gaps. Let's see it in practice.

10 Again, position the playhead between the words "so" and "in this film."

11 Choose Trim > Trim Start, or press Shift-[(left bracket).

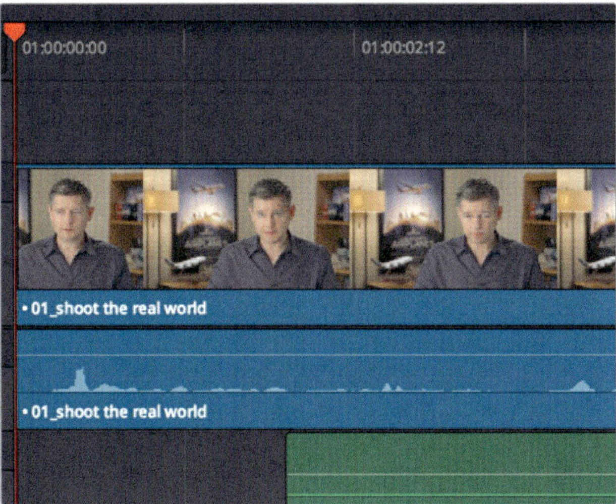

The same beginning frames are trimmed from the clip, but now the clip starts at the beginning of the timeline and the change ripples through the rest of the timeline, thereby shortening the overall duration. Let's look at the end of this clip.

12 Position the playhead at the start of the timeline, and play over the first clip.

A more powerful ending to this clip would be after the statement, "Shoot the real world."

13 Position the playhead at the end of the statement, "Shoot the real world," while ensuring that you do not include any of the following sentence.

TIP You can press the Left and Right Arrow keys to nudge the playhead one frame forward or backward to position it more precisely.

Instead of using the trim start function, you can use the trim end function to remove frames from the end of the clip.

Because you have the trim edit mode tool selected, removing all the frames from the playhead to the end of the clip will ripple the timeline, as did the ripple delete function you used in the previous lesson. But just as you disabled auto select in the previous lesson for the music track so you didn't delete a portion of the music, you also need to disable auto select here so you don't trim the end of the music.

14 Click the auto select button to disable it on Audio 2.

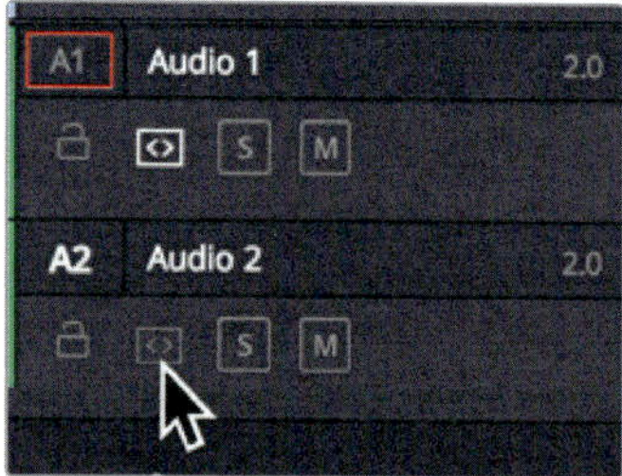

15 Choose Trim > Trim End, or press Shift-] (right bracket).

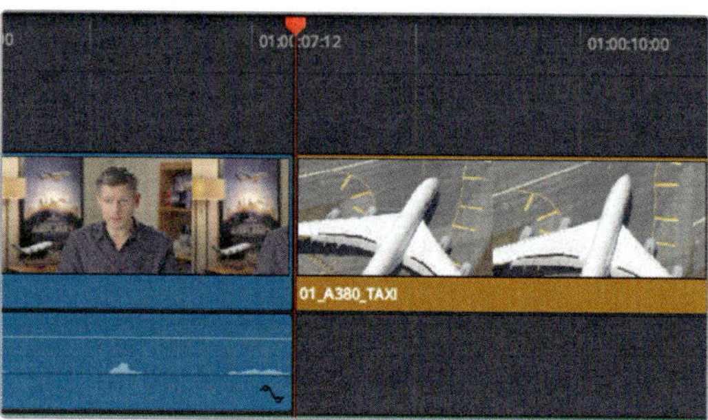

The end of the clips on Video 1 and Audio 1 are trimmed to the playhead but the music track remains unchanged. The remaining clips in the timeline are shifted to the left by the same number of frames that you just removed.

16 In the toolbar, select the selection mode tool, or press A.

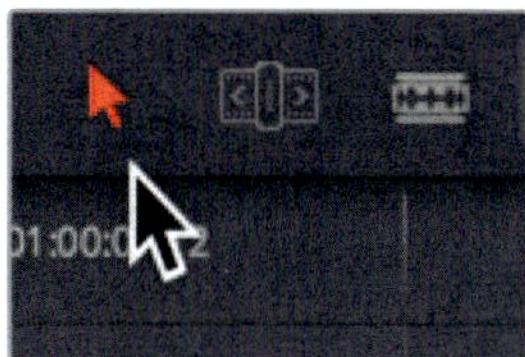

17 Position the playhead at the start of timeline, and start playback to review your top and tail trims.

You should have two take-aways from this exercise. The first (and obvious) one is that trim start and trim end are two very quick ways to tighten your edits through an entire timeline. The second and more fundamental take-away is that the selection mode tool opens gaps while the trim edit mode tool ripples the timeline.

Ripple trimming

The multiple ways you can use the trim edit mode tool makes it fast, precise, and flexible. Let's look at another clip to decide how trimming might improve it.

1 Position the playhead at the end of the interview clip you just trimmed.

2 Play the timeline to view the next three clips, until the interview subject returns to the screen.

The shot of the Maldives starts too late. The water plane is already overhead when the clip begins. You need to add more frames to the beginning of the Maldives shot so it starts with the plane out of the frame. Let's zoom in and center that clip in the timeline window.

3 Position the playhead at the start of the 10_MALDIVES clip.

4 Below the timeline, drag the scroll bar to the right to center the playhead.

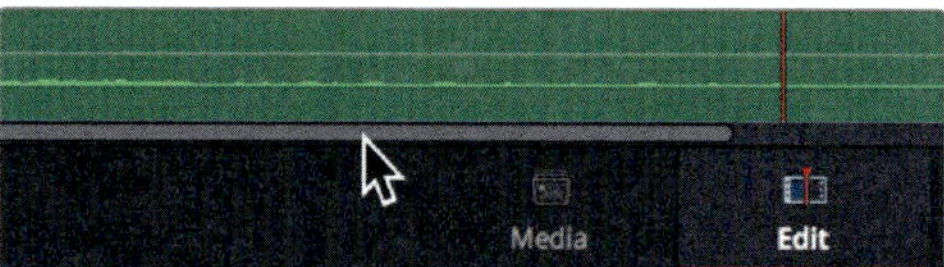

5 In the toolbar, drag the zoom slider to the right to increase the size of the 10 MALDIVES clip in the timeline.

You now have two choices. You can use the selection mode tool to add frames to the beginning of the Maldives, and thereby overwrite some of the end frames on the A380 clips that comes before it. Or, you can use the trim edit mode tool and ripple the timeline, thereby leaving the A380 clip unchanged.

Because the A380 clip is fine as it is, let's choose to ripple trim.

6 In the toolbar, click the trim edit mode button, or press T.

With the trim edit mode selected, you can no longer use the cursor to select clips and move them in the timeline. Now, the primary purpose of the cursor is to select a cut point and the side of that cut point you want to trim.

7 Position the trim edit mode cursor at the end of the 02_A380_TAKEOFF clip.

The cursor changes to the ripple trim cursor that you use to lengthen or shorten a clip's duration. When the cursor is to the left of the cut, it will trim the end, or tail, of the cut.

8 Position the trim edit mode cursor over the beginning, or head, of the 10_MALDIVES clip.

When the cursor is on the right side of the cut, you can trim the head of the clip. This is where you want to trim.

9 Drag the head of the cut slightly to the left and hold the cursor there for a moment.

As you drag, the timeline viewer splits to show you the last frame of the outgoing clip on the left and the first frame of the incoming clip on the right. This two-frame side-by-side display is designed to show how the action and framing from the two sides of a cut will match up (or not).

10 Continue dragging left until the **10 MALDIVES** plane is off the screen.

TIP When dragging to trim, the cut point may snap to the playhead and make it difficult to precisely position the trim. When this happens, tap the N key to disable snapping as you trim.

When removing frames, a ripple trim pulls in all the clips after the trim point to close the gap (much like the ripple delete that you performed in the previous lesson.) When you add frames using the ripple cursor, you not only lengthen the clip, but you also lengthen the overall duration of your program.

11 To review the trim point, choose Playback > Play Around/To > Play Around Current Frame, or press / (slash).

TIP The number of seconds played before and after any play around command is determined by the pre-roll and post-roll settings in the editing user preferences.

Adding frames during trimming requires that you have access to additional frames of the captured clip that were not included within its in and out points as the clip was edited into the timeline. Those unused portions of each clip placed in the timeline are known as **handles**.

If you edit the entire length of a clip into the timeline, you will not have handles available to extend that clip, so you will be able only to remove frames when you trim.

Trimming using numbers

When you are a trimming specific number of frames, instead of dragging the cut point using a visual guide, it is easier to use the keyboard to enter the exact number of frames you want to move, or to nudge the trim one frame forward or backward.

1 Make sure the start of the 10_MALDIVES clip is still selected with the trim edit mode tool.

   From the previous review of this trim, you can see that the plane still comes in too abruptly. Let's add one second more to the start of this clip. Instead of trying to drag one second precisely, you can enter the number using the keypad.

2 Type **-1.** (minus, one, period), and press Return (Mac) or Enter (Windows).

   One second is added to the start of the 10_MALDIVES clip. Using a negative number to add frames may seem a bit counterintuitive, but the positive and negative values are based on the timeline direction. Moving a clip or cut point to the left is a negative move, whereas moving to the right is a positive move.

3 To review the cut, choose Playback > Play Around/To > Play Around Current Frame, or press / (slash).

   You can continue to enter seconds and frames to refine the precise starting point for the Maldives clip. But you can also quickly nudge the cut one frame at a time by pressing the , (comma) and . (period) keys.

4 To remove three frames from the start of the Maldives clip, press the . (period) key three times.

TIP Pressing Shift- , (comma) or Shift-. (period) trims in five-frame increments.

5 To review the trim point, choose Playback > Play Around/To > Play Around Current Frame, or press / (slash).

6 To return one frame to the start of the Maldives clip, press the , (comma) key once.

> **TIP** Choosing Playback > Loop and then playing around the current frame allows you to use the . (period) and , (comma) keys on-the-fly as you loop over the transition.

Whether you use the number pad or drag to trim is really your choice. Although using the number pad may be faster, it is also less visual. When trimming by dragging you are better able to see the frames, but you sacrifice work speed. The right choice for you is whichever method you feel most comfortable with in any given situation.

Selecting tracks to trim

So far in this lesson you've trimmed only video clips. What happens when you select a clip that includes an audio track?

1 Press Shift-Z to view the entire timeline, and then position the playhead between the 06_MILKYWAY and 02_If it was possible clips.

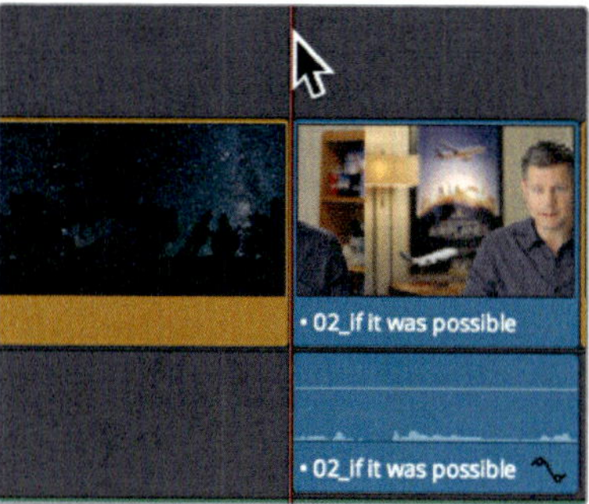

2 In the toolbar, drag the zoom slider to zoom in on these two clips.

> **TIP** In Mac, press Cmd-+ (plus sign) or Cmd- - (minus sign) to incrementally zoom in and out of the timeline. In Windows, press Ctrl-+ (plus sign) or Ctrl- - (minus sign) to incrementally zoom in and out of the timeline.

3 Click the trim edit mode button, or press T, to enter trim edit mode, if necessary.

4 Place the pointer over the right side of the cut point, over the start of the 02_If it was possible clip.

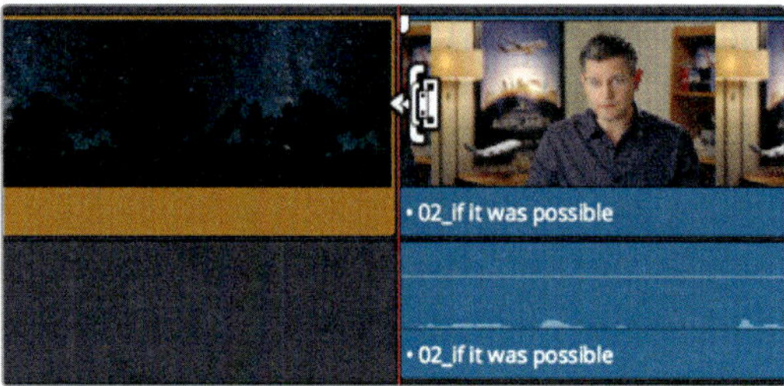

5 Click to select the start of the 02_If it was possible clip for ripple trimming.

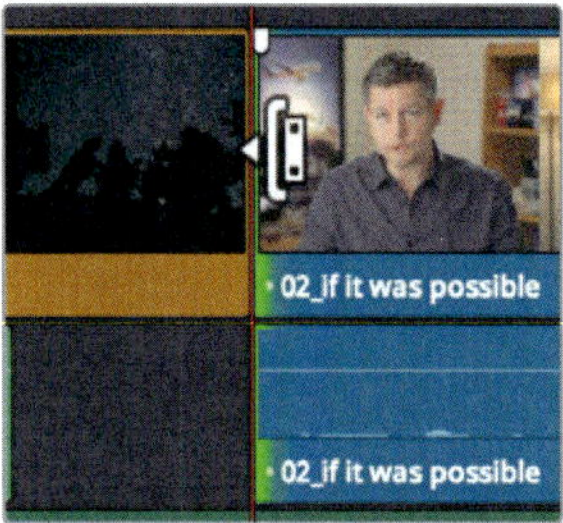

Even though you selected only the video track, both the video and audio tracks are selected. The audio and video are from the same interview clip, so they are linked. This behavior is similar to moving clips in the timeline with the linked selection button enabled.

6 In the empty timeline area above the video track, click to deselect the edit point.

7 Click the linked selection button to disable it.

TIP You can also hold down the Option key (Mac) or the Alt key (Windows) to temporarily select the video edit point without disabling the linked selection button.

8 Once again, select the start of the 02_If it was possible clip for ripple trimming.

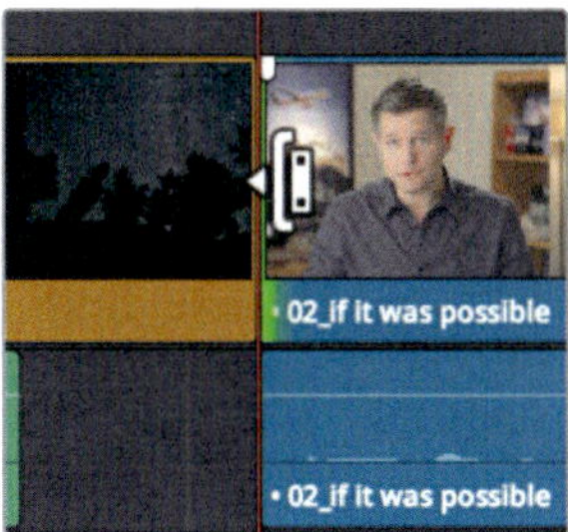

With the linked selection function disabled, only the video edit point is selected.

9 To review the selected cut, choose Playback > Play Around/To > Play Around Current Frame, or press / (slash).

Offsetting the video from the audio so that one is seen or heard sooner than the other is a technique used in editing to improve program flow. Often called **J-cuts** and **L-cuts**, these edits are most commonly used in dialog scenes, but they can be used here to lead you into the next shot with more continuity. To further increase continuity, let's start his voice a few seconds before the picture.

10 Drag the cut to the right about one second until you are between the statements "It was really important" and "Every shot was original." Use the audio waveform in the audio track as a guide to position the video cut in the gap between the sentences.

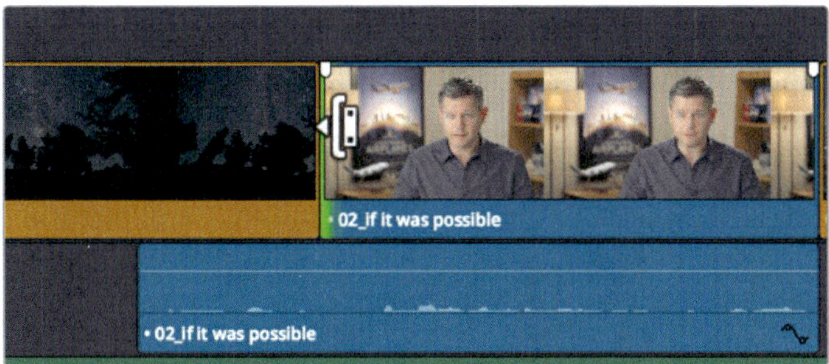

TIP Enabling Trim > Dynamic Trim allows you to use the J-K-L keys for trimming.

Before you review that cut, it is worth taking a short detour and asking why the audio track moved under the 06_MILKYWAY clip when you trimmed the video track? That has a lot to do with the state of the auto select button that you used in Lesson 4 when you were deleting a range within a clip.

Here, the auto select function was trying to keep your timeline in sync while you were trimming. To more fully understand how this works, let's trim a bit more, this time with auto select disabled for the audio track.

11 On Audio 1, click the auto select button to disable it.

12 Again, select the start of the 02_If it was possible clip and drag it to the right for a few frames.

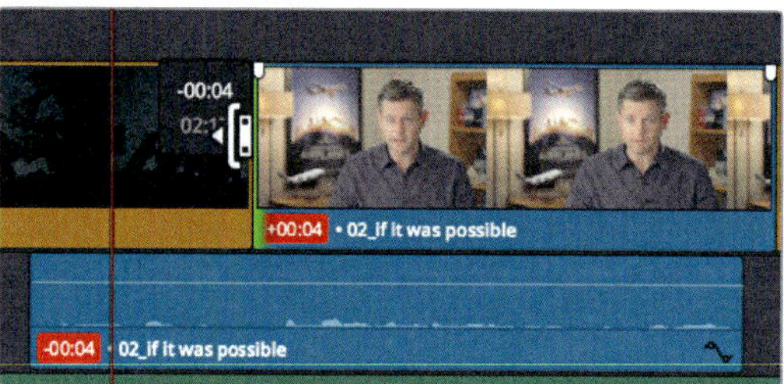

Because you disabled the auto select button for the Audio 1 track, it didn't adjust when you trimmed frames from the video track. Consequently, the audio went out of sync with the now-short video track. The red badges indicate that the number of frames in the audio and video tracks are out of sync. With more understanding of auto select, let's undo that action and finish up the lesson.

13 Choose Edit > Undo, or press Cmd-Z (Mac) or Ctrl-Z (Windows), to undo the previous trim.

14 On Audio 1, enable the auto select button.

15 Click the linked selection button to enable linked clips.

16 In the toolbar, select the selection mode tool, or press A.

Finally, let's review the trim you made.

17 Position the playhead at the start of 06_MILKYWAY clip, and review your J-cut.

Keeping audio and video in sync is always a concern (and a chore) for editors. The linked selection function is invaluable in assisting you with that effort on a clip-by-clip basis and the auto select buttons are invaluable on a timeline basis. Although it's necessary to disable both in some situations, it's good practice to enable them most of the time.

Using roll trimming

While a ripple trim alters a single side of a cut point, Roll trims simultaneously trim both the end of the outgoing clip and the start of the incoming clip. These types of trims are useful when you want to retain the overall timeline duration or you are trying to match action.

1 In the timeline, position the playhead at the end of the 06_MILKYWAY clip.

2 Press the Spacebar to play through the end of the 05_BAY AREA_LIGHTS clip.

For this trim, you will need to trim only the video track because the audio in this interview ends where you want it, when he says, "Everything was real." However, the two shots would flow better if you heard the word "real" while the Bay Area shot was onscreen. Just a small overlap is all you need.

3 In the toolbar, click the linked selection button to disable it.

A rolling trim can be performed using either the selection mode tool or the trim edit mode tool. The behavior for a roll trim is exactly the same no matter which tool you use. So, you can leave your selection mode tool selected.

4 Center the mouse pointer over the video cut point.

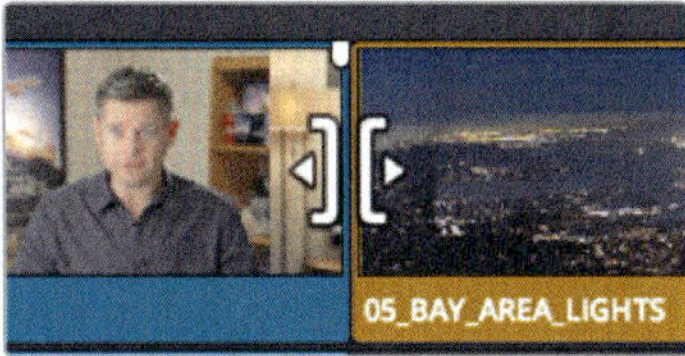

When the mouse pointer is centered over the cut, the cursor changes to a roll trim cursor.

5 On the video track, click the cut point to select both the end of the 02_if_it_was_possible interview clip and the start of the 05_BAY AREA_LIGHTS clip.

TIP When a cut point is selected with the ripple or roll cursor, pressing the U key toggles between each side of the edit to select that side for trimming.

With both sides of the cut selected, any adjustments will be made equally to both sides of the cut. Removing frames from the end of the interview clip will add frames to the start of the Bay Area clip and vice versa.

6 Drag the cut to the left until the playhead is just before the words "was real." In the audio track, use the audio waveform as a guide to position the video cut just before those words.

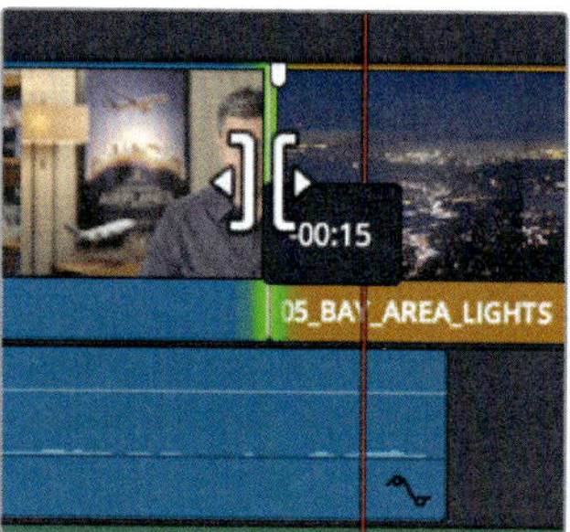

Let's play the transition to see your results.

7 To review the cut, choose Playback > Play Around/To > Play Around Current Frame, or press / (slash).

You can continue to refine the cut by pressing the , (comma) and . (period) keys to nudge it into the exact spot you want.

Slipping a clip

As you refine your timeline, making a clip longer or shorter isn't always going to be the best refinement. Occasionally, you'll want to use a portion of the source clip that is either earlier or later than what is currently in the timeline. In other words, you want to shift the range without changing the clip's duration or position in the timeline. This trim is called **slipping** a clip.

1 Choose View > Zoom > Zoom to Fit, or press Shift-Z, to see the entire timeline.

2 Position the playhead at the start of the 01_A380_TAXI clip.

3 Press the Spacebar to play the timeline until you see the 10_MALDIVES clip.

The 02_A380_TAKEOFF clip starts a bit slow and ends before the plane is out of the frame. Could you trim both ends of the cut to fix this? Sure, but you could perform the same correction more quickly using the slip tool.

4 Position the playhead over the 02_A380_TAKEOFF clip, and in the toolbar, drag the zoom slider until you fill the timeline window with the A380 clip and the clips on either side of it.

5 Click the trim edit mode button, or press T.

Just as when you chose between ripple or roll trims, the placement of the mouse pointer is important when choosing the slip cursor.

6 Place the mouse pointer over the upper-middle region of the 02_A320_TAKEOFF clip.

The cursor changes to a slip cursor. With the slip cursor in place, you're ready to slip the clip.

7 Drag to the left to slip the clip until you see the plane leave the frame in the upper right of the viewer.

As you drag, the viewer changes to a 4-up display that allows you to compare all relevant outgoing and incoming frames. The upper two frames show the starting and ending frames of the clip being slipped. The lower-left frame shows the previous clip's unchanging last frame, and the lower-right frame shows the next clip's unchanging first frame. The 4-up display enables you to compare and match the action of all three clips: the clip you are slipping and the two on either side of that clip.

NOTE Depending upon the window configuration and the size of your display, you may need to stop dragging a clip, reposition the mouse, and then resume dragging.

Having the plane leave the frame completes the action of the take-off and makes for a more pleasing clip.

8. Position the playhead before the slipped clip, and then play the timeline to review your edit.

   This looks good but it could be improved if you didn't cut to the Maldives shot exactly as the plane leaves the frame. You need to add five frames of the plane out of frame before you cut.

   Once you click a clip with the slip tool, you can use keyboard shortcuts to slip left or right in one- or five-frame increments.

9. Press Shift-, (comma) to slip the clip five more frames.

10. Position the playhead before the slipped clip, and play the timeline to review your edit.

    If that looks good to you, then leave it as it is. If you feel the clip needs more or less "breathing room" before you cut to the next clip, use the keyboard shortcuts to refine it.

11. When you are done, in the toolbar, select the selection mode tool, or press A.

Slipping a clip is most often used more subtly than you have done here. You'll find that you frequently will slip clips just a few frames to get a perfect match with the surrounding clips.

Opening gaps using the selection tool

You can also use the default selection mode tool for trimming. It does not ripple trim; instead, it opens gaps when a clip is made shorter or overwrites the next clip when a clip is extended.

1 Choose View > Zoom > Zoom to Fit, or press Shift-Z, to see the entire timeline.

2 Position the playhead at the start of the 08_SOUTH_POLE clip.

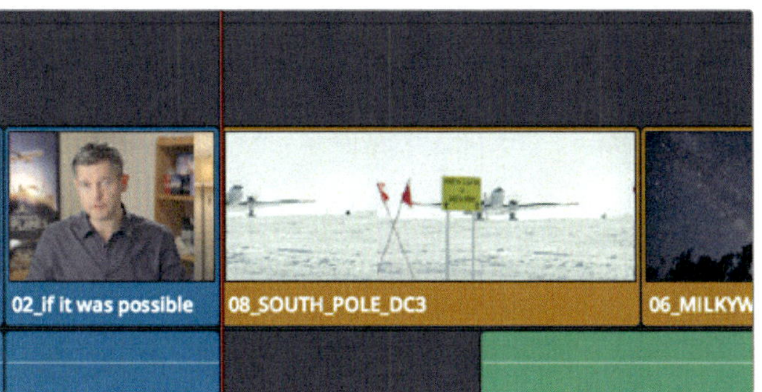

3 Press the Spacebar to play the timeline until you see the 05_BAY_AREA_LIGHTS clip.

06_MILKYWAY is a long clip. It would be better to shorten it. You could just overwrite a new clip to cover up part of the 06_MILKYWAY clip, but in many cases you can't do that. For instance, maybe you haven't yet imported the clip or you haven't decided which clip you want to use. Still, you may want to shorten the clip and leave the gap as a placeholding reminder. For whatever reason, opening gaps is a valuable trimming function that you can apply with the selection mode tool.

4 Position the playhead over the center of the 06_MILKYWAY clip.

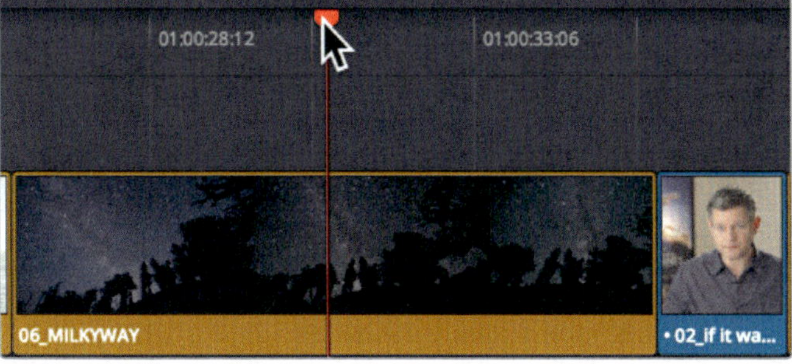

5 In the toolbar, drag the zoom slider to increase the size of the clip in the window.

6 Hover the mouse pointer over the start of the 06_MILKYWAY clip.

A Resize Trim cursor appears that is similar to the ripple trim cursor. However, the results are different when compared to using the Trim tool.

7 Drag the start of the 06_MILKYWAY clip to the right until the tool tip displays +4:00.

Unlike using the ripple trim tool, when using the trim edit tool, trimming one side of an edit using the selection mode tool leaves a gap.

Marking a Gap

You can't leave the gap in your program so you will have to fill it with a clip. Let's look in the B-Roll bin for a clip to fill this gap.

1 Choose Workspace > Reset UI Layout to unhide the Media Pool, and click the Single viewer mode button to display the source viewer.

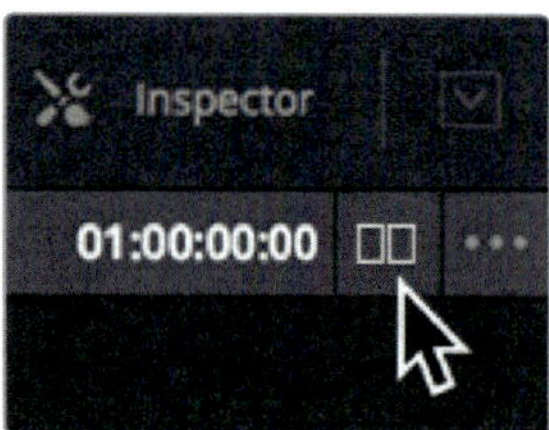

2 In the B-Roll bin, double-click the 07_Kenya clip to load it into the viewer, and press the Spacebar to play it.

The 07_KENYA clip is the only clip in the bin that you haven't used, so it is a good candidate for filling the gap.

3 In the viewer, drag the jog bar back until the plane's reflection in the water points straight up.

You'll mark your In point here.

4 Press I to mark an in point on the source clip.

The gap is the range in the timeline you are trying to fill, so you needn't add an out point on the source clip; but you will need to mark in and out points in the timeline based on the duration of the gap.

5 Drag the timeline playhead into the center of the gap.

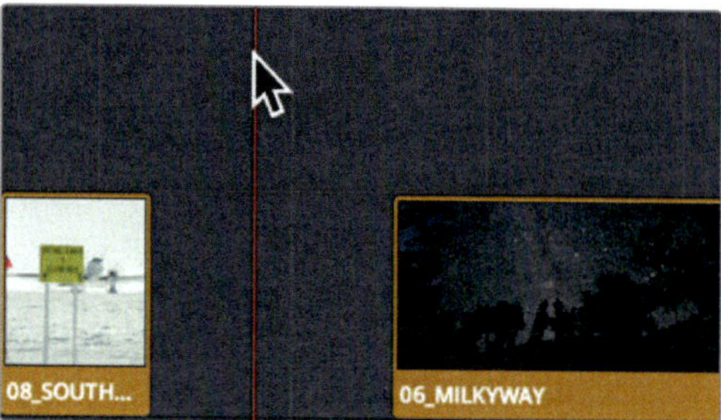

Instead of moving the playhead, marking an In point, moving the playhead again, and marking an out point, you can use the Mark Clip command to mark the clip under the playhead.

6 Choose Mark > Mark Clip, or press X.

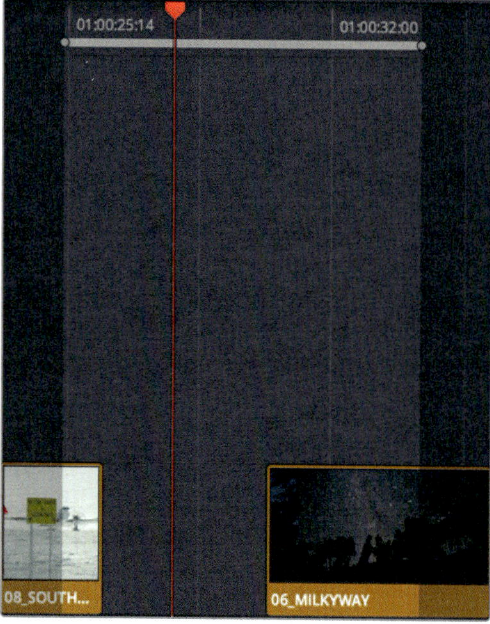

The Mark Clip command added in and out points in the timeline, but it used the audio track on A1 as the duration to mark. You really want to ignore the audio track and use the gap. This is another use case for auto select. It helps keep tracks in sync when trimming and identifies the tracks when deleting a range (as you did in Lesson 4).

Here, auto select will allow you to target the track you want the Mark Clip command to use when setting in and out points.

7 In the timeline header, disable the auto select buttons for Audio 1 and Audio 2, leaving only Video 1 with an auto select button enabled.

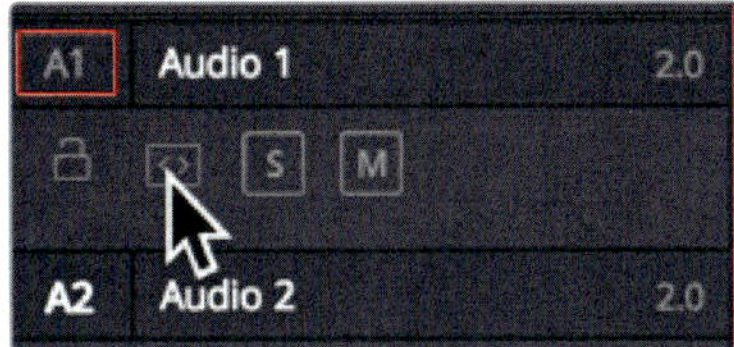

8 Choose Mark > Mark Clip, or press X.

TIP When multiple video tracks have Auto Select enabled, the lowest numbered video track is the target track.

With auto select enabled on Video 1, the Mark Clip command correctly used the gap duration to set the in and out points. All that is left to do is make the edit.

9 In the toolbar, click the overwrite button to edit-in the 07_KENYA clip.

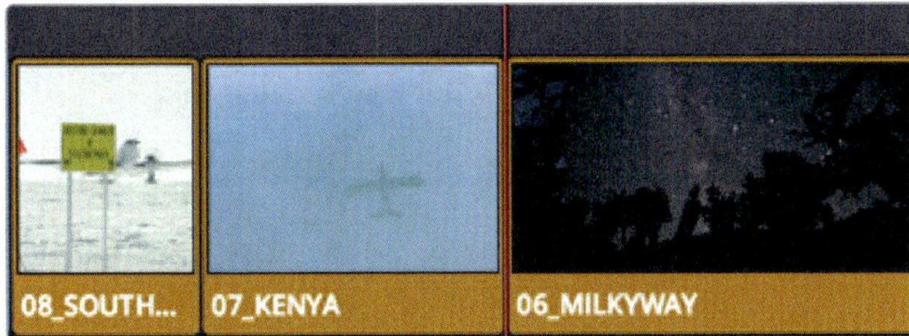

This is the same behavior you saw on the very first exercise in this lesson when using the trim start function opened a gap between the start of the timeline and the first clip. It is a fundamental behavior to keep in mind.

Trimming in the source viewer

At any time—not just while trimming— you may want to view the available handles for a clip in the timeline to see if the unseen content is worth using. Instead of searching through all of your clips in all of your bins, you can open the clip from the timeline into the source viewer.

1 In the timeline, position the playhead at the start of the 08_SOUTH_POLE DC3 clip.

2 Press the Spacebar to play over the newly added 07 KENYA clip and the 06 MILKYWAY clip.

The yellow plane just enters the top of the frame before you cut. Instead of blindly trimming, you should check to see what the rest of this source clip includes.

3 In the timeline, double-click the 07_KENYA clip to open it in the source viewer.

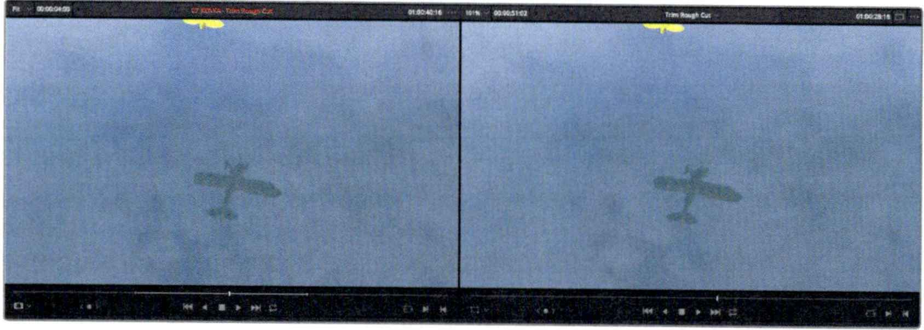

The Kenya clip opens in the source viewer with in and out points marked to indicate what is currently used in the timeline.

4 Press Spacebar to play until the end of the 07 KENYA clip in the source viewer.

This shot only gets better the longer you play it. You should extend it by at least another second in the timeline. But here is the really important part to understand. When you double-click a clip in the timeline, you are not opening that clip from the bin, you are opening a live version of the clip in the timeline. That means if you modify those in and out points in the source viewer, it modifies the range used in the timeline.

TIP When you open a timeline clip into the source viewer, the name above the source viewer appends the name of the timeline to the clip name.

5. To change the out point of the clip in the timeline, drag the jog bar in the source viewer until the flamingos are halfway down the right side of the frame.

6. Press O to mark an out point.

   The clip in the timeline is extended based on the new out point you set in the source viewer. Also, because the selection mode tool was the selected tool in the toolbar, the Milkyway clip was overwritten by the extended 07_KENYA clip.

7. In the timeline, position the playhead at the start of the 08_SOUTH_POLE DC3 clip.

8. Press Spacebar to play the newly extended 07 KENYA clip and the 06 MILKYWAY clip.

With so many options and methods available, trimming may initially seem a daunting phase of your workflow. Editors tend to use the trim edit mode most often, unless they need to open a gap in the timeline. In those situations, trimming with the selection tool is the better choice.

Sliding a clip

Let's look at one last option with the trim edit mode tool. Sliding a clip is almost the opposite of slipping. Sliding retains a clip's displayed content in the timeline but moves it to change only its position in the timeline. It is most similar to moving a clip along the timeline; but whereas moving a clip will overwrite the clip in the direction that you drag, sliding a clip trims the adjacent clips without leaving any gaps.

1 In the toolbar, click the trim edit mode tool, or press T.

2 Position the mouse pointer over the brown clip label region of the 07_KENYA clip.

When the pointer is over the bottom of the clip, it changes to a slide cursor.

3 Drag the clip to the right to slide it until it the tool tip displays +1:00.

Dragging to slide a clip displays a 4-up viewer arrangement similar to the one you saw when using the slip tool. The difference is that the upper frames stay the same because they represent the first and last frames of the clip that you are sliding. The lower frames change to show you the last frame of the previous clip and the first frame of the following clip.

4 Play the timeline to review your edit.

5 Click the selection mode tool, or press A, to exit trim edit mode.

Trimming sub-frame audio

All the video trimming techniques you used in the previous exercises also can be used when you are just trimming audio. However, audio does have one additional trimming feature that is not available when trimming video. Audio trimming allows you to trim less than a single video frame.

1 Press Shift-Z to see the entire timeline.
2 Position the playhead in the middle of the 02_A380_TAKEOFF clip.

You are going to trim the narration, so it will be easier to do so if you do not hear the music on Audio 2.

3 In the Timeline track header, click the mute button on Audio 2.

The mute button is used to temporarily silence all audio clips on a particular track.

4 Play the timeline to hear the end of the audio narration, "If it was possible to shoot it, we wanted to go shoot it."

This piece of narration slightly cuts off the start of the word "if." You need to add to the start of the word "if" so it sounds right.

5 Position the playhead at the start of the audio segment.

6 In the toolbar, drag the Zoom slider all the way to the right to zoom in to the audio clip.

You can increase the vertical size of the audio track using the Timeline View Options panel, thereby making the waveform easier to see.

7 In the toolbar, click the timeline view options menu.

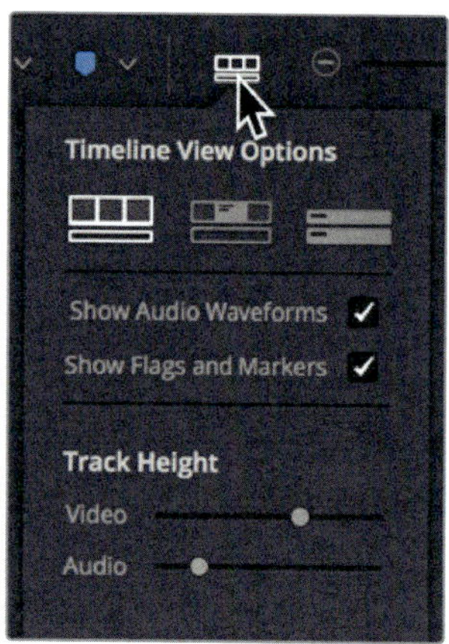

The Timeline View Options panel includes a number of ways to change the timeline's appearance. Although you'll look at most of them in upcoming lessons, you'll now drag just the Audio Track Height slider to increase the vertical size of the audio tracks.

8 In the Timeline View Options, drag the Audio Track Height slider to the right until it is about halfway to the end.

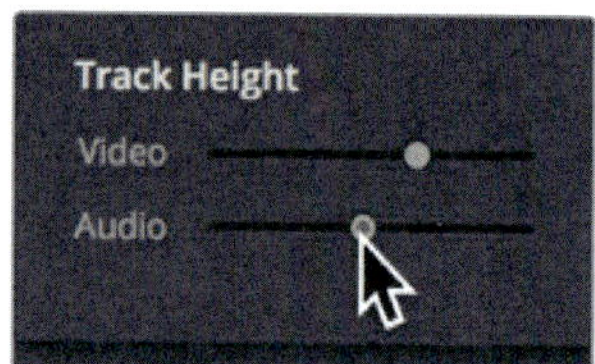

This one change will help you see soundwaves more clearly in the clip's audio waveform. Now you'll add one frame to the start of the audio clip.

9 Option-click (Mac) or Alt-click (Windows) the right side of the audio cut with the selection mode tool, and press the , (comma) key once to add one frame.

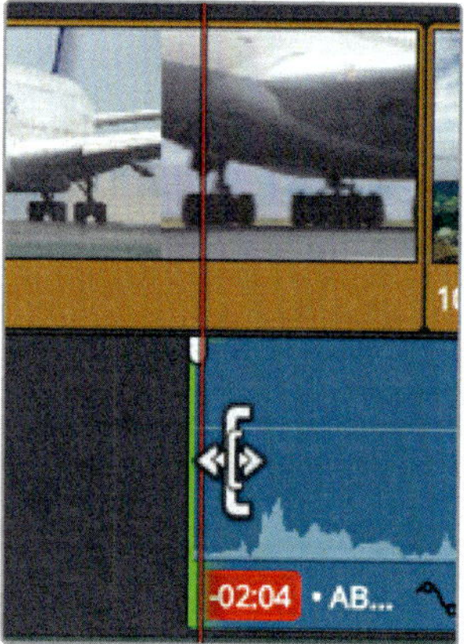

10 Play over the area to hear the one-frame trim.

You now hear the ending of the word prior to "if." The correct relocation for this cut is somewhere less than one frame. You can trim on a sub-frame boundary by turning off the snapping function.

11 In the toolbar, click the snapping button and the linked selection button. Press the Right Arrow key to move the playhead forward one frame.

12 Using the selection mode tool, drag the right side of the audio clip halfway toward the playhead.

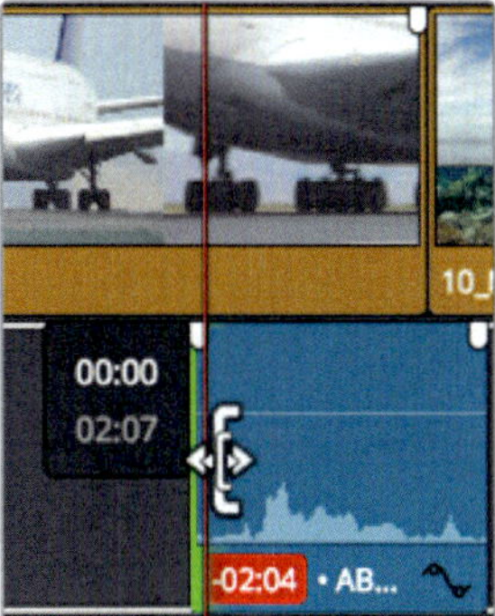

13 Play over the area to hear the sub-frame trim.

The word ""if" now sounds perfect.

14 In the toolbar, click the Snapping button and the linked selection button to reenable them.

Moving and trimming audio at the sub-frame level allows you to eliminate short stutters and breaths in narration that otherwise could detract from the message you are trying to convey.

DaVinci Resolve's trimming functions have a lot of flexibility and capability loaded into them. Mastering their power will help elevate your story telling because they are the primary way you will control the rhythm and pacing of your program.

Lesson 6

Applying Transitions and Effects

Once you have the basic structure of a scene, you can begin to open up a whole new avenue of creativity by adding graphics and effects. It doesn't matter what type of program you are creating; these additions can be as simple as a cross-fade or as fantastic as a morphing transition. Many effects are subtle, even hidden, while others are meant to be attention grabbers. In this lesson, you'll work with a sampling of DaVinci Resolve 14's tools for creating visual interest; but more importantly, you'll learn the concepts of adding transitions and effects so you can confidently continue to explore on your own.

Time

This lesson takes approximately 50 minutes to complete.

Goals

Fading clips in and out

Many programs start with a fade transition (fade-in) and end with one (fade-out). When you fade, you are mixing two elements. One element is a video clip and the other is a completely black frame, or in DaVinci Resolve, an empty part of the timeline.

1 Open the Age of Airplanes project, if necessary.

2 In the Rough Cuts bin, double-click the Transitions and FX rough cut.

Because you will be altering the video track in this lesson, you can modify the layout again to suit your needs.

3 Choose Workspace > Layout Presets > Big Trim > Load Preset to use your previously created layout.

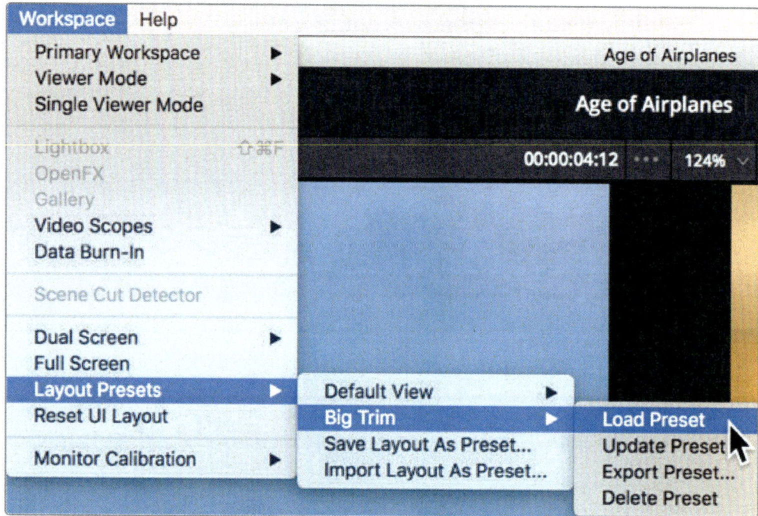

4 Click the single viewer mode button to enlarge the viewer.

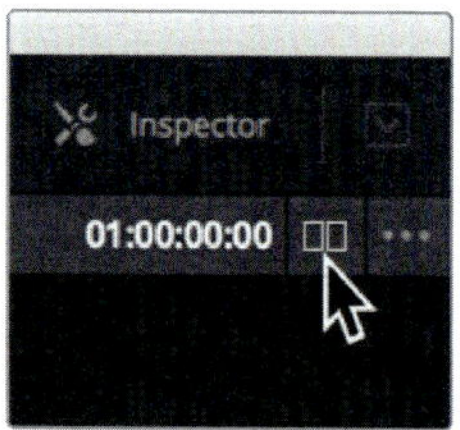

You'll also customize the timeline view to make the audio tracks smaller and the video track larger.

5 Click the Timeline View Options menu.

6 Drag the Audio Track Height slider all the way to the left.

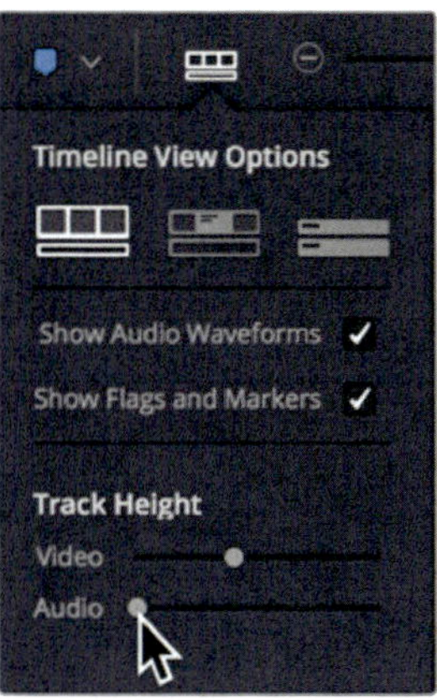

7 Drag the Video Track Height slider about mid-way, and click the Timeline View Options button again to hide the options.

8 Position the playhead at the start of the timeline.

9 To begin this trailer, you'll add a very quick fade-in.

10 In the timeline, place the pointer over the 01_shoot the real world clip.

As with audio clips, two fade handles—video fade handles, in this case—appear in the upper-left and upper-right corners of the clip.

11 At the start of the clip, drag the handle toward the center of the clip until the tool tip reads +00:10.

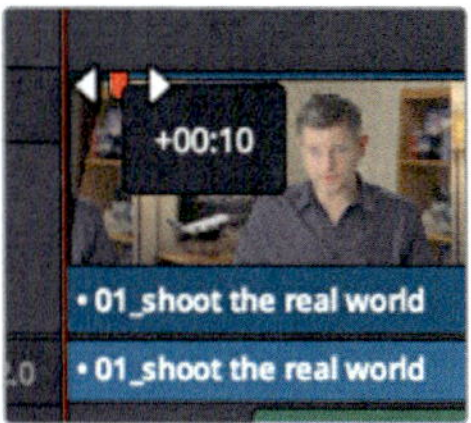

You've added a 10-frame fade-in to the start of the scene.

12 Play the start of the timeline to watch the fade.

Adjusting fade handles is a fast and easily accessed method for placing and refining fades-in and -out.

Adding cross dissolves

A transition is a gradual transformation from the ending of one shot to the start of the next. You've seen them many times on TV and in films, even if you didn't consciously make note of them. The most common video transition is the cross-dissolve, which is an overlapping fade between the end of one clip and the beginning of the next. The quickest way to add a cross-dissolve is to create it directly in the timeline.

1 In the timeline, go to the first red marker.

2 Play over 08_SOUTH_POLE_DC3 and the next clip, 07_KENYA.

You would like to blend these two clips softly into each other. To achieve this, you'll add the default cross dissolve transition.

3 Drag the zoom slider to zoom in to the clips you'll be working on.

TIP You can hold down the middle mouse button to pan in the timeline.

As if you were going to perform a rolling trim, locate the mouse pointer directly over the edit between 08_SOUTH_POLE_DC3 and 07_KENYA.

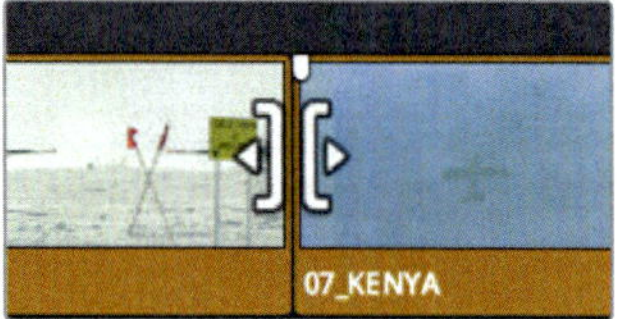

4 When the pointer changes to a rolling trim cursor, click the cut to select it.

The green rolling trim handles appear on both sides of the edit point.

5 Choose Timeline > Add Transition, or press Cmd-T (Mac) or Ctrl-T (Windows).

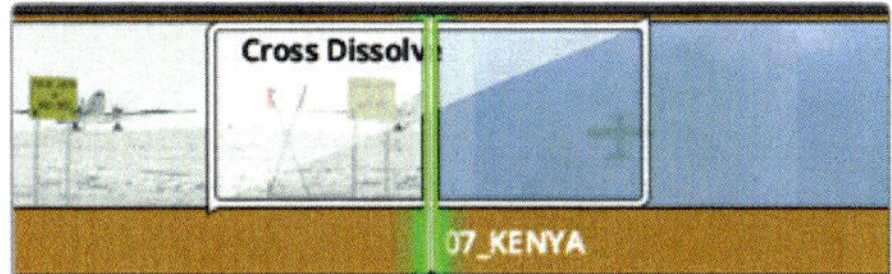

A dissolve with a one-second duration is added to the edit point.

TIP You can change the default transition duration in the editing category of the user preferences window.

When creating a transition, frames from the two clips will overlap. That is, half of the transition frames are taken from the unused portions of the outgoing clip and half from the incoming clip. These video handles, which you used for trimming in Lesson 5, are now used to extend clips to support the transition.

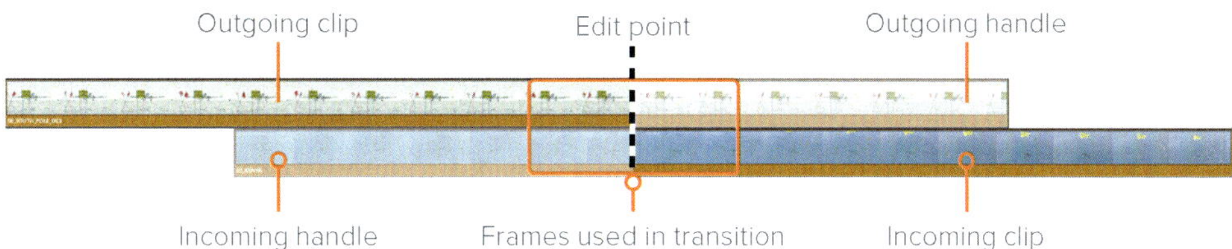

6 Play the two clips to view the dissolve.

TIP To remove a transition, zoom in and select it using the selection tool, and press Delete.

Shortening and lengthening transitions

It would be great if the default one-second dissolve duration fit every scene perfectly, but it won't. It may cut off some of the action or extend a clip too far so that you see a previously hidden camera shake. Whatever the reason, at some point you will want to change the dissolve duration. The easiest and most straightforward way to do so is by dragging in the timeline.

1 Place the mouse pointer over the right edge of the dissolve between 08_SOUTH_POLE_DC3 and 07_KENYA.

2 Drag the right edge in towards the edit until the tool tip reads -00:06.

As you drag, the transition is shortened by six frames on both sides of the edit for a total decrease of 12 frames. This dissolve transition is aligned to the center of the cut, so the transition will remove the same number of frames on each side of the cut regardless of how short you make it. How long can you make a transition? That depends upon the length of the two source clips in the bin.

3 Drag the right edge away from the edit until the selected edit will not extend any further.

The transition can extend only so far because you will eventually run out of handles on one of the clips, which means that no more media is available to create a longer transition.

Customizing transitions

Each transition has several adjustments that you can use to customize its appearance. Some of the simpler transitions, such as the cross dissolve, have fewer parameters than specialty transitions such as wipes. In every case, customization controls appear in the Inspector.

1 Double-click the Cross Dissolve transition to open it in the Inspector.

> **TIP** If you are finding it difficult to select a transition, drag the zoom slider to zoom in on the timeline until you can select the transition icon without accidentally selecting the cut point.

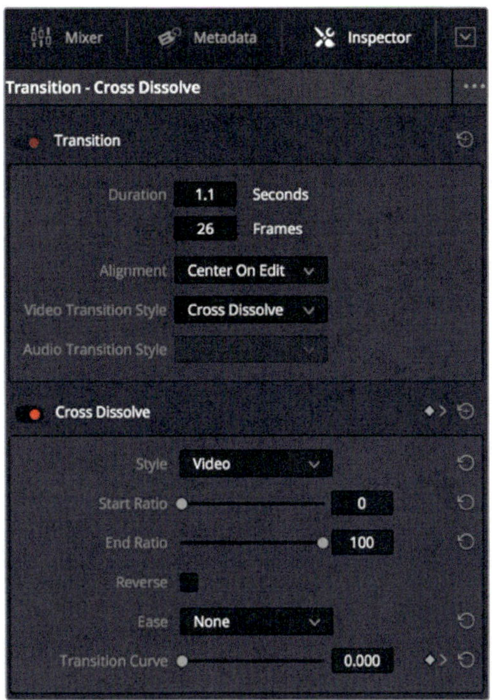

The Inspector displays the transition parameters. The upper half of the Inspector has parameters that are common to all transitions. These include Duration, Alignment, and Transition Style. The lower half has parameters specific to the current transition.

2 In the Alignment pop-up menu, choose Start On Edit.

Previously, half of the dissolve's duration started before the edit point; now it begins at the edit point.

3 In the Cross Dissolve Style menu, choose Film.

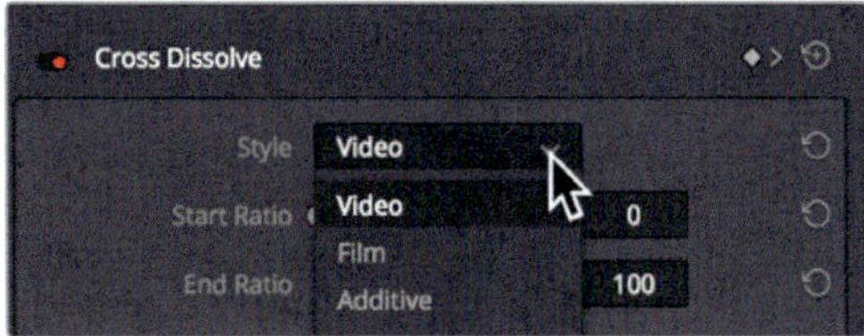

The Style menu allows you to choose from a variety of Cross Dissolve styles. The Film dissolve mimics the subtle luminance and acceleration response of a dissolve optically generated for film. You can further increase the acceleration curve by choosing an Ease In & Ease Out acceleration.

4 In the Ease menu, choose In & Out to create a smoother transition.

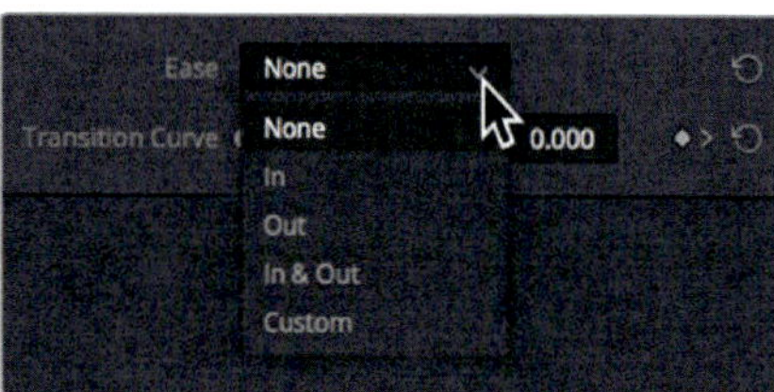

5 Play the modified Film Dissolve to view your changes.

Now that you have customized a transition, you'll learn how to save it for repeated use.

Saving custom presets

After customizing a transition, you can save that transition and its customizations into the Effects Library for use in future projects.

1 In the timeline, right-click the customized Cross Dissolve transition.

2 In the pop-up menu, choose Create Transition Preset.

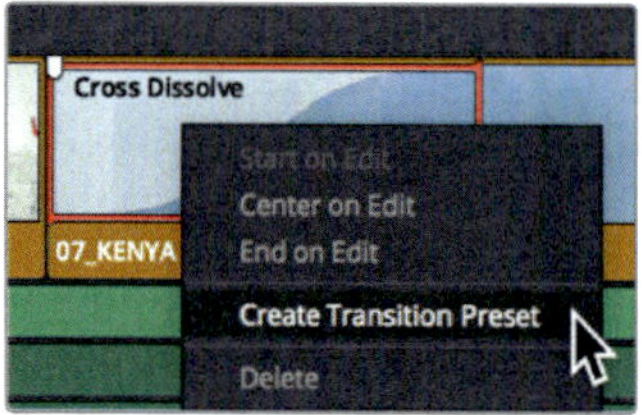

3 In the Transition Preset dialog, enter the name **Filmic Dissolve**, and click OK.

Saved presets are saved into the Effects Library.

4 In the upper-left area of the Interface toolbar, click the Effects Library button.

The Effects Library contains all transitions, titles, and filter effects. The categories along the left side make it easy to find the effect type you are looking for.

5 In the sidebar, select the Video Transitions category.

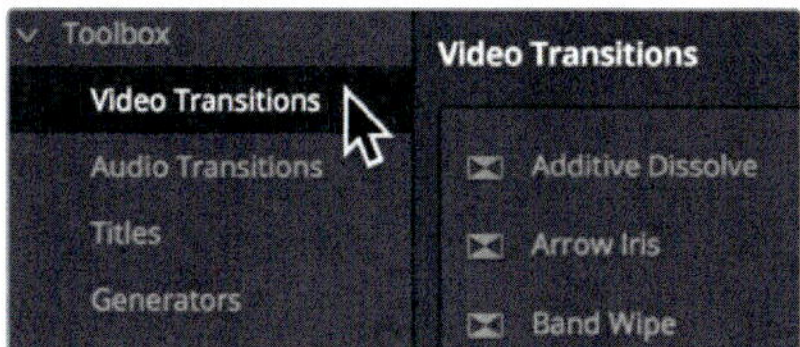

All custom presets have a yellow icon to the left of their names to make them easier to locate.

6. Scroll through the Effects Library to locate your saved preset.

   If your saved preset is something you plan on using as a signature transition throughout a program, you may want to save it as the standard transition. The standard transition has the benefit of being the transition that can be added from the Timeline > Add Transition menu or a keyboard shortcut.

7. In the Effects Library, right-click your saved preset, and choose Set As Standard Transition.

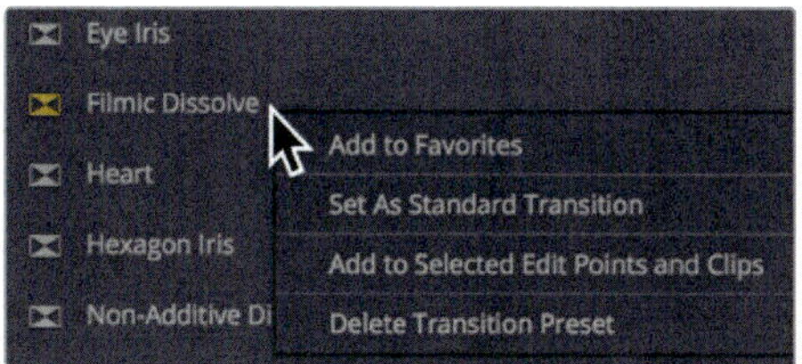

The new standard transition has a red tag to the left of its name to identify it as the standard transition. It is now the standard transition for all projects you create on this system.

Applying transitions and filters from the Effects Library

DaVinci Resolve includes many types of transitions, each with a unique visual style. Other transitions may not be as useful storytelling tools as the dissolve, but they can be handy in specific situations. Because other transitions are not as commonly used, you add them directly from the Effects Library and not by using a keyboard shortcut.

1. In the second interview clip, position the playhead over the jump cut.

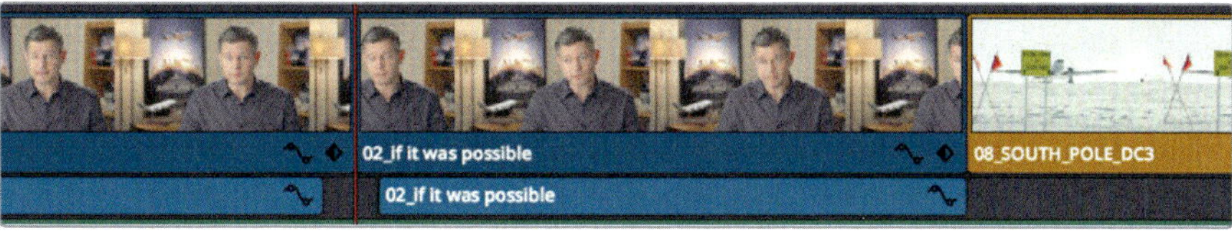

2 Press the / (slash) key to play around the edit point.

This is the jump cut in his interview that you inadvertently created previously. It is time to try to fix this problem using a unique transition called Smooth Cut.

3 At the top of the Effects Library, click the search button.

Instead of scrolling through the long list of transitions or filters, you can search for the name of the effect you want.

4 In the search field, type **Smooth**, and press Return to locate the Smooth Cut transition.

5 From the Effects Library, drag the Smooth Cut transition over the jump cut in the timeline.

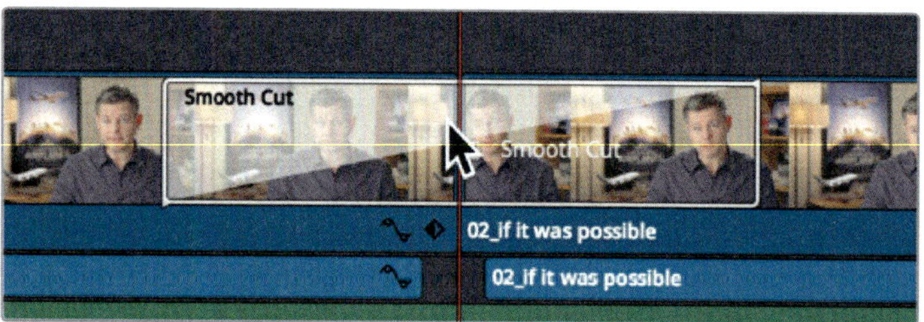

6 Release the mouse button when the transition cursor is centered over the cut.

The Smooth Cut is a sophisticated transition that warps the two sides of a cut, so they seamlessly blend together. If the jump cut is subtle, the results of the Smooth Cut can make it seem like one continuous shot. It works best when it is used for a very short duration of around three or four frames.

7 Zoom into the timeline until you are clearly able to see the name, Smooth Cut.

8 Place the mouse pointer over the right edge of the highlighted Smooth Cut box.

9 Drag in toward the cut until the tooltip reads -00:09 and 00:04.

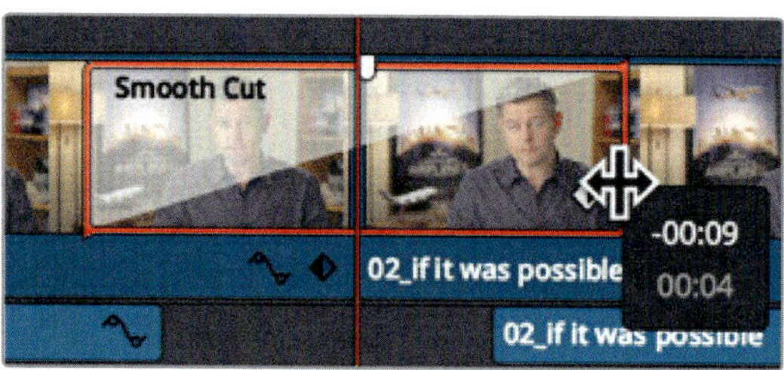

10 Play the timeline to view the repaired jump cut.

Are there ethical questions about making an edited interview clip appear to be one continuous shot? This aviation trailer doesn't really create a moral dilemma, but you should still decide for each of your projects.

Applying filter effects

The Effects Library also holds filter effects that you can use to create unique visual effects or to solve common visual problems.

1 Press Shift-Z to see the entire timeline.

2 Press Shift-Down Arrow until you are on the last red marker on the 05_BAY_AREA_LIGHTS clip.

This clip appears a little flat and would look nicer if it were shot with a tilt-shift lens. Luckily, DaVinci Resolve has a built-in Tilt-Shift Blur filter to simulate just such a lens.

3 In the Effects Library, clear the search field by clicking the clear button to the right of the search field.

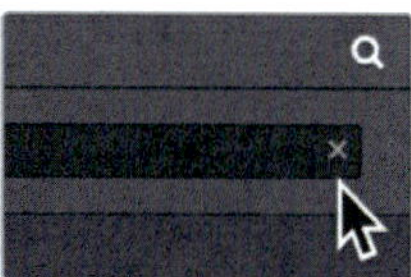

4 In the sidebar category list, select OpenFX > Filters, and in the search field, type **Tilt**.

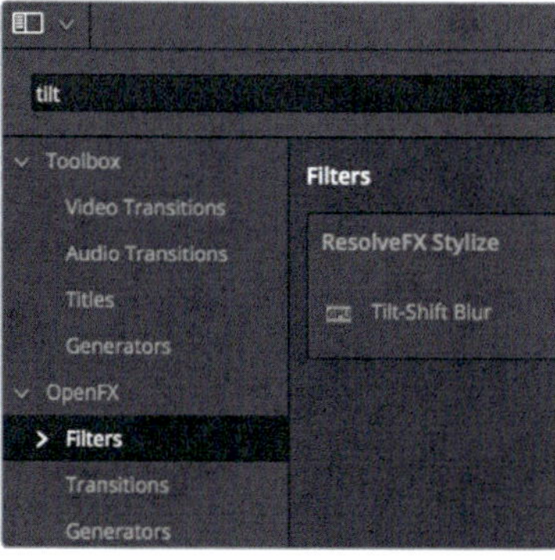

5 Drag the Tilt-Shift Blur filter onto the 05_BAY AREA LIGHTS clip in the timeline.

The Tilt-Shift Blur filter is added; but if you are using the free version of DaVinci Resolve, a dialog box indicates that this effect is available only in the Studio version.

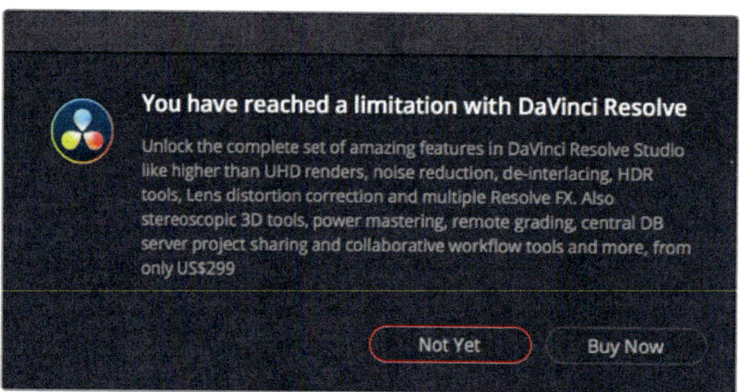

However, you can still try it out without upgrading, although the results will include a watermark.

6 Click Not Yet to continue. You have many parameters available in the Inspector to customize its look.

7 In the timeline, select the 05 BAY AREA LIGHTS clip, and then click the Inspector button in the upper-right corner of the interface.

When a filter effect is applied to a clip, the Inspector will display an OpenFX tab next to the standard Video tab.

8 At the top of the Inspector, click the OpenFX tab.

TIP To reset a single parameter or the entire effect, click the circular Reset button to the right of the Inspector.

You can quickly compare the filtered shot to the original by disabling the effect in the Inspector.

9 At the top of the Inspector, click the Disable button to the left of the filter effect's name. Click it again to enable the effect.

If you are using the free version of DaVinci Resolve, you can remove this effect from the timeline.

10 Click the trash can icon at the top of the Inspector.

Effect choices are neither right or wrong. You can use as many or as few as you think you need to tell your story. The only potential wrong is the end result. You must always ask yourself if an effect helps set the right mood or atmosphere for the story. Does it help to direct audience attention in the right direction? If the answer is yes, then try it. Ultimately, it is your own aesthetic that will help you decide.

Using third-party effect plug-ins

OpenFX is a cross-platform visual effects plug-in standard used by DaVinci Resolve and Fusion software as well as other applications. Popular plug-in packages—such as GenArts' Sapphire, RE:vision Effects' ReelSmart Motion blur, and BorisFX Continuum Complete—can be added to DaVinci Resolve to perform many stylized operations that would be otherwise difficult or impossible to perform using only DaVinci Resolve's standard tools. These plug-ins can be applied in exactly the same ways you would apply other items from the Effects Library.

The installation and licensing of plug-ins is managed by each vendor's installer. Once installed, OpenFX plug-ins appear in the Effects Library and in the OpenFX category.

Reframing shots

Most of the time, you're going to be editing high-definition (HD) or maybe ultra high-definition (UHD) projects that use a 16×9 aspect ratio. But sometimes you'll need (or want) to edit and view your program using a different aspect ratio. You can do so using the output blanking menu.

1 Choose Timeline > Output Blanking > 2.39.

The aspect ratio of 2.39:1, which is used for 35mm theatrical widescreen, is applied in the viewer.

TIP The controls used by the output blanking menu to create masks are located on the Color page in the sizing palette's output sizing mode.

When you apply output blanking, it does not remove the hidden part of the clip, it just covers it up which allows you to reframe clips to better fit the new aspect ratio.

2. Move the playhead over the first interview shot, and in the timeline, select the clip.

   The Inspector now shows the video tab controls for the selected clip. These controls include typical parameters for Zoom size, Position, Rotation Angle, and Crop.

   You can move your subject down in the viewer to give him more head room and make a more pleasing composition.

3. Hover the mouse pointer over the Position Y numeric field.

   Any numeric field can be used as a virtual slider just by dragging within the field.

4. In the Y Position numeric field, drag to the left until the value shows -70.00.

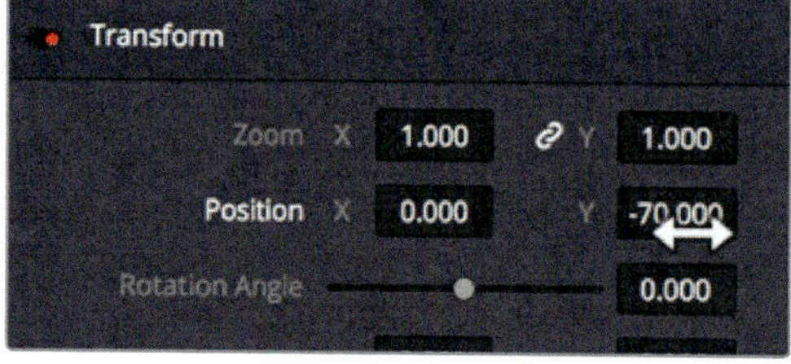

   This is better positioning for the subject. You'll have to copy this to the remaining interview shots.

5. Choose Edit > Copy, or press Cmd-C (Mac) or Ctrl-C (Windows).

6 In the timeline, select the next interview clip to the right. Then Cmd-click (Mac) or Ctrl-click (Windows) the two remaining interview clips in the timeline to select them.

7 Choose Edit > Paste Attributes, or press Option-V (Mac) or Alt-V (Windows), to open the Paste Attributes window.

8 In the Video Attributes category, select the Position checkbox, and click Apply.

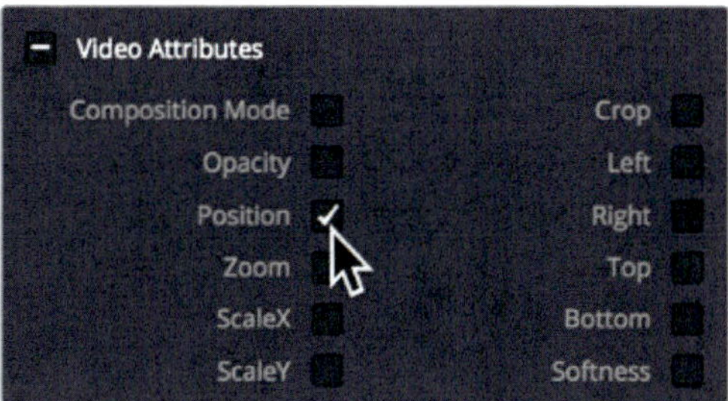

This reposition value applied to the first interview clip is pasted onto the selected interview clips.

Using onscreen controls

You can apply common transform operations such as position, scale, and rotation using onscreen controls in the viewer, rather than dragging virtual sliders in the Inspector. Let's reframe a shot by resizing it in the viewer.

1 Press Shift-Down Arrow to go to the last red marker on the 05_BAY_AREA LIGHTS clip.

This is the last clip before your credits, so to indicate that you are bringing this trailer to a close, let's simulate the camera pulling back and away from the scene. First, you'll use the onscreen control to resize the clip a bit larger.

2 In the timeline, select the first half of the 05 BAY AREA LIGHTS clip, and move the playhead over it to see the clip in the viewer.

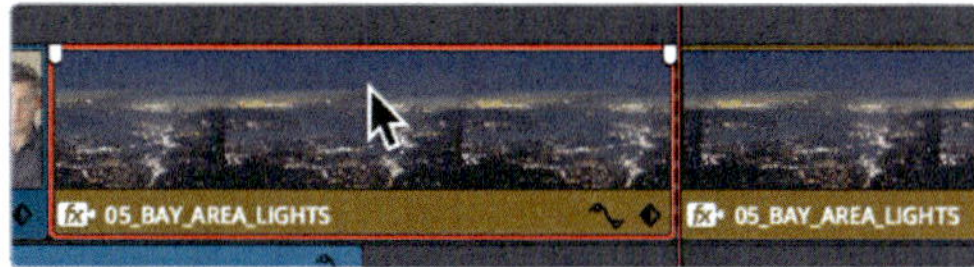

3 In the lower-left corner of the viewer, click the timeline viewer pop-up menu, and choose Transform.

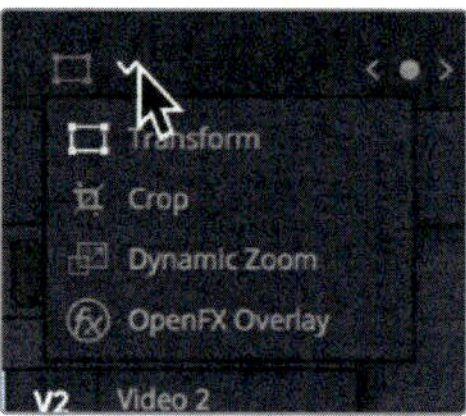

A bounding rectangle appears in the viewer to show the size and position of the frame.

4 To scale the clip, drag a handle in any corner of the bounding box away from the center until the zoom value shows 1.200.

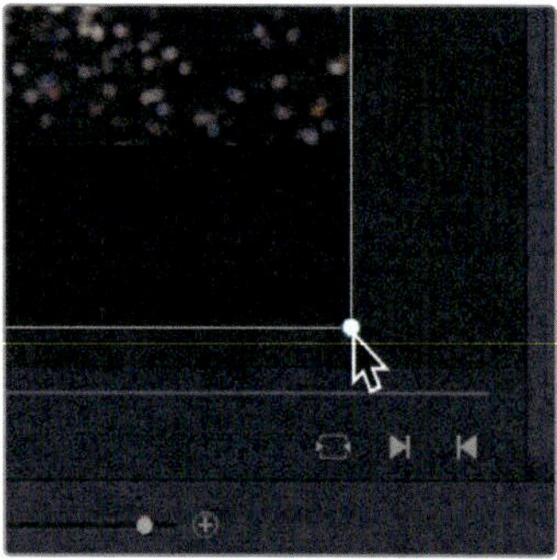

By default, the zoom X and Y values are linked together so the clip is resized equally in X and Y, thereby keeping its normal aspect ratio.

TIP In the magnification menu (in the upper-left of the viewer) you can change how the frame is sized in the viewer.

5 In the lower-left corner of the viewer, click the transform rectangle to hide the onscreen controls.

Animating a resize

Almost every parameter displayed in the Inspector can be animated over time by setting two keyframes. You'll use the current resized frame as the starting keyframe, and then set a new keyframe at the end of the clip.

1. Move the playhead to the start of the 05 BAY AREA LIGHTS clip.
2. Select the clip to display the controls in the Inspector

   Animating the parameters of an image over time requires that you set keyframes. Keyframes allow you to assign different parameter values to specific frames. DaVinci Resolve then interpolates between the two values to create a smooth animation.

   You have already scaled the clip up for your starting position, so all you need to do in this first frame is add a keyframe.

3. In the Inspector, click the zoom keyframe button to the right of the Zoom X and Y numeric fields.

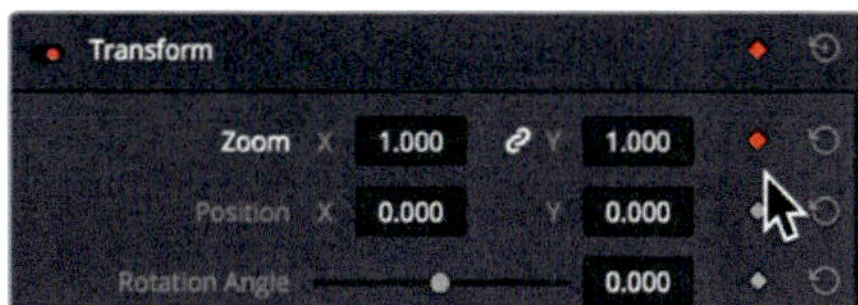

   Animating a parameter requires that you set at least two values for it on separate frames. The first value at the start of the clip is now set. You'll set the second value where you want the image to appear at the end of the clip.

4. In the timeline, position the playhead at the end of the 05 BAY AREA LIGHTS clip.

   DaVinci Resolve uses an auto-keyframe model. When a parameter already has one keyframe set, changing the position of the playhead, and then modifying the parameter, automatically adds a second keyframe.

5. Place the cursor over the Zoom X numeric field, and drag to the right until the value is reset to 1.0.

6. Press the / (slash) key to play around the currently selected clip.

   You can further control the smoothness of keyframe-controlled effects using the keyframe curve editor in the timeline.

7. Zoom into the timeline to see the 05 BAY AREA LIGHTS clip clearly.
8. With the clip still selected, in the lower-right corner of the timeline segment, click the sine wave-shaped curve editor button.

The curve editor appears beneath the clip.

In the curve editor, white dots represent the keyframes. Buttons along the top of the curve editor allow you to change the interpolation of a selected keyframe.

9 Click the first keyframe on the left side of the curve editor.

10 Click the first interpolation button on the left to place an ease-out interpolation.

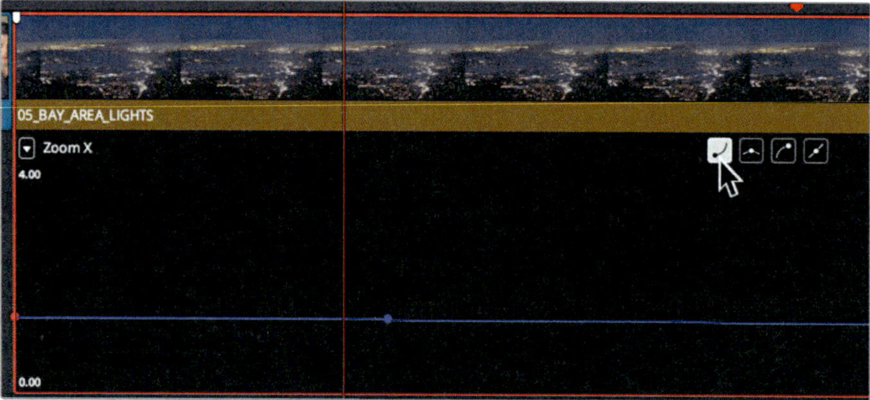

11 Click the last keyframe on the right side of the curve editor.

12 Click the second-from-last Interpolation button on the right to place an ease-in interpolation.

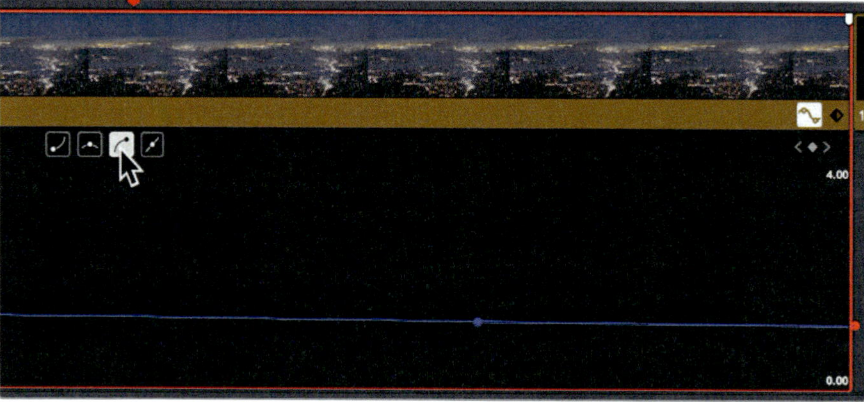

13 Click the sine wave-shaped curve editor button to close the curve editor.

14 Press the / (slash) key to play around the currently selected clip.

Rendering and background caching

Depending on the speed of your computer and disk drives, as well as the media file types you are using, all effects may not play back smoothly on your computer. The fps (frames per second) indicator above the timeline viewer shows the actual playback framerate that your computer is achieving. If the number has a red dot next to it, your system is playing your project slower than the actual project frame rate.

To optimize playback performance for complex effects, DaVinci Resolve can automatically render and cache such effects to your disk drive. Although DaVinci Resolve can use three distinct caching systems to render files, this exercise will focus on smart caching for the Edit page.

The first task is to make sure caching is turned on.

1 Choose Playback > Render Cache> Smart.

   The smart cache operates on timeline-specific effects such as transitions, opacity adjustments, and composite mode superimpositions. Regions of the timeline that require caching have a red bar over them, whereas regions that are already cached have a blue bar over them.

   All of this is easy enough, but you have additional settings to customize the caching operation.

2 Choose File > Project Settings > Master Settings.

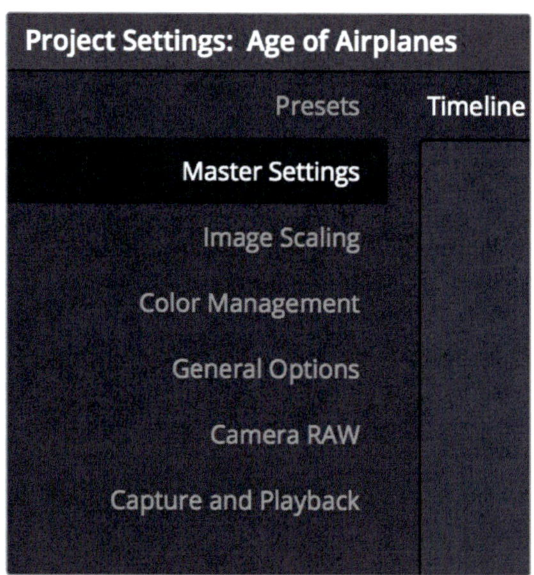

Master Settings include a few Optimized Media settings related to caching. The "Cache frames in" menu sets the compression format that DaVinci Resolve uses to save the rendered files. The choices common to Mac and Windows include uncompressed 10-bit and 8-bit formats, and Avid's DNxHR formats. In Mac, DaVinci Resolve also includes Apple's ProRes compression format.

The default setting creates a high-quality 10-bit file that will look good in your final output. If you are temporarily working on a portable or a laptop with a slow disk drive, you may want to opt for a marginally lower-quality 8-bit format such as Avid DNxHR HQ or ProRes 422 to enable faster processing. For now, you'll leave this setting at the default value and move on to background processing.

3 Make sure "Enable background caching" is selected.

   When background caching is enabled, effects rendering begins based on the length of time your computer sits idle.

4 In the "Enable background caching" numeric box, enter **3**.

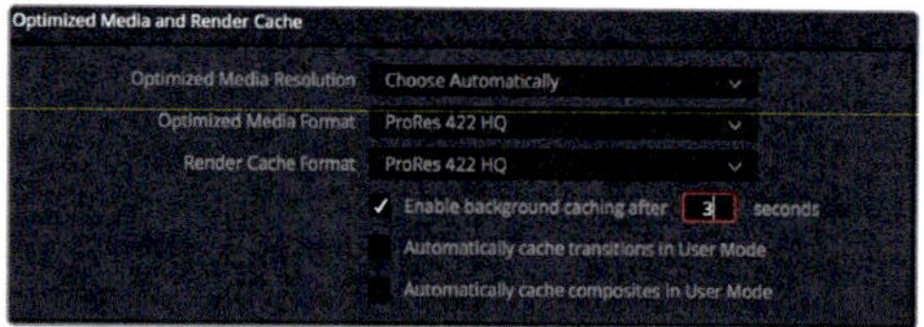

   Background caching will now begin to render effects after your computer sits idle for three seconds.

TIP To delete all of the rendered cache files for the current project, choose Playback > Delete Render Cache > All.

Now that your smart caching is set, DaVinci Resolve will automatically render everything necessary to optimize playback for your effects. When you reopen a project, cached clips are still cached; but when you change any cached effect, it will need to be re-cached.

Creating a constant speed change

Changing the playback speed of a clip is done for a variety of reasons. Sometimes it's used to accentuate dramatic action, and sometimes it's used so the timing of a clip fits into a scene. In almost every production genre, you'll have a need to speed up clips, slow them down, and even stop and hold on a frame for a few seconds.

The most common type of speed change is a constant speed change. It uniformly alters the playback of a clip in the timeline to turn it into a slow-motion clip or a fast-motion clip of a single framerate.

1 In the timeline, position the playhead at the start of the 06_MILKYWAY clip.

2 Play over the clip (always a good idea before you change its speed.)

  This clip goes on a bit too fast for a shot of the nighttime sky. You want to slow the speed but not change its duration in the timeline.

3 Select 06_MILKYWAY. Right-click the clip, and in the pop-up menu, choose Change Clip Speed.

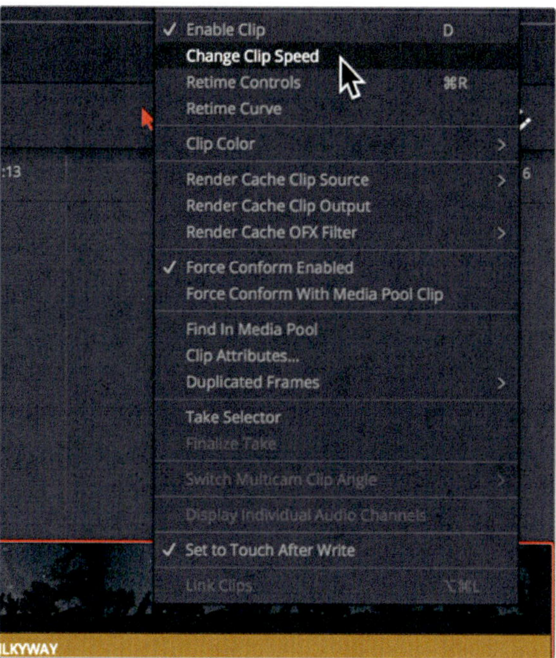

The Change Clip Speed dialog appears with a number of controls for clip playback.

4 In the Speed numeric field, enter **50**, and click Change to close the dialog.

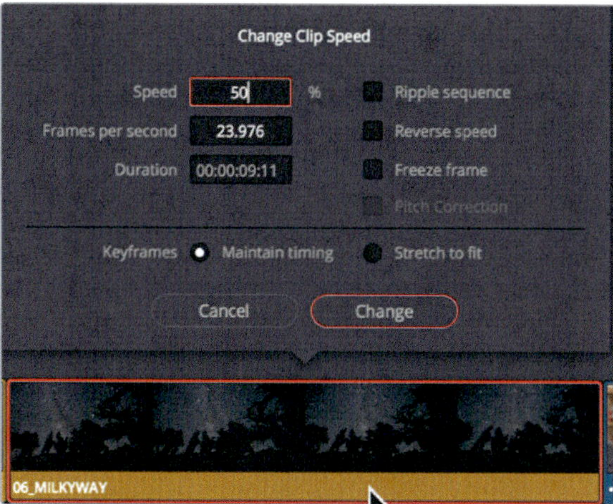

Setting this value to 50% means that the clip will play at half the timeline frame rate—in this case, 12 frames per second. To indicate that the clip's playback speed has been changed, a small speed change icon is displayed next to the clip's name in the timeline.

5 Play the clip to see the speed change results.

When creating a slow-motion clip, the default settings in the Change Clip Speed dialog do not change the overall duration of the clip or the timeline. However, DaVinci Resolve does have tools that allow you to modify the speed of a clip by changing its duration.

Retiming using the selection and trim tools

While the Change Clip Speed dialog creates constant speed changes and retains the clip duration, the retime controls stretch and shrink a clip's duration by slowing it down and speeding it up.

1 Right-click the 06_MILKYWAY clip, and in the pop-up menu, choose Retime Controls, or press Cmd-R (Mac) or Ctrl-R (Windows).

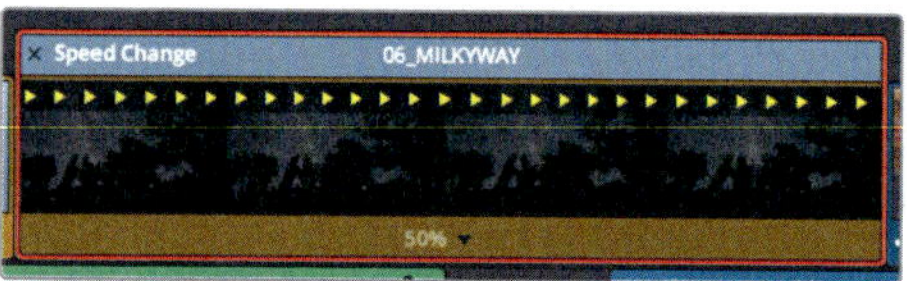

The Speed Change bar appears at the top of the clip in the timeline. The current speed of the clip is listed along the bottom. The speed of a clip is altered by trimming the Speed Change bar in the timeline.

2 In the toolbar, select the trim tool, or press T.

3 Move the pointer to the right edge of the Speed Change bar.

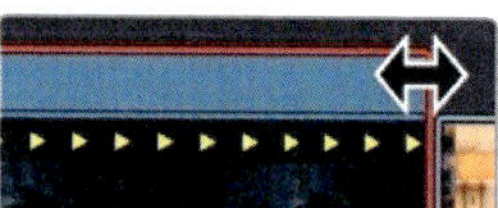

The pointer turns into a double-arrow cursor.

4 Drag the edge of the Speed Change bar to the right to lengthen the clip until the speed display at the bottom of the clip reads 40%.

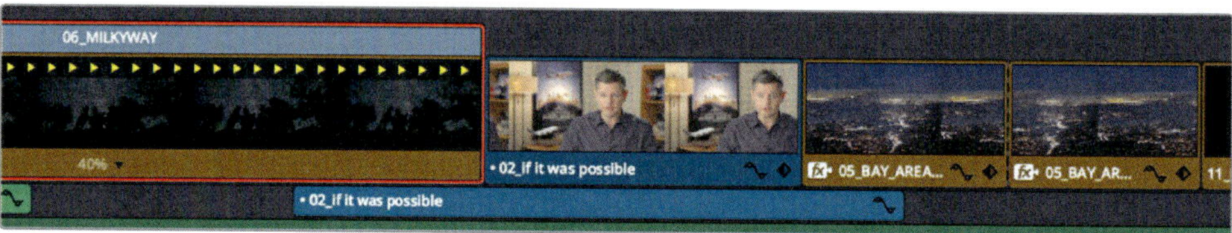

Dragging the Speed Change bar to the left extends the duration of the clip by slowing down its playback speed. When the trim tool is selected, the Speed Change bar ripples the remaining clips in the timeline. However, if you undo that last step, you can see how the same trimming feature can change depending upon which tool is selected in the toolbar.

5 Press Cmd-Z (Mac) or Ctrl-S (Windows) to undo the previous speed change.

6 In the toolbar, click the Selection mode tool, or press A.

7 Move the pointer to the right edge of the Speed Change bar, and drag to the right until the clip completely overwrites the incoming interview clip.

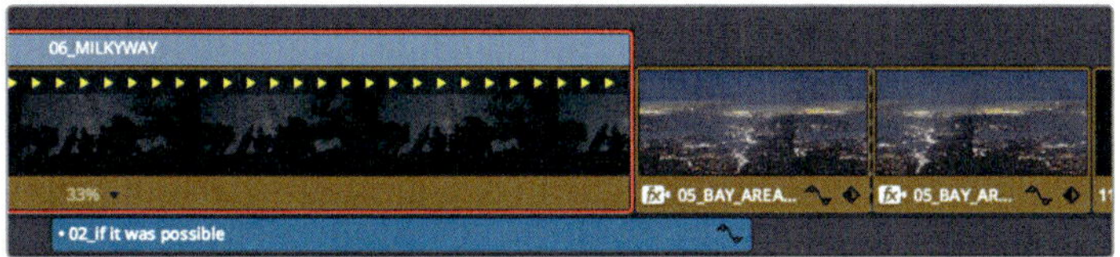

8 Play the retimed clip to see the results.

TIP To return a clip to its original speed, click the clip speed pop-up menu at the bottom of the clip, and choose “Reset to 100%”.

With the trim tool selected, the timeline is rippled, thereby pushing the rest of the clips to the right. The clip’s duration and the overall timeline duration are extended.

Reversing a clip

When the retime controls are displayed, you have the option of using a pop-up menu at the bottom of the clip to change the speed instead of trimming it using the Speed Change bar.

1 At the bottom of the 06_MILKYWAY clip, click the retime pop-up menu.

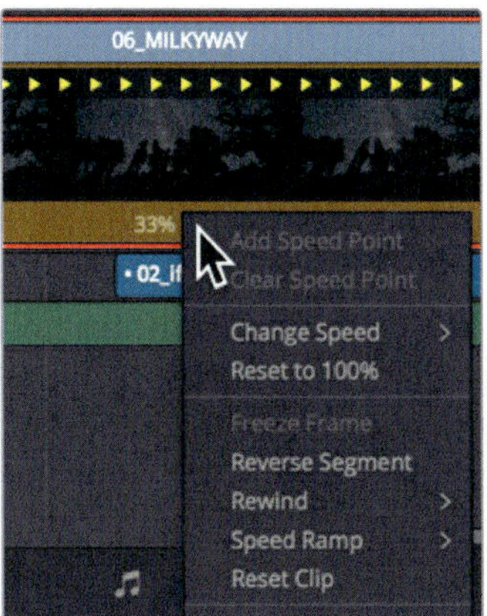

This menu includes options for common slow and fast speeds, as well as the ability to set speed points for variable retiming. Still other options allow you to reset or reverse the playback.

2 In the menu, choose Reverse Segment.

3 Play the retimed clip to see the results.

The clip now plays in the reverse direction at the same speed.

Changing retime processing

You can change the way a retimed clip is processed, and thereby change the way it looks. The various retime processing settings located in the Inspector trade speed for quality. Which you choose will depend on the type of movement in your shot and how much processing time you are willing to allow.

1 Select the retimed 06_MILKYWAY clip.

2 Open the Inspector.

3 Scroll to the bottom of the Inspector.

4 Click the Retime Process menu to open it.

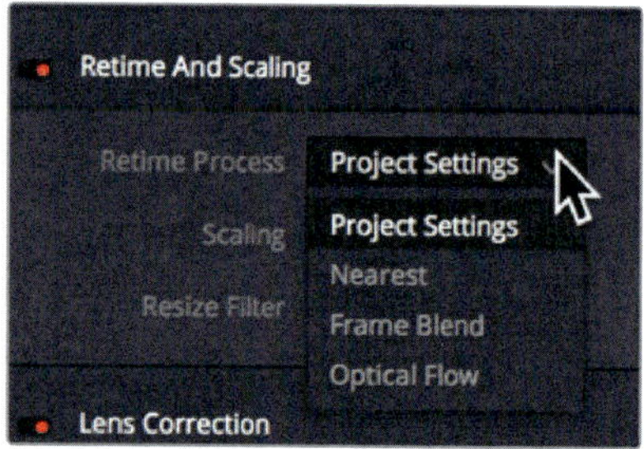

You have three options for processing clip retiming: Nearest, Frame Blend, and Optical Flow.

- Nearest is the fastest processing option but delivers the lowest-quality results. This simple operation duplicates frames to create slow motion which often causes stepping artifacts even in clips that have just a moderate amount of movement. Nearest is the default option set on the master settings of the project settings.

- Frame Blend is a slightly more processor-intensive option that delivers better-looking results. It also duplicates frames to create slow motion, but then blends them to produce smoother motion. This is the most reliable option and generally delivers acceptable results.

- Optical Flow is the most processor-intensive and highest-quality process. It uses motion estimation and warping techniques to generate new frames from the original source frames. The results can be exceptionally smooth when the motion in a clip is unobstructed. However, if two moving elements cross while moving in different directions (such as legs crossing when walking) or the camera movement is erratic, optical flow processing can cause stretching and tearing artifacts.

TIP Additional motion estimation controls in the Frame Interpolation section of the project settings can sometimes improve small tears or stretching artifacts.

5 In the menu, choose Optical Flow.

  When Optical Flow is selected, a red bar appears above the clip to indicate that the process needs to cache. If smart caching is enabled, the rendering will be performed in the background and you'll be able to see the results in a few seconds.

6 When the clip is cached, play over the optical flow results to see the smoother motion.

It's a good idea to first try Optical Flow processing to see if it produces acceptable results, and then revert to Frame Blend, as necessary. Optical Flow, unlike the other retime processing types, **requires** that you cache the result.

An Introduction to Audio Post and Sound Design

Chances are you've heard the adages "Seeing is believing" and "A picture is worth a thousand words." However, when it comes to motion pictures, both the visuals and soundtrack are equally important. In fact, a great soundtrack sells the onscreen illusion, manipulates emotions, transports the audience into the scene and captivates their imagination. A lousy soundtrack, on the other hand, keeps the audience at a distance, distracts from the story, and draws attention to production flaws, performance issues, and plot holes.

Audio post production is much more than simply adjusting volume levels and mixing tracks. Transforming production sound into a powerful soundtrack requires time, technical skill, creative vision and execution, as well as a full set of professional audio tools. The good news is that DaVinci Resolve 14 includes the tools to create a professional soundtrack from start to finish. Before you dive into the following audio chapters, it's a good idea to understand the audio post production process and workflow.

Keep in mind that many elements affect the workflow you'll use: the type of project, budget, format, length, deliverables and distribution methods often dictate the size of the post audio team, amount of time, and tools available to get the job done. For this introduction, let's focus on the fundamental post production audio processes necessary for both narrative and documentary style projects. Although the following pages explain the different jobs and stages in audio post production, having the Fairlight page built into DaVinci Resolve means you can perform the same steps on your projects with no additional crew or budget.

What is audio post production?

Let's start with a few basic terms. **Audio post production** refers to the process of making a soundtrack for moving images. Notice the use of "moving images," which encompasses all projects great and small from movie theaters to streaming videos and everything in-between. A **soundtrack** is simply the audio that accompanies a finished project.

How your audience experiences the finished project is greatly influenced by the soundtrack. In fact, a well executed soundtrack may go unnoticed for hours by the audience while it is immersed in show. On the other hand, it takes only a few seconds of an amateurish or sloppy soundtrack to lose the audience not only from the story, but possibly from the theater or to a different channel.

If you've ever recorded or watched a home movie, especially one shot at an exciting public place such as the beach or an amusement park, then you've got first hand experience with some of the inherent challenges in both recording and listening to natural production sound. All those excess environmental sounds and distractions create a need for audio post production to transform raw sound into successful soundtracks with clear dialogue, realistic effects, and lush acoustic soundscapes wrapped in an emotionally powerful score.

What is the audio post production workflow?

Since the advent of synced sound in motion pictures, the first rule of audio post has been, "Never start working on audio until the picture is locked." Locked suggests that there will be no more changes to the picture edit from this point forward.

In reality, changes always happen. Why does this matter? Because, soundtracks need to maintain a frame accurate relationship with the picture to stay in sync. If they are off by as little as one or two frames, the sight and sound will be noticeably out of sync, a situation that is distracting, unprofessional and likely to lose your audience.

In a traditional post production workflow, changes to locked picture have a cascading snowball effect on audio post. But when you're working with DaVinci Resolve, which is the only professional editing software that includes a full digital audio workstation (DAW), no matter what editing changes are made, you can update your project immediately and efficiently. This gives you tremendous creative flexibility if you are working on your own, because you can go back and forth between editing picture, audio work and color correction as often as needed.

For larger productions, DaVinci Resolve solves the issue of updating, transferring files to other systems, and conforming projects between editorial and audio post because editing and audio post production are done in the same project without ever leaving the application. Best of all, audio post production can start on the exact same timeline that the editor used so you have zero chance of losing frames or going out of sync. Once audio post begins, the editor can use a duplicate timeline to make any new changes. Then the audio editor can easily merge the changes between timelines with DaVinci Resolve's powerful timeline comparison tool.

DaVinci Resolve has the audio tools needed for the highest quality audio post production, and is ideal for small projects yet powerful enough for big Hollywood studios and broadcast productions to use as well. Whether you are working on your own or with a large post production team, you can easily migrate projects to a large facility for experienced audio sound designers and engineers to mix and master the soundtrack.

Now let's break down the different phases and jobs in a traditional audio post production workflow. With DaVinci Resolve you can perform all of these steps as needed by yourself or with a team of audio professionals on your own projects.

Spotting the soundtrack

A **spotting session** is when the supervising sound editor and the sound designer (often the same person) sit down with the director, editor and composer to look for soundtrack elements that need to be added, fixed or re-recorded. Notes from a spotting session are combined into a spotting list that details music cues, important sound effects, dialogue fixes, and additional audio notes.

DaVinci Resolve has simplified these spotting sessions with the timeline markers that you can use in either the Edit page or Fairlight page. The marker index in the Fairlight page serves as an interactive spotting list, that not only includes information and a thumbnail for each marker, but also moves the playhead to the selected markers position in the timeline.

Production dialogue editing

Dialogue editing is the tedious behind the scenes task of splitting dialogue into separate tracks, removing unwanted sounds, replacing individual words or phrases for clarity and balancing separate clip audio levels for consistency. Why go to all that trouble? Because spoken words are as important to a soundtrack as the lead vocals in a hit song. Keep in mind that dialogue editors are responsible for all spoken words including dialogue, narration, and voice over.

Production dialogue editing starts with creating separate tracks for each character, then moving all of those dialogue clips into a specific track. This crucial step is necessary because each voice in a production is different and, therefore, needs to be processed individually with volume normalization, equalization, and effects specific to that voice.

Next, the dialogue editor cleans up the tracks and removes any unwanted human sounds (like tongue clicks and lip smacks). If a distracting sound can be physically cut out, this is the time to do it. Plug-ins and effects can help eliminate unwanted clicks, pops, and noise automatically; but be aware that any processing you add to a clip, can affect a voice, as well.

After the dialogue is cleaned up, the volume levels are balanced to be consistent on each dialogue track. If dialogue can't be used because it is damaged, noisy, or unclear, it must be replaced with audio from other takes or re-recorded. The process of re-recording production dialogue is called **automatic dialogue replacemen**t (ADR) or **looping**.

Dialogue editing can be time consuming and laborious. Once again, DaVinci Resolve includes easy navigation, precision editing tools and shortcuts that can simplify and speed up the process.

Sound design and sound effects editing

Once the dialogue editing is finished, the creative process begins! The sound designer's creative input to the soundtrack is similar to that of the director of photography for the picture. Sound designers are responsible for the overall acoustic experience for the audience. They also oversee the many individual tracks of sound and music that comprise the soundtrack. These audio tracks include dialogue, ambience, hard sound effects, and foley sounds.

Not only do sound designers determine the aural illusion and mood of the soundtrack, they also create, record, and enhance sound elements that only exist in their imaginations. After all, many projects need sound effects that don't exist in the real world. Where do you go to record dragons, aliens, or zombies? Those sounds must be created or designed from scratch using a combination of real sounds, simulated sounds, and a lot of processing and effects.

While the sound designer determines the depth and detail of the sound effects tracks, the sound effects editor places each sound effect in corresponding tracks. Sound effects fall into four main categories:

Natural sound, also known as **Nat sound** or **production sound**, is anything other than dialogue recorded by a microphone on location during the shoot.

Ambience, or ambient sound, is the realistic conglomerate of sounds that establish a location, such as waves rhythmically crashing and sea birds chattering for remote seaside ambience.

Hard sound effects are so named because they need to be physically synced to picture and are necessary for the story or scene.

Foley sound consists of any character-driven sound effects caused by characters interacting with their onscreen environments. Foley sounds are named after Jack Foley, a legendary sound editor at Universal Studios, who originally developed the technique of recording reenactments on a stage. Foley sound replaces the original production audio for everything from fist fights to footsteps and clothing movement.

Audio editing tools in DaVinci Resolve's Fairlight page are designed specifically for the precision editing and placement required when editing sound effects. And DaVinci Resolve's clip speed changes are perfect for advanced sound design and pitch effects.

Music editing

Music editing involves placing different music elements into the soundtrack to enhance the mood or story. All soundtrack music falls into one of two categories: music occurring within the scene that the characters can hear, so-called source or **diegetic music**; and **non-diegetic music** that is added in post for the benefit of the audience, the **background score**.

Diegetic music needs special attention to make sure that the volume levels, placement, effects and presence fit the context of the scene.

Non-diegetic music added in post production for emotional effect or impact includes the score, stingers, and stabs. Stingers are singular notes or chords that build tension and suspense. Stabs are quick bursts of music that work like an exclamation point to draw attention to something or someone in the story or narration.

Enhancing and sweetening tracks

Once the dialogue tracks are edited and the sound effects and music added, it's time make subtle improvements to the sound of each track so that they work in context with the other tracks in the mix. The tools used to improve the sound in a track are similar in many ways to the tools colorists use to improve individual shots within a scene. Because you are learning to use DaVinci Resolve, and color correction is an integral part of the post production process, it seems fitting to show the similarities between adjusting audio and color.

For all intents and purposes this process could be called audio correction. You manipulate four fundamental elements to enhance or "sweeten" audio tracks so they work together as intended in the final mix: volume level, dynamics, equalization and pan. DaVinci Resolve controls all four of these elements on every track without the need for additional plug-ins or patching.

Volume controls are used to adjust the loudness of a track on a decibel scale, and are similar to luminance (brightness) because both volume and luminance have strict broadcast standards, and are usually the first thing the audience notices in each scene. Volume levels can be adjusted on each clip, track, and the main output, just as luminance (black and white levels) can be adjusted on individual clips, scenes, and output. In DaVinci Resolve, you can change the volume level of a clip in the timeline or Inspector. Track volume is controlled by faders in the mixer. You can also change the volume levels over time using automation.

Dynamics controls adjust the **dynamic range**, which is difference between the loudest and quietest peaks in a track. A track's dynamic range is very similar to contrast within a shot. A track with a high dynamic range has very loud and quiet elements within the track, such as a character whispering and then screaming in the same scene. A low dynamic range would be fairly flat, such as a commercial voiceover in which the volume level of the talent is very even from start to finish. If you have ever worked with a Waveform or Parade scope in the Color page, controlling a track's dynamics is very similar to adjusting the white and black levels of a clip. Just think of white as the loudest you can get (-3db) and black as the quietest.

The Fairlight page mixer includes the four most common dynamics controls in one easy-to-use panel. The compressor is used to narrow the dynamic range by lowering the loudest peaks and bringing them closer to the lowest peaks. The expander, in contrast, expands the dynamic range to increase the difference between the loudest and quietest peaks. The limiter and gate both work as acoustic "brick walls" to limit sound from exceeding a target level (limiter), and to prevent sounds lower than a set threshold from being heard (gate).

Pan controls place the sound of a track within a panoramic stereo field. These controls are used to compose the acoustic experience just as a cinematographer composes the visuals of a shot. Tracks can be precisely located anywhere from left to right to sound as if they come from an offscreen source, or somewhere within the frame. DaVinci Resolve includes advanced pan controls in both the Edit page and Fairlight page with both 2D (stereo) and 3D sound placement for surround sound systems.

Equalization (EQ) controls manipulate specific frequencies to enhance the overall sound, and are just like working with color, saturation, and hue in color correction. For example: the human voice is based on a fundamental frequency shared by millions, the additional frequencies add tonal qualities to "color" the voice and make it unique and recognizable. The primary function of equalization is to lower frequencies that detract from the voice and boost the positive frequencies to improve the overall sound. The Fairlight page mixer includes a six-band **parametric equalizer** on each track which is the perfect tool for enhancing and "sweetening" audio tracks.

Mixing and Mastering

The last step of audio post is mixing the tracks and mastering the output. Assuming that all of the other steps were completed prior to the mix, the process is fairly straightforward. The goal of mixing and mastering is to balance the levels coming from each track so they sound good as a whole. This is accomplished by making subtle changes to the track levels, or combining similar tracks into submixes to make them easier to control with one fader. The final master needs to sound great and meet delivery standards for loudness. Fortunately, the Fairlight page includes everything you need to mix tracks and loudness meters to make sure the levels are right on target.

Now that you understand some of the technical steps and creative tools that are essential in an audio post production workflow, you can dive in to the upcoming lessons and start putting them to use on your own projects!

Lesson 7

Working with Audio on the Edit Page

Depending on the size of your project, you may be responsible for the basic audio tracks or for the entire sound design. Even if you plan to hand off your final mix to an audio editor, you still need to give your client a sense of what that final audio mix may sound like.

On the Edit page, DaVinci Resolve 14 includes audio editing and mixing tools to help you mix your soundtrack. DaVinci Resolve also includes all of the capabilities of a full-blown digital audio workstation (DAW) on its Fairlight page. In this lesson, you will work on the Edit page to create additional audio tracks for sound effects, and then set the audio levels to create a balanced mix.

Time

This lesson takes approximately 45 minutes to complete.

Goals

Working with markers

You will create your mix using a similar version of the current timeline. Because that timeline already has narration and music in place, you'll begin by identifying areas that could benefit from sound effects. You can use markers to annotate clips in the source, identify a specific time in the timeline, or label a range of time. Such markers are often used as reminders for a task that you want to do later. In this timeline, you'll add markers in the timeline to identify the four areas that need additional sound effects.

1 Open the Age of Airplanes project, if necessary, and choose Workspace > Reset UI Layout.

2 Select the Rough Cuts bin, and double-click the Audio Rough Cut timeline to open it into the timeline viewer.

3 In the timeline, position the playhead in the middle of the 02_A380_TAKEOFF clip.

This clip could use a nice big roaring take-off sound. To place a marker here, you must first select the clip.

4 Using the selection mode tool, click the clip in the timeline.

You can add markers using the marker button in the toolbar.

5 Click the marker button.

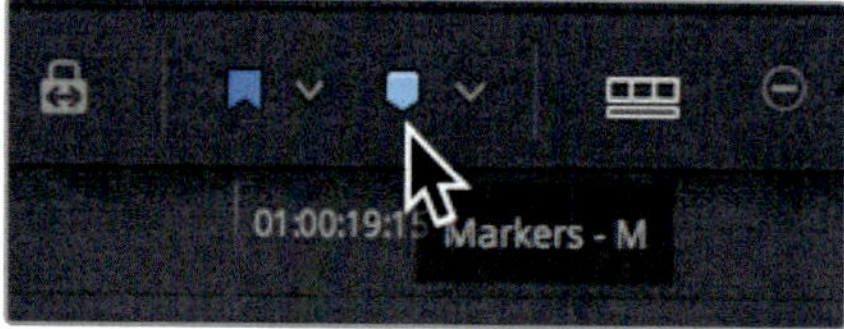

TIP You can turn the marker visibility on and off in Timeline View Options > Show Flags and Markers.

A blue marker is added to the selected clip in the timeline at the current playhead position.

6 On the clip, double-click the blue marker; or select the marker, and press Shift-M to open the marker dialog.

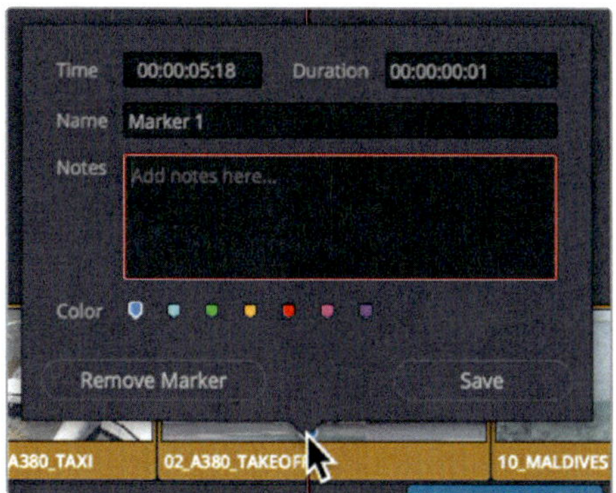

You can change the color of markers to further organize your work. For instance, you could add green markers where graphics were needed and purple markers where effects were needed. You could also add notes to markers that are more descriptive of your changes.

7 Click the red color swatch, and in the Name field, type **SFX**. In the Notes field, type **Add Roaring Take Off**. Click Save.

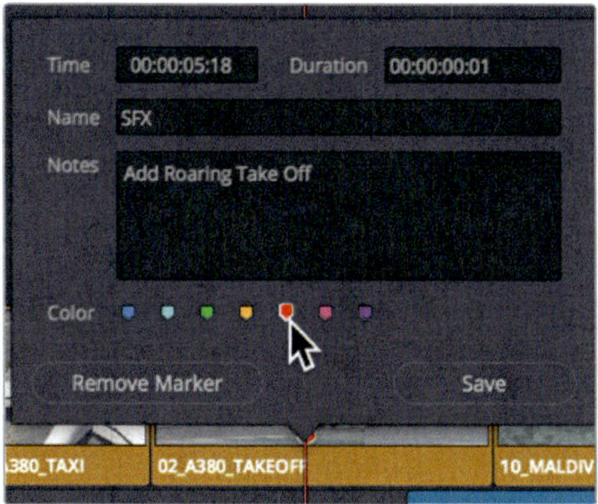

With one clip completed, you have two remaining timeline locations that need markers and notes.

8 In the timeline, position the playhead in the middle of the 10_MALDIVES clip and select it.

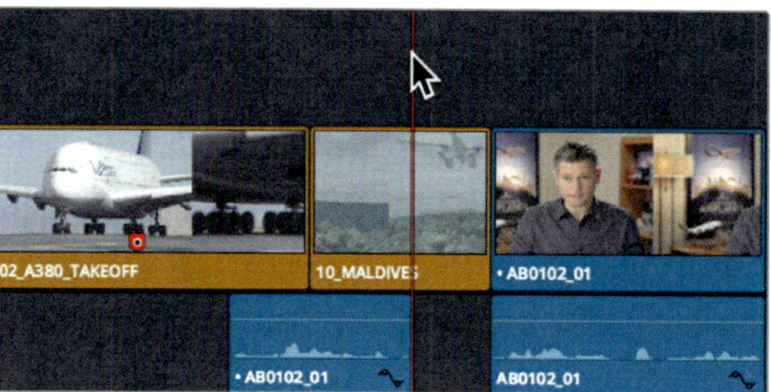

For this clip, you'll use the fastest method, a keyboard shortcut.

9 Press the M key twice.

The first M key press adds the marker; the second press opens the dialog.

10 In the dialog Name field, type **SFX**; and in the Notes field, type **Add Overhead Sea Plane**. Click Save.

> **TIP** To delete a marker, open the Marker dialog, and click the Remove Marker; or select the marker, and press the Delete key.

11 Next, position the timeline playhead in the middle of the 08 South_Pole_DC3 clip, and click in the gray area above the clip so nothing is selected in the timeline.

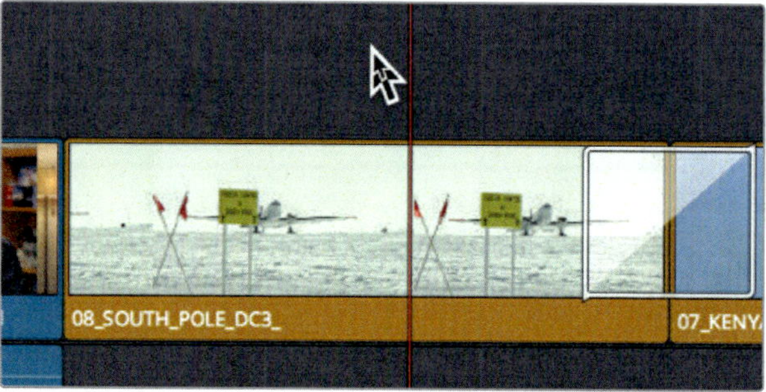

For this clip, you'll use the fastest method of all, a keyboard shortcut.

12 Press the M key.

The marker is added to the timeline ruler under the playhead location. Unlike the other markers applied to the clip, if you were to drag or cut and paste the South Pole clip to a new location in the timeline, the marker would not move with it. With that exception, the marker works exactly the same as the clip markers.

13 Press M again to open the dialog, and in the Name field, type **SFX**. In the Notes field, type **Add Loud Prop Plane**. Click Save.

> TIP Ripple trimming a clip in the timeline will move a marker in the Timeline Ruler to the same duration as the trim.

All your red color markers are added into the program. Now let's look at how we can add markers to a source clip.

Adding markers in the source viewer

You'll also find markers invaluable when applied to source clips. Markers on a source clip can add notes as in the timeline. You also can use markers to identify multiple areas of a clip that you may want to use in your timeline. In this way, markers can be used as placeholders for multiple in and out points.

1 In the Audio bin, double-click the Sound FX clip to open it in the source viewer.

This is a typical sound effects clip that contains many sounds. Let's play it and listen for a sound effect that you might want to use somewhere in your timeline.

> NOTE Although the images in this book show bins in specific configurations, it is not critical that your setup match them exactly. Your bins may be in icon view or list view depending on your preference.

2 Play the first five seconds of the Sound FX clip.

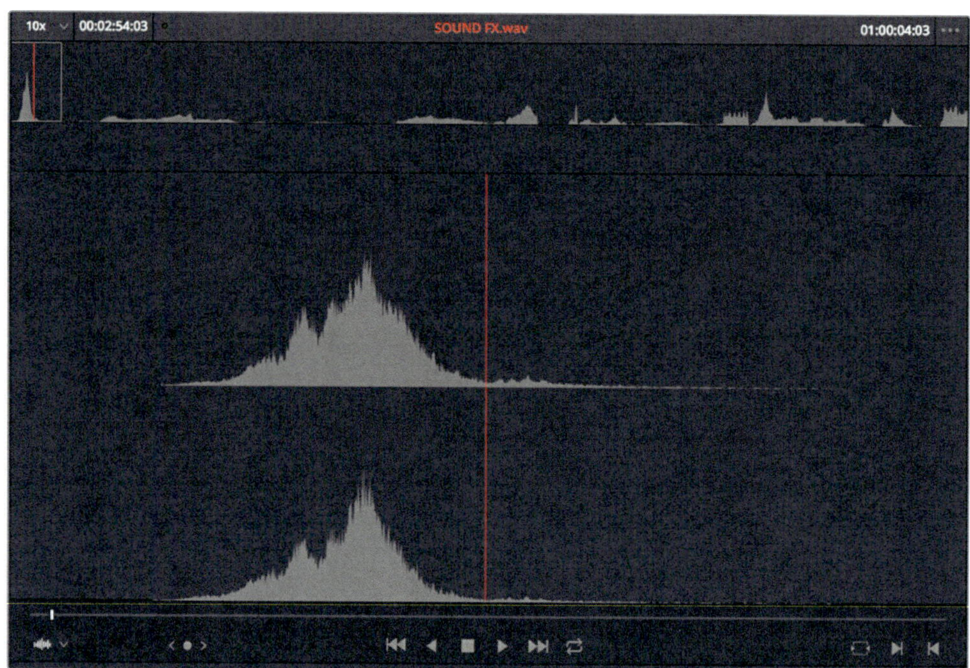

Fortunately, the first sound effect in this clip is a prop plane buzzing. To add a marker to the source clip in the viewer, you can use the same keyboard shortcut.

3 Position the source viewer's jog bar over the peak of the sound effect (the highest part of the waveform).

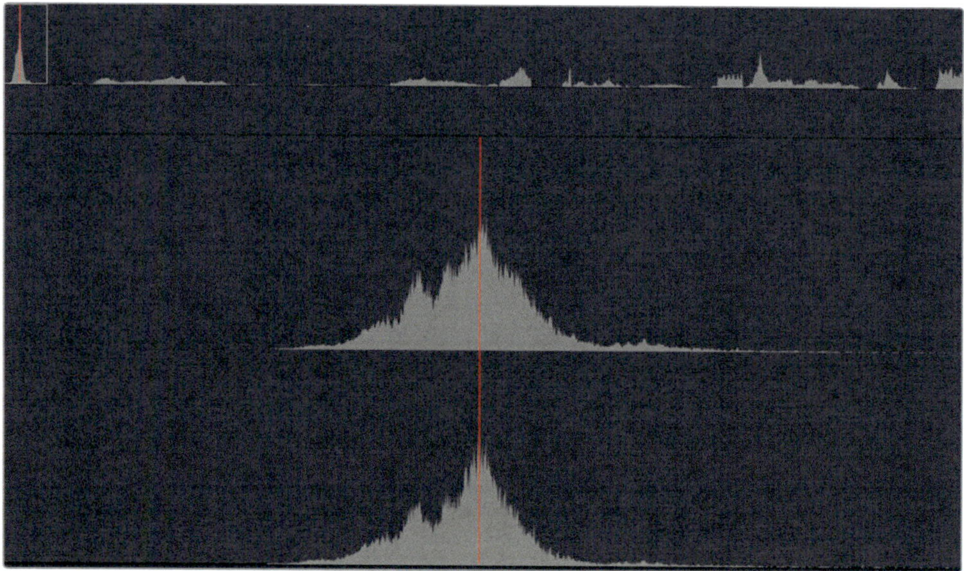

4 Press the M key.

A marker is added to the source clip under the source viewer's jog bar. You also can add notes to source clip markers.

5 Press M again to open the dialog, and in the Name field, type **Plane Overhead**. Click Save.

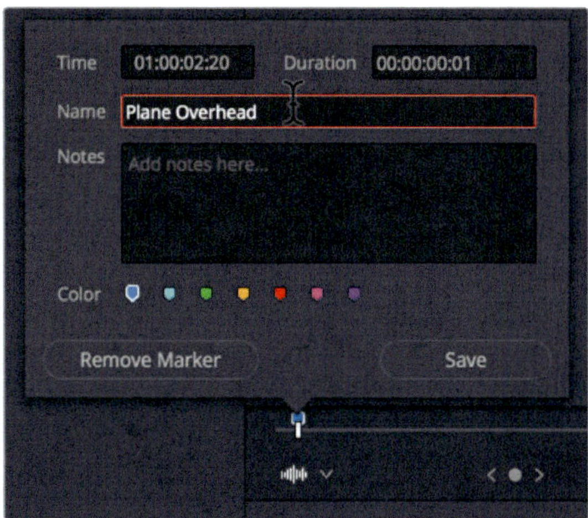

Your first sound effect is marked and identified. The marker will help you locate this effect later when you want to begin editing.

Marking a range of frames

In the source viewer and the timeline, you can also spread markers across several frames. Doing so is useful when you want to identify a specific section that will need attention, or to simulate multiple in and out points on a clip. You can use this function in a long sound effects clip that may contain multiple affects you want to use. Considering the previous marker that you added in the source viewer, you need to locate two more sounds for your program. First, you'll look for a loud prop plane sound for the South Pole shot.

1 In the source viewer, continue playing the Sound FX clip for about 30 seconds.

Somewhere in that 30 seconds is a loud, good-sounding prop plane take-off. Before you decide to use this sound effect, you can use markers to identity a potential range for the South Pole clip. Then, you can continue to search the source clip for possible alternatives.

In the source viewer, position the jog bar where the prop plane sound effect begins.

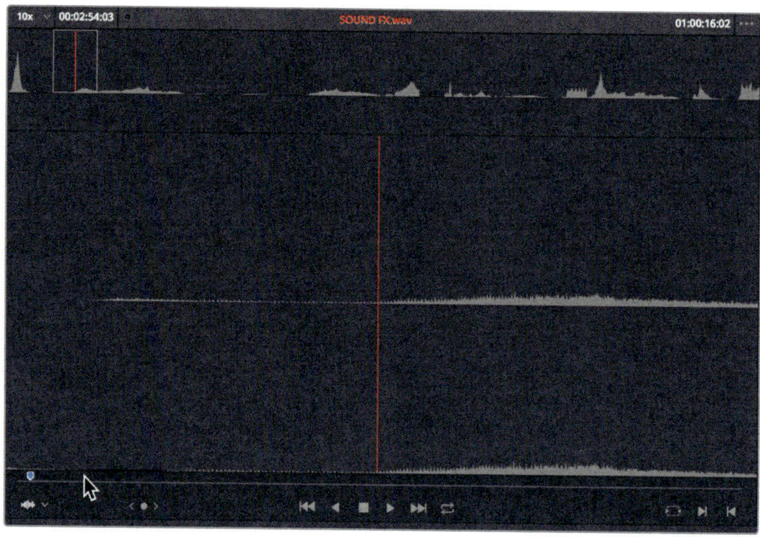

The initial sound of the prop plane engine would not fit the visual of the South Pole plane. You'll need to play a little farther into the sound effect and place a marker when the engine really gets going.

2. Play the clip to find a location where the engine sound is revving loudly.
3. Press M to add a marker in the source viewer.
4. Press M a second time to open the dialog. In the Name field, type **Loud Prop Plane**.

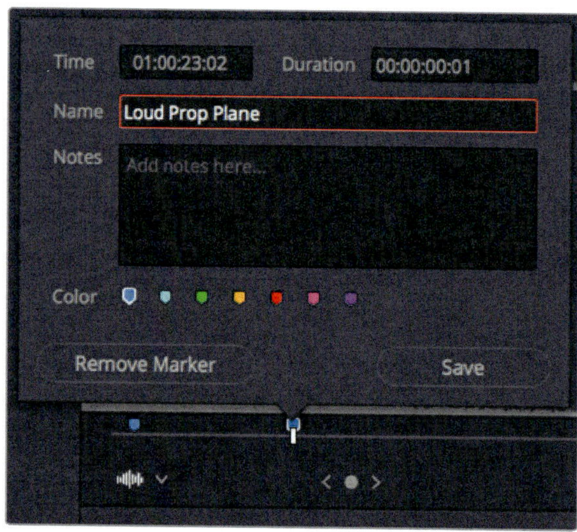

5. To extend the range of the marker, click at the end of the Duration field, and delete the last three digits.
6. Type **300** to create a duration of three seconds, press Enter on the keyboard and click Save.

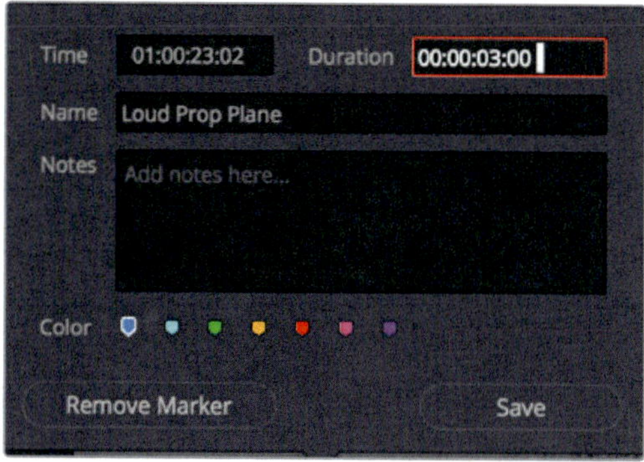

TIP Instead of entering a duration for the marker range, you could Option-drag (Mac) or Alt-drag (Windows) the marker to extend the range.

This time you'll add in and out points around the area you are interested in and then convert them to markers.

This will be a roaring jet taking off for the A380 Takeoff clip.

7 Drag the jog bar slowly across the clip until you hear the loud jet take-off, somewhere near the middle of the clip. (Hint: A medium-sized peak can be found in the audio waveform where the jet sound is located.)

Once you locate the start of the jet sound, you'll use an in and out point to mark the range.

8 In the source viewer, position the jog bar where you hear the jet take-off start.

9 Mark an in point, and then locate the end of the sound and mark an out point.

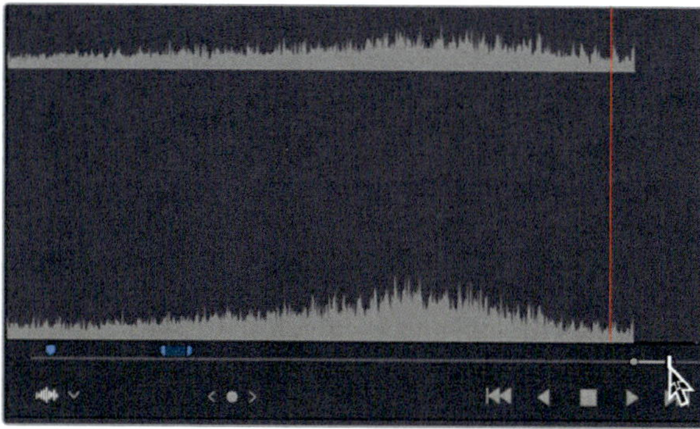

10 Right-click between the marked in and out points, and in the pop-up menu, choose “Convert In and Out to Duration Marker”.

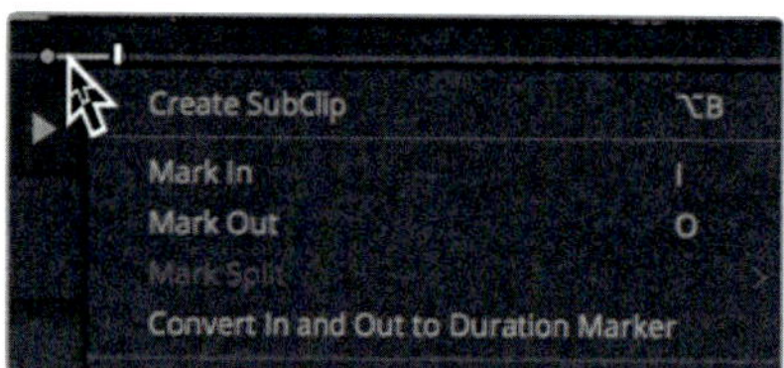

The Duration marker is added to identify the same frames as the in and out points. You can use the same pop-up menu to open the dialog.

11 Right-click in the marker range, and in the pop-up menu, choose Modify Marker.

The dialog opens.

12 In the Name field, type **Roaring Jet Take Off** and click Done.

You've identified sound effects that fill the requirements of the markers in the timeline. Now, you'll edit them into your trailer to enhance the cinematic experience.

Customizing the interface for audio

Throughout this lesson you'll work on the audio tracks, so it makes sense to customize the timeline view by increasing the audio track heights so you can more clearly see the audio waveform displays. Doing so will help you locate specific sounds and evaluate the audio more effectively. Furthermore, those waveforms are a handy visual reference to determine the volume of the audio.

To open up more room in the timeline for audio tracks, you can push the tracks up, thereby removing some of the headroom that exists above the video track.

1 Locate the mouse pointer between the video track and the audio tracks in the timeline.

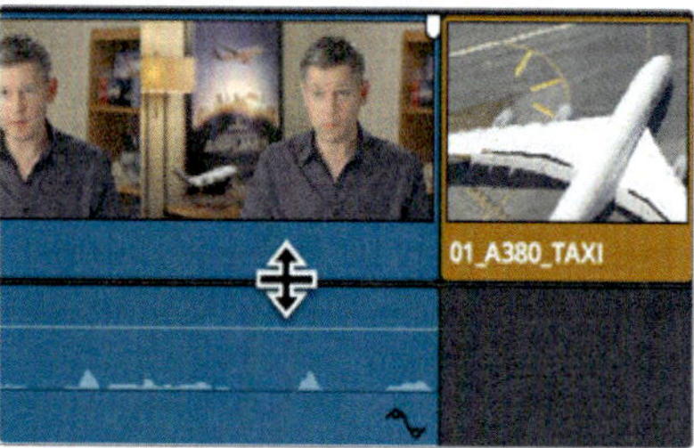

The pointer changes to a resize cursor, indicating that you can drag up or down to assign more or less room to audio or video tracks.

2 Drag up on the horizontal divider that separates the audio and video tracks until the video track is at the top of the timeline.

As you did in the previous lesson, you can change the appearance of the tracks in the Timeline View Options menu.

3 In the toolbar, click the Timeline View Options button to open the pop-up menu.

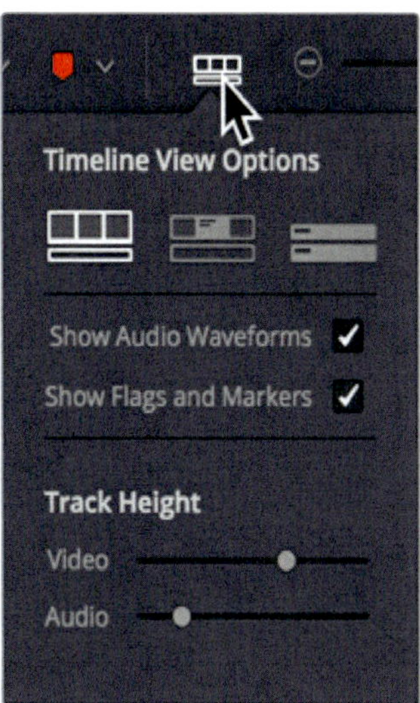

Here you will change the height of the video and audio tracks to suit your upcoming tasks.

4 Click the third track appearance button to the right to collapse the audio and video tracks.

5 Select the Show Audio Waveforms checkbox to turn on the audio waveform display.

6 Drag the video and audio track height sliders all the way to the right to increase the track sizes.

7 Click the Timeline View Options button to hide the pop-up menu.

Now your timeline is better configured for audio editing and mixing. You'll later add other interface elements to help control and monitor the audio; but for now, displaying larger waveforms in the timeline will help you edit in the sound effects.

Adding and patching tracks

Your timeline is fairly well organized considering the small amount of audio that it contains. Audio 1 is devoted to the interview and Audio 2 is devoted to music. To keep those tracks organized, you'll insert an empty audio track to accommodate the new sound effects.

1 In the timeline header, right-click the Audio 2 label to open the pop-up menu.

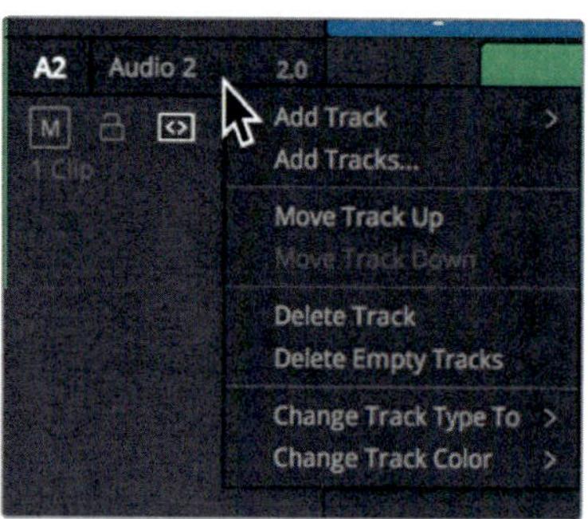

This menu has options to add, move, and delete tracks. Your stereo sound effects need to be edited into a new stereo audio track.

2 In the pop-up menu, choose Add Track > Stereo.

> **TIP** In the timeline, audio clips with multiple channels are displayed as a single audio clip on a single track. When you first create a new audio track, you choose its audio format (mono, stereo, 5.1 surround, or multi-channel). You can, however, change it later.

An Audio 3 track is added to the timeline. Now you need to patch the track so that the audio content in the source clip is edited onto this new Audio 3 track.

3 Drag down the A1 destination control to Audio 3 to align the A1 track in the source viewer with the Audio 3 track in the timeline.

Now that the tracks are patched, you can edit your first sound effect into the timeline. Let's locate that first sound effect.

4 In the source viewer, navigate to the first marker in the clip by choosing Playback > Previous Marker three times, or pressing Shift-Up Arrow three times.

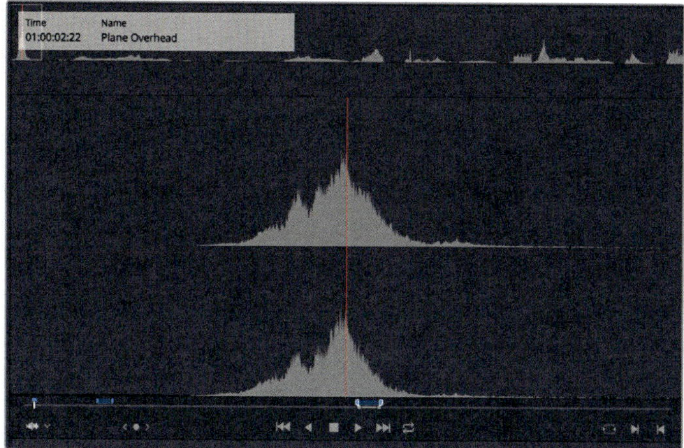

The source viewer should be located at the first marker. The marker overlay in the viewer shows that this is an overhead prop plane. This may work perfectly for the Maldives shot where the plane flies in at the top of the frame.

5 Click in the timeline viewer, or press the Q key, to activate the timeline viewer.

6 Navigate to the marker over the Maldives clip by choosing Playback > Previous Marker, or pressing Shift-Up Arrow.

One of the best ways to edit sound effects is to use the replace edit that you used previously. When adding sound effects, you typically are trying to match the action in a frame with a precise sync point in the sound effect. Most often these sync points are not at the start or end of a clip, so you need to use another method for aligning the clips. With a replace edit, you can position the jog bar over the sync point of the sound effect, and then position the timeline playhead on the video frame where you want the sound effect sync point to be heard. The source clip is already located on the loudest part of the buzzing overhead plane. That point should line up with the sea plane just entering the top of the frame.

7 Position the playhead on the frame where the nose of the plane and the floats have just entered the frame.

TIP If the playhead is difficult to move precisely, press the N key, or click the snapping button in the toolbar, to disable snapping. The playhead will no longer snap to the marker.

Unlike using a replace edit to replace a clip on the same track, you must set in and out points when replacing a clip into an empty track as you are doing here in Audio 3.

8 Choose Mark > Mark Clip, or press X, to mark in and out points for the duration of the Maldives clip.

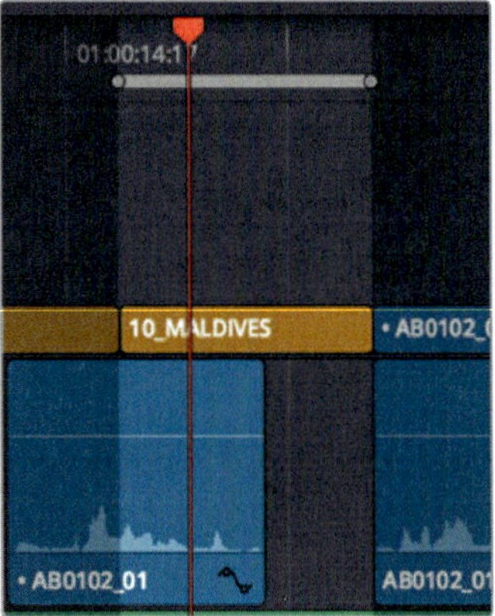

9 In the toolbar, click the replace button, or press F11.

10 Play over the sound effect to hear it synchronized with the picture.

Now you have an organized track layout that places the subject's voice on Audio 1, music on Audio 2, and sound effects on Audio 3. Maintaining media organization throughout your editing stage is always important, but may be even more important when editing audio because you'll often find yourself managing a dozen tracks or more.

Color coding tracks

You can organize your tracks even further by adding another layer of color organization. You already color coded the video clips, but you can also color code tracks in the timeline. To make it easier to parse your timeline, let's color code the music track with orange and the sound effects track with green.

1 Right-click the Audio 2 timeline header.

2 In the pop-up menu, choose Change Track Color > Orange.

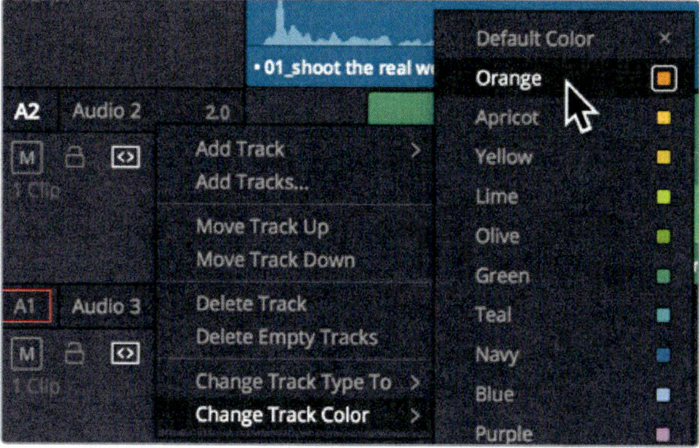

The music clip on Audio 2 changes to the orange color.

3. Right-click the Audio 3 timeline header.
4. In the pop-up menu, choose Change Track Color > Olive.

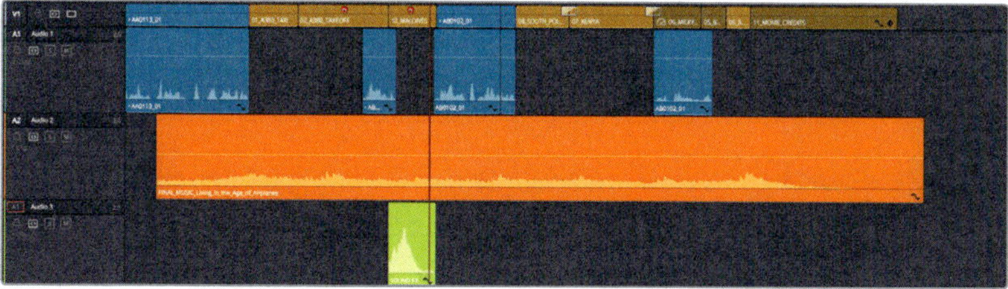

The sound effect clip changes to an olive color. Any new sound effect that you edit onto that track will also assume the olive color. However, when you have assigned a color to a clip in the bin, that color will override the timeline track color.

Finding markers using the Edit Index

In a short timeline such as the one you have here, navigating to each marker isn't much of a challenge. On more involved projects, however, you'll need a quick way to locate one specific marker among dozens of other markers. The Edit Index is a list view of all the editing events (clips and markers) in the current timeline.

1. At the top of the DaVinci Resolve window, click the Edit Index button.

The Edit Index opens below the Media Pool, showing all of the editing events and columns of metadata. This is too much information to digest even in a short timeline.

2 In the upper-right corner of the Edit Index, click the options menu.

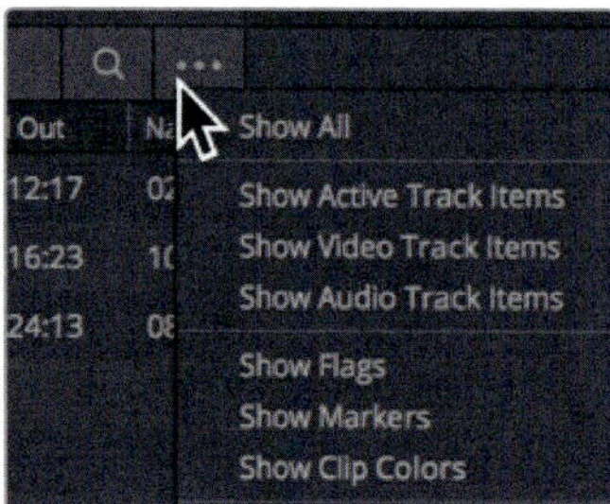

3 Choose Show Markers > All.

The Edit Index changes to show only those markers in the timeline, making it much easier to review the list. But you can make it even easier. As with the bin's list view, you can choose to show or hide columns of information.

4 Right-click a column header to open the menu of columns.

5 In this menu, deselect each column except for Number, Color, and Notes.

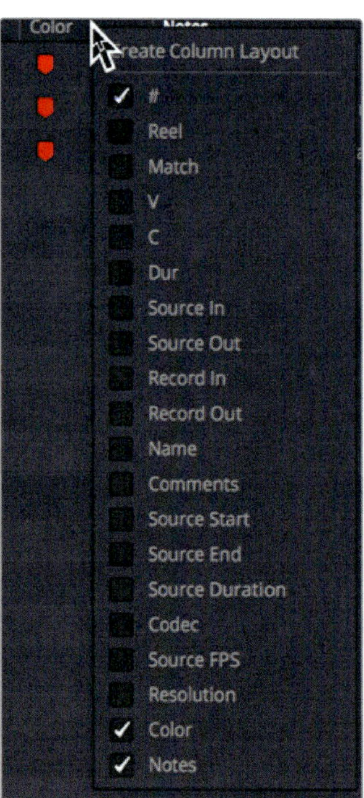

Now you can identify each marker by its color, and the note you entered.

6 Click the red marker that contains the note, "Add Roaring Take Off."

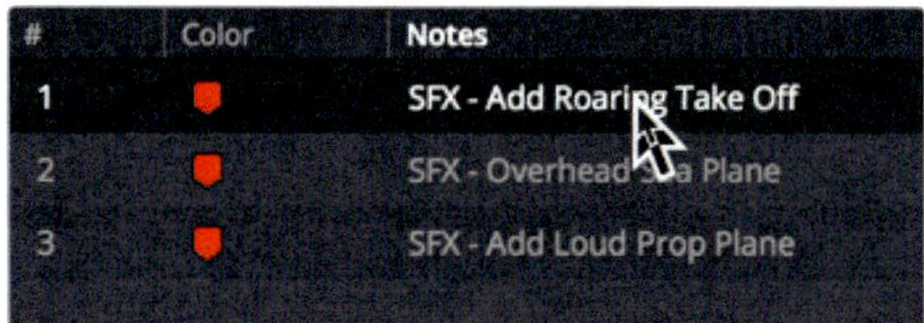

The timeline jumps to the marker on the 02_A380_Takeoff clip. This is the next clip to which you'll add sound effects. Instead of trying to find the marker in the Source viewer, DaVinci Resolve has an easier way that you'll use next.

Viewing markers in a bin

Markers added to source clips can be viewed in the list view of a bin. If the markers have a duration, you can use them like multiple in and out points and edit them directly into the timeline.

1 Above the Media Pool, click the list view button.

The sound effects clip has a disclosure arrow to the left of its name. Clicking this disclosure arrow will display the markers you added to the clip.

2 Click the disclosure arrow next to the Sound FX clip.

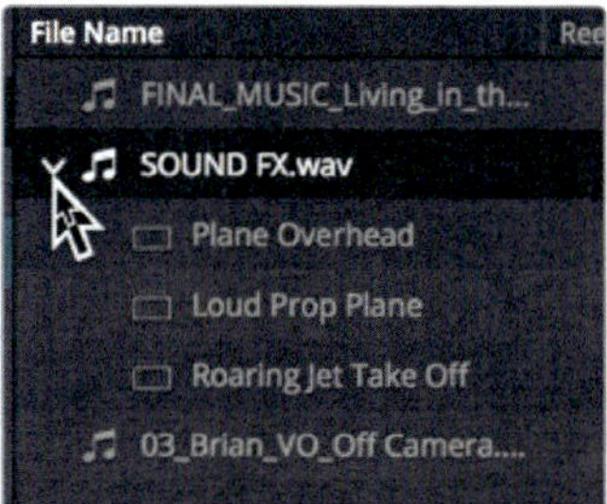

Each marker on the source clip is displayed under the clip's file name. If these markers have duration, you can just drag them into the timeline, and the duration markers will be used as in and out points for the clip.

3 From the Audio bin, drag the Roaring Jet Take off marker to the Audio 3 track so it aligns with the start of the 02_A380_TAKEOFF clip.

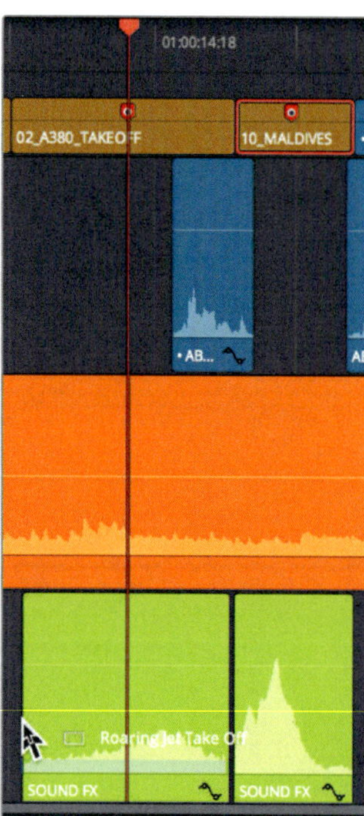

One more sound effect to add.

TIP Double-clicking the marker in the bin's list view will open the clip into the source viewer with the jog bar placed at the marker.

4 In the Edit Index, click the SFX - Add Loud Prop Plane entry to relocate the timeline playhead to that marker position.

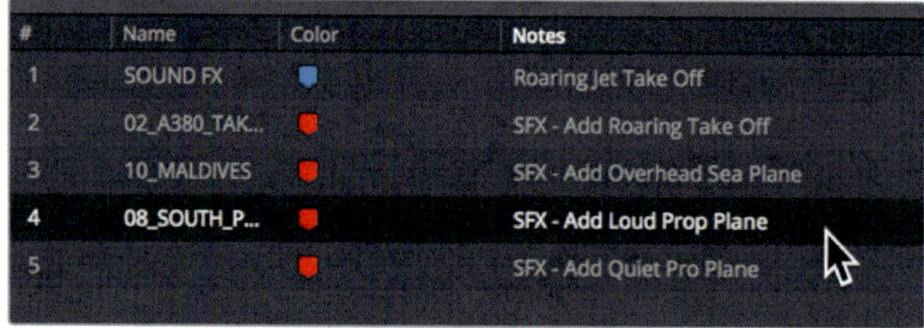

#	Name	Color	Notes
1	SOUND FX		Roaring Jet Take Off
2	02_A380_TAK...		SFX - Add Roaring Take Off
3	10_MALDIVES		SFX - Add Overhead Sea Plane
4	08_SOUTH_P...		SFX - Add Loud Prop Plane
5			SFX - Add Quiet Pro Plane

5 From the Audio bin, drag the Loud Prop Plane marker to the Audio 3 track so it aligns with the start of the 08_SOUTH_POLE_DC3 clip.

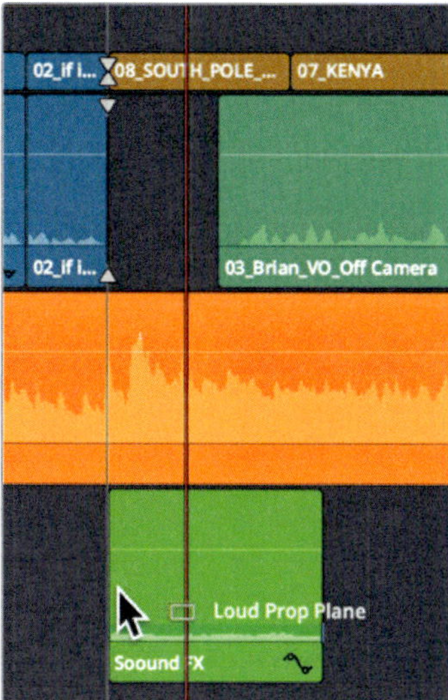

6 Move the playhead to the start of the timeline, and play over your newly added sound effects.

All of your audio tracks, music, narration, and sound effects are now in place.

Linking clips

Clips that contain audio and video captured together are linked automatically in DaVinci Resolve. That behavior makes it easy to move and trim them together so they stay in sync. However, that is not the case with your newly added sound effects. They have no link to the video they were placed with. However, you can create a sync relationship between them so that when you move one of the video clips that has a sound effect under it, they are attached and move together.

1 In the timeline, select the 02_A380_TAKEOFF clip.

2 In Audio 3, Cmd-click (Mac) or Ctrl-click (Windows) the sound effect clip directly under 02_A380_TAKEOFF.

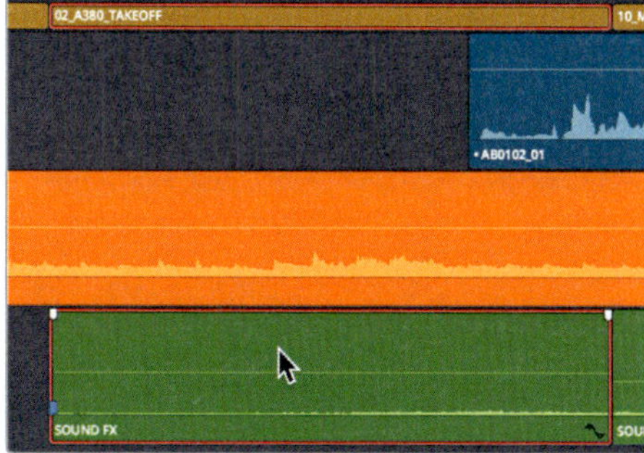

These two clips should stay together if you later want to move the A380 clip or even delete it. To do so, you need to link them together.

3 Right-click the selected SOUND FX clip, and in the pop-up menu, choose Link Clips.

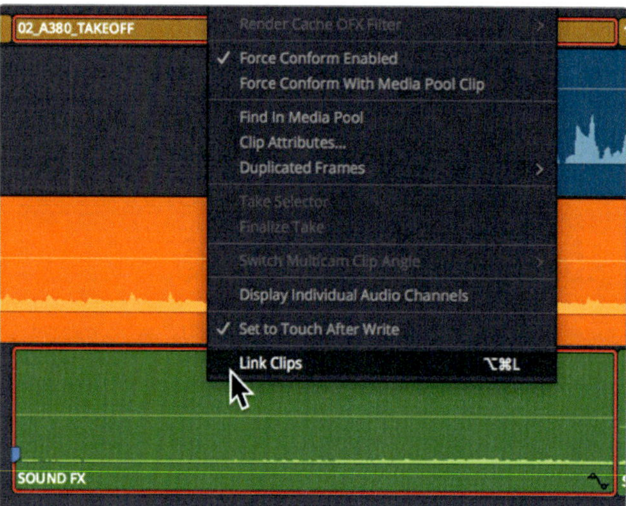

A small white dot appears in the timeline before the clip name. The dot signifies that this audio clip is linked to another clip. If you selected one in the timeline, the other would also be selected just as if they were recorded together.

4 Go through the timeline and link the remaining three sound effects with their respective video clips.

The linked selection button in the timeline toolbar works for these linked clips just as it does for linked clips that were recorded together.

Monitoring, soloing, and muting audio

When you start to edit your audio, the first thing you need to do is sit back and listen. Just play the tracks to hear them in the context of the picture.

1 Press the Home key, and then press the Spacebar to play the timeline to its end.

> **TIP** Some Mac keyboards do not have Home and End keys. In their place, press Fn-Left Arrow to move the playhead to the start of the timeline and Fn-Right Arrow to move to the end.

To get a sense of the contents of each audio track, you can listen to each track independently by soloing it.

2 Press the Home key to move the playhead back to the start of the timeline.

3 On Audio 1, click the solo button to temporarily silence the other audio tracks.

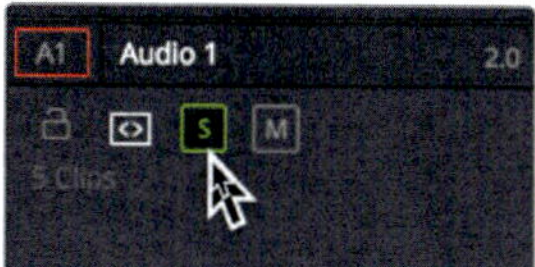

4 Press the Spacebar to play the timeline.

Only the interview is heard because the other two tracks are not soloed.

5 While the timeline plays, click the solo button on Audio 3 to hear the sound effects.

You can solo multiple tracks to hear only the track you select.

To accurately monitor the levels of your audio, you can use the audio meters.

6 In the upper-right corner of the DaVinci Resolve interface, click the Mixer button to open the audio mixer.

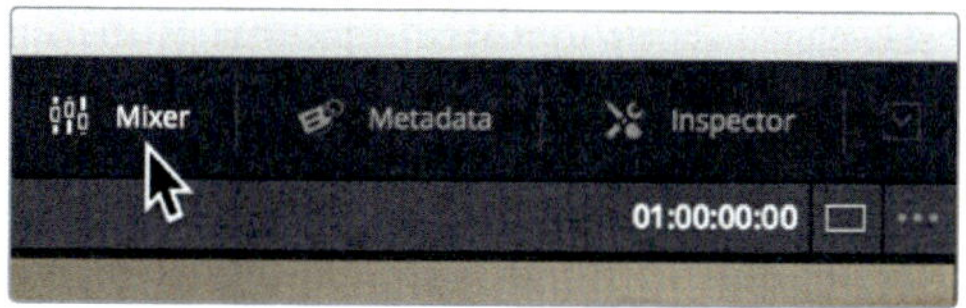

You will use the Mixer panel to monitor each track's audio level while watching the audio meters. Meters are a visual guide for making audio volume and processing adjustments. In DaVinci Resolve, the meters repeatedly generate an average (mean) level with a thin one-pixel line showing you the peak. But here in the Edit page, the goal is really just to create consistency amongst the clips on the same track, so all your narration is at the same level and all your similar sound effects are at a consistent level.

7 Play the timeline to hear the soloed tracks as you watch the meters.

You can see each track's audio displayed in the meters and set the appropriate level for each track.

Setting levels in the Inspector

With all video and audio edited into your timeline, you're ready to set the relative audio levels for your program. You'll start by examining the tracks that contain the primary audio—in this case, it's the interview clips on Audio 1—and setting those clips to their maximum volume levels.

1 Position the playhead at the start of the timeline.

2 Click the solo button for Audio 3, so that only Audio 1 has solo enabled.

3 In the toolbar, drag the zoom slider to zoom in on the first audio clip.

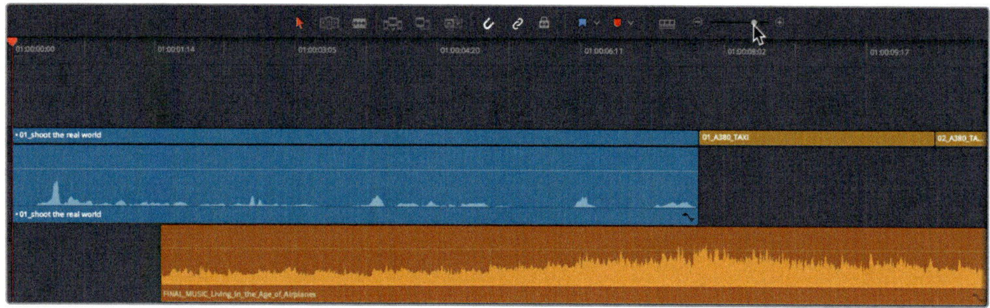

4 Select the first audio clip on Audio 1.

5 In the upper-right corner of the DaVinci Resolve window, click the Inspector button, or press Cmd-9 (Mac) or Ctrl-9 (Windows). Then, click the Audio tab, if necessary.

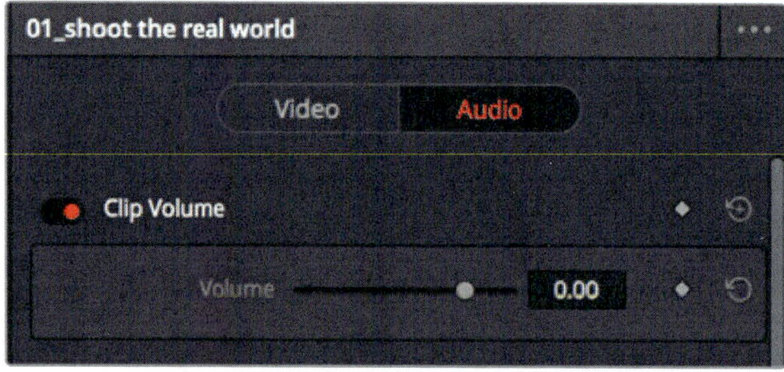

When an audio clip is selected in the timeline, Level and EQ parameters are displayed in the Inspector.

6 Drag the Volume slider to the right to increase the clip's volume. Keep dragging until the tops of the peaks in the waveform are displayed in a lighter shade of blue.

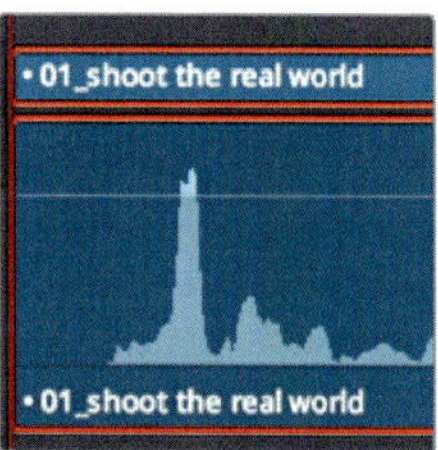

TIP To increase or decrease the volume of a selected clip, press Cmd-Option-+ (plus sign) and Cmd-Option- - (minus sign) in Mac, or Ctrl-Alt-+ (plus sign) and Ctrl-Alt- - (minus sign) in Windows.

The lighter blue highlight identifies an audio peak that is above the -20 dB reference level. For your primary audio clips, a decent starting point is to set your narration primarily around -12 dBFS, and make sure your peaks are nowhere near 0 dBFS.

7 Play through the clip and watch the meters. If the meters show levels going consistently above -12 dB., drag the volume slider to the left.

It is important to get close to that -12 db threshold, otherwise you might be setting your primary clips too low and not taking advantage of the full dynamic range of digital audio recording.

Copying and pasting audio attributes

You adjusted the level for one of the four narration clips in the timeline. Three additional clips could use adjustments similar to the one you just performed. But instead of adjusting each clip individually, you can copy the volume adjustment from one clip and paste it onto the others.

1 Select the first narration clip on Audio 1, and choose Edit > Copy, or press Cmd-C (Mac) or Ctrl-C (Windows).

2 Press Shift-Z (Mac) or Shift-Z (Windows) to view the entire timeline.

3 Select the audio under the 02_A380_TAKEOFF clip.

4 Shift-click the last audio clip on Audio 1 to select all of the remaining clips on that track.

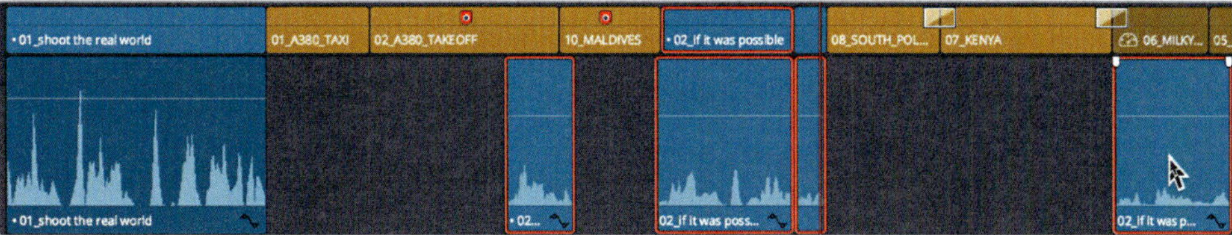

5 Choose Edit > Paste Attributes, or press Option-V (Mac) or Alt-V (Windows).

The paste attributes window includes clip attributes that you can copy and paste from one clip to another.

6 Select the Audio Attributes Volume checkbox, and click Apply.

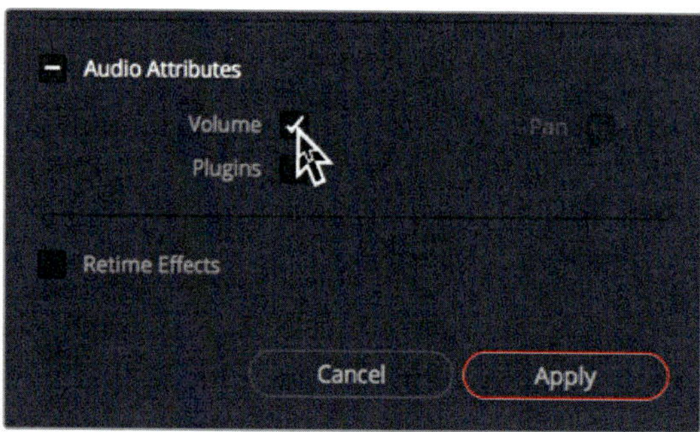

The volume attributes from the first interview clip are pasted onto the selected clips.

7 Play over the timeline. Drag the Volume slider to make any additional volume adjustments that seem appropriate.

Copying and pasting attributes to set clips at similar levels can save you a lot of time when working with a complex timeline. You can also paste other attributes, including video attributes.

Setting levels in the timeline

For quick audio level adjustments, you can change the audio level of a clip graphically in the timeline using the volume curves.

1 In the track header for Audio 3, click the solo button to hear only the sound effects.

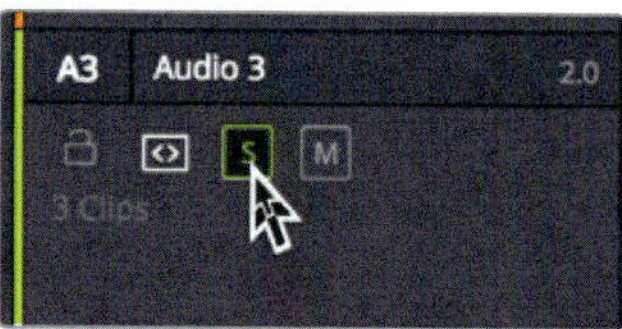

2 Position the timeline playhead at the start of the first sound effect on Audio 3.

3 Press Shift-Z (Mac) or Shift-Z (Windows) to return to the previous zoom level.

4 If necessary, drag the scroll bar at the bottom of the timeline so that you can see the first two sound effects in the Timeline window.

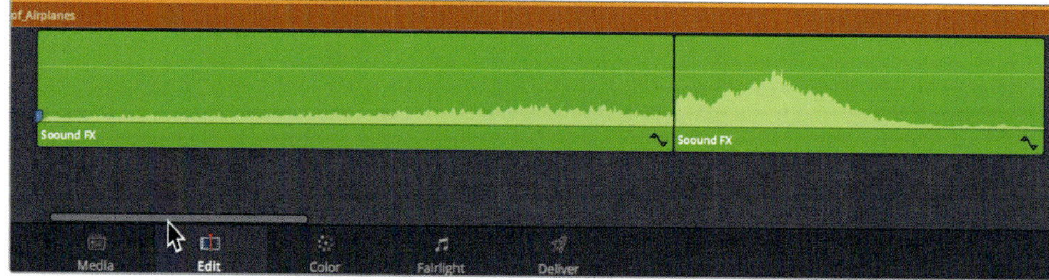

5 Play these two sound effects and watch A3 on the meters to monitor their levels.

Both sound effects are too loud. They are not meant to be at the same level as a direct hit from a photon torpedo! Because you are dealing with digital audio, any sound that reaches a level of 0 dBFS will be clipped and distorted. Unlike analog audio, digital audio has no headroom, so peaks in your audio that reach or exceed 0 dBFS will sound horrible.

As a guide, try to set sound effects to fall somewhere between -10 and -30 db. Of course, no hard and fast rule exists because the final and best setting comes down to creative intent and your ears, but you can use the -10 to -30 dB range as a rule-of-thumb target as you set levels.

6 Hover your mouse pointer over the thin white line that runs through the first sound effect audio clip on Audio 3.

The thin white line in the audio clip is the volume curve and represents the volume level of the clip. Dragging down the volume curve (line) will lower the level just as it would if you dragged the Volume slider to the left in the Inspector.

7 When your mouse pointer changes to an up and down arrow pointer, drag the volume curve down until the tool tip reads roughly -14 or -15 db.

8 Because the second sound effect is a bit louder, drag its volume curve down until the tool tip reads roughly -20 db.

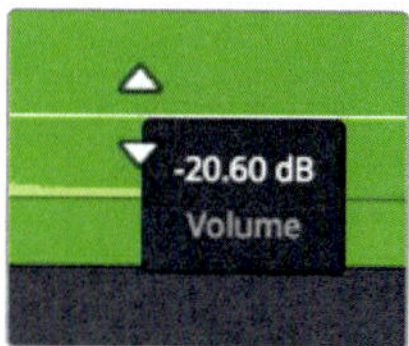

9 Position the timeline playhead at the start of the first sound effect on Audio 3.

10 Click the solo button on Audio 1 so you can hear the narration with the sound effects.

11 Press the Spacebar to listen to the first two sound effects and the narration.

You'll learn how to adjust the remaining clip in the next lesson when using the Fairlight page. For now, let's add a yellow marker to the timeline with a note to remind yourself that you need to set levels for the final sound effects.

12 In Audio 3, position the playhead over the red marker above last sound effect.

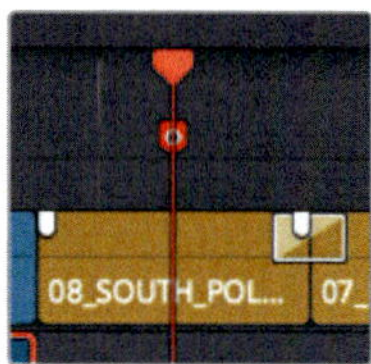

Because you already have a red marker here, you can modify it for your current purpose.

13 In the timeline ruler, double-click the red marker to open the dialog.

14 Change the Name field to **Fix Level on A3**, and remove the text in the Note field.

15 Click the yellow color tag at the bottom of the dialog. Click Save.

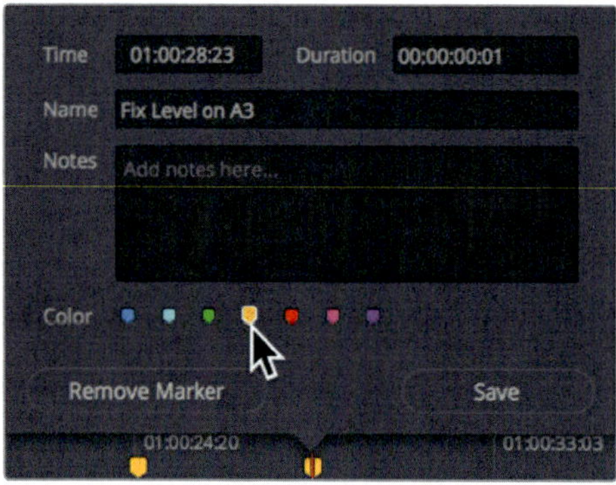

The marker will carry over into the Fairlight page, so you can easily locate these clips when you want to continue the audio mixing.

Changing a level within a clip

The music track is the final track that you'll integrate into your mix. Level setting here is slightly more involved than with the other tracks because you really want to set two different levels within this one music clip. The music should be at a quiet level as it plays under the interview portion of the timeline, and then gradually increase in volume when the interview stops. You can set multiple levels within a clip in different ways but on the Edit page you'll use keyframes.

1 In the track header, disable the solo buttons for A1 and A2.

All three audio tracks should now be audible.

Initially, you'll set a low volume level for the music as the interview begins. You'll want it low enough so that it doesn't interfere with the spoken words but loud enough to add atmosphere to the scene. You can set the level by dragging the volume line directly in the timeline. Again, you will use a general rule of thumb that music should fall somewhere between -20 and -30 dBFS.

2 For the music clip, drag the volume line down until the tool tip reads roughly -20 dB.

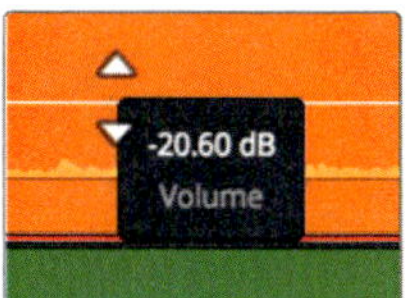

3 Play the timeline with the audio mixed in.

Dragging the entire level curve has set the clip's overall level. This level should suit the interview portion of the timeline, but it's much too low to underscore the plane shots. By adding keyframes to manipulate the volume line, you can change a clip's volume over time.

4 Position the playhead at the end of the first interview clip.

Here is where you want the audio to be loud so you will add a keyframe to the music clip.

5 On the music clip, Option-click (Mac) or Alt-click (Windows) the volume line under the playhead position to add a keyframe.

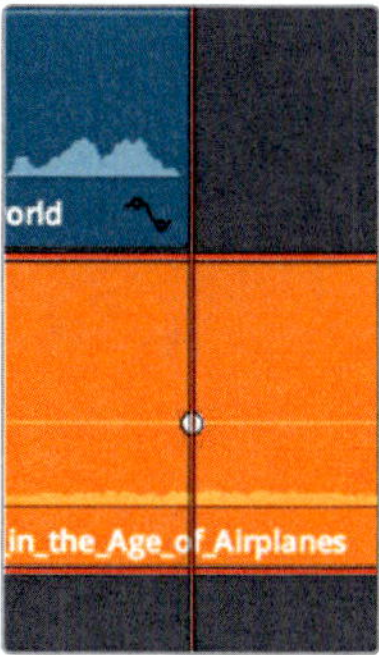

A white dot, the keyframe, is added to the curve. (You may have to move the playhead slightly to see it.) This keyframe marks the point where the music volume should be loud. You now need to set a keyframe before this point to identify where the music should start to crescendo, or ramp up.

6 Click in an empty gray area of the timeline to ensure that no clips are selected.

7 Type **-12**, and press Enter/Return to move the playhead back half a second.

8 Option-click (Mac) or Alt-click (Windows) the level curve under the playhead position to add a keyframe.

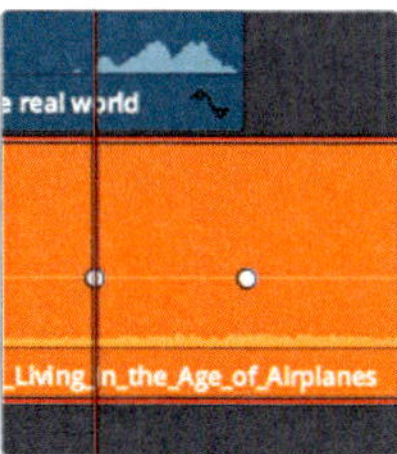

You've now set the duration that the gradual volume change, or ramp, will occur. Changing the vertical position of the second keyframe will cause a ramp up in volume.

9 Position the pointer over the second keyframe on the music clip.

   The pointer changes to a move pointer to indicate that it is over a keyframe.

10 Drag the keyframe up until the tool tip reads roughly -10.00 db.

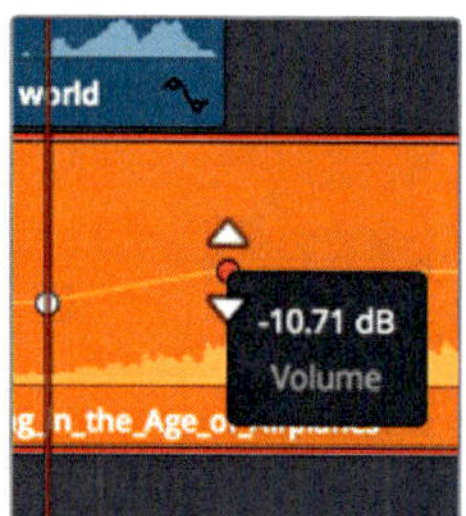

11 Play the first part of the timeline until you hear the mix with the new music ramp.

   You've only set the music volume levels under the first interview section. It is now too loud for the rest of the timeline. You could continue to add keyframes and adjust levels for the entire timeline, but let's hold off and complete the adjustments in the Fairlight page in the next lesson.

Adding audio fades

Most audio fades are added as a corrective process to soften the incoming or outgoing audio clip. Subtle fades are often applied to narration and dialogue when plosives, particularly those that start with P and B, are too harsh to leave unchanged. You'll also find that you'll use the obvious fade-in and fade-out on music.

1 Play over the start of the timeline until the music begins.

   No matter how low you set this music, it always comes in abruptly. It is meant to build during the interview and hit a crescendo when the first image of the plane appears. You can achieve this slow build up with a reasonably long fade-in.

2 Press Shift-Z to see the entire timeline.

3 In the timeline, place the pointer over the music clip.

Audio fade handles appear in the upper-left and upper-right corners of the clip.

4 Drag the left handle in toward the center of the clip until the tool tip reads +2:00

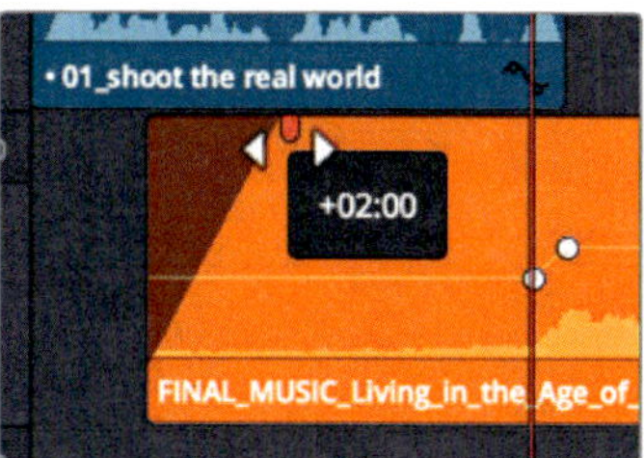

You added a two-second fade-in to the start of the clip.

5 Play over the start of the timeline to hear your new fade in.

You can use any combination of fade handles, level curves, and Inspector tweaks that you feel most comfortable with to refine your audio tracks; but these tools and techniques are ultimately just a starting point. To build a true multitrack, cinematic soundtrack, in the next lesson, you'll use DaVinci Resolve's Fairlight page.

Lesson 8

Mixing Sound in Fairlight

If filmmaking is primarily a visual art, why do professional productions spend so much of time, talent, and (yes) money to create perfect soundtracks? The answer is that high-quality sound is essential to visual storytelling. It often makes the difference between an amateur production and a professional one.

The Fairlight page in DaVinci Resolve 14 is designed specifically for realizing cinematic-quality sound in your film and video productions. Most importantly, it is built into your editing and color-grading application, which means you can refine the edit, color, and sound mix, right up until the time you master your final delivery. That's what makes the Fairlight page, and DaVinci Resolve, a game changer for filmmakers.

In this lesson, you'll navigate the Fairlight page, and rework the mix you developed on the Edit page using Fairlight's simple bussing configuration window.

Time

This lesson takes approximately 30 minutes to complete.

Goals

Exploring the interface

Let's talk about workflow. You are almost finished with your edit, and your project is now more or less complete. At some point, you have to move on to audio finishing. Traditionally, all audio post-production was done separately from editing using a standalone audio software. As a result, many media conversions and duplication of work went on just to successfully and accurately share the edited timeline in the audio post application. You are about to experience how that workflow is accelerated in DaVinci Resolve 14.

1 Open DaVinci Resolve, if necessary; and in the project manager, open the Age of Airplanes project.

2 In the Rough Cuts bin, double-click the Fairlight Rough Cut to open it in the timeline.

  This timeline is slightly different from the one you used in the previous lesson. Although it has all the volume changes you made, along with the keyframing and fade-up on the music track, it has checkerboarded the narration so that every other cut is on A1 and then A2. The sound effects have been checkerboarded onto A3 and A4.

  This style of checkerboarding and keeping similar audio elements together is a fairly standard way to organize your audio tracks as they grow.

3 At the bottom of the DaVinci Resolve interface, click the Fairlight button.

  That's the only workflow step required to move from editing to audio post!

  While the interface may appear to be one large timeline, the Fairlight page is actually divided into four simple sections.

The Interface toolbar shows and hides various palettes such as mixers and meters.

The transport controls include standard and audio-specific transport functions.

The toolbar includes commands, as well as modes buttons.

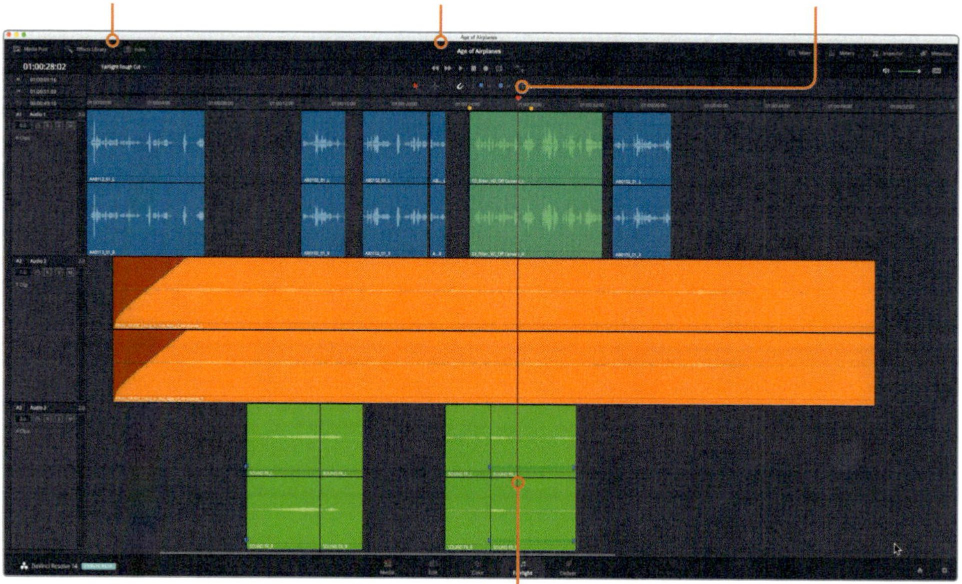

The timeline is optimized for audio mixing and uses single channel or multichannel tracks.

Your current timeline—with the same color-coded tracks, markers, and audio levels that you were viewing in the Edit page—is now available in the Fairlight page. It looks different because the Fairlight page is optimized for multichannel audio recording, editing, mixing, and sweetening. However, many of the playback techniques you learned when using the Edit page will also apply here.

4 Drag the playhead to the start of the timeline, or press the Home key.
5 Press the Spacebar to play your project.
6 Press the K key to pause, and then press L twice to fast forward.
7 Press J to play backward.
8 In the transport controls above the timeline, click the Stop button.

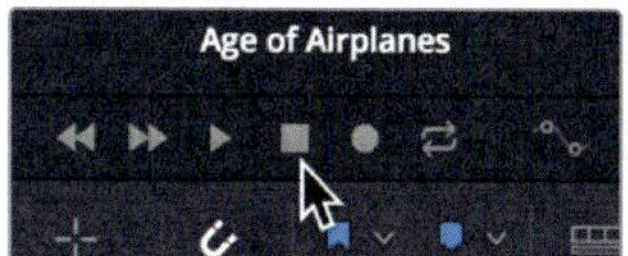

Other standard transport controls such as play, rewind, and fast forward are also included in the strip of transport controls along with audio-specific transport controls such as Loop and Record.

TIP You can record audio, such as a voiceover, directly in the Fairlight page. Just click the Input pop-up menu at the top of each channel strip in the mixer to open the Patch Input/Output window. Select your input recording device and output track. Once a track has been patched, it can be armed for recording by clicking the R button, in the track header. Finally, click the record button in the transport controls to start recording at the playhead position.

Previewing video

In the Interface toolbar, you can open panels as you need them. By default, the viewer and the meters are hidden.

1 In the upper-right corner of the Interface toolbar, click the Meters button to open the Monitoring panel.

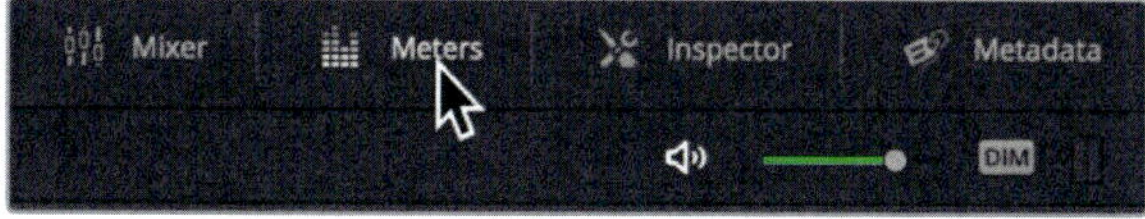

2 Press the Spacebar, and watch the meters and viewer during playback.

3 Press the Spacebar to stop playback.

The meters are a handy place to see all of your track and output levels. Because you have only a few audio tracks for this project, let's expand the viewer as a separate window and hide the meters.

4 In the viewer, click the expand button to turn it into a floating window.

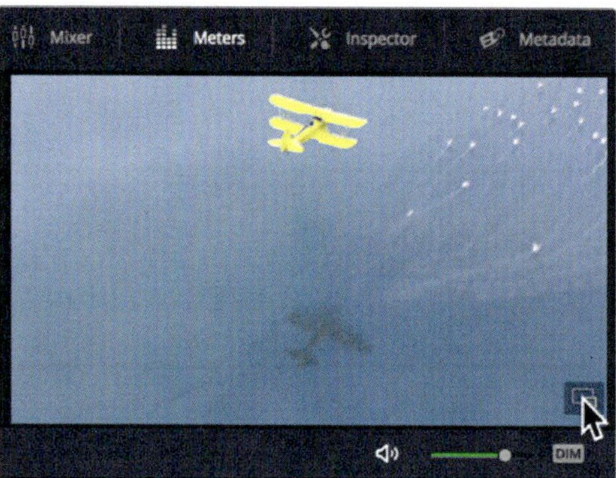

The viewer can be resized and positioned anywhere in the user interface.

5 In the viewer, drag out the lower-right corner to increase the size of the viewer.

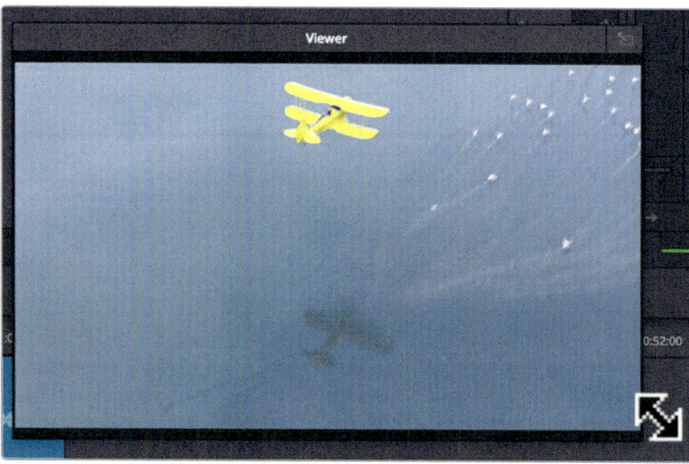

If you have a Blackmagic DeckLink card or UltraStudio, you also can view the video on a second video display.

Renaming and color coding tracks

Organization becomes even more important as you expand the number of audio tracks and want to navigate through them quickly. Instead of trying to remember the track number that you used for different types of tracks, you can rename them as you like.

1 In the track Header, click the Audio 1 track header to select it.

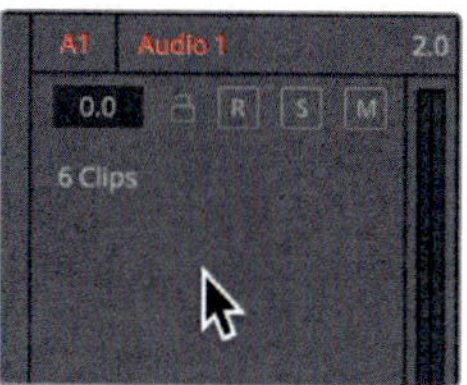

Dealing with a handful of tracks is easy, but let's extrapolate a bit to imagine fifty tracks or more. Suddenly, you're going to want more than color to find specific tracks. Luckily, you can name the audio tracks right in the header.

2 On the A1 Track header, double-click the name "Audio 1". Type **Narration 01** and press Return.

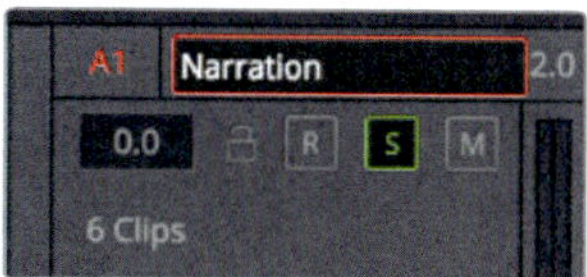

3 Because Audio 2 is also narration, double-click the Audio 2 name, type **Narration 02**, and press Return.

4 Change the names of Audio 3 and Audio 4 to **SFX 01** and **SFX 02**, respectively.

5 Click the Audio 5 track and rename the track to **Music**.

As you did in the Edit page, you can also better organize by color coding the new tracks here in the Fairlight page.

6 Right-click the A4 SFX 02 track header, and choose Change Track Color > Lime.

7 On the A5 Music track, right-click the header, and choose Change Track Color > Yellow.

Again, when you're working with only a few tracks, naming and color coding those tracks may seem like overkill; but when your track list starts to grow, you will be glad you started the process in an organized fashion.

Viewing a spotting list

Communication is essential in post-production because one person seldom does all the work. As a result, it is important to document all of the elements of your project. Even when you do plan to do all editing and audio mixing yourself, you will find that you will still need to keep track of the many, many details of your project.

Among those details are the basic decisions about where music and sound effects should be placed. Those choices are documented in a **spotting list** which is typically created when the director, the sound editor, and the composer watch the program together and produce a list identifying, **spotting**, all the places that sound effects and music are needed.

Even when you are one person wearing all those hats, you will still need that spotting list of all the audio elements you'll want to add. The Index is a panel that takes all the work out of creating a spotting list by leveraging those markers you added on the Edit page along with new markers you can add in the Fairlight page.

1 In the upper-left of the Interface toolbar, click the Index button.

The Index is divided into two tabs. The first tab lists all the tracks in the current timeline. The second tab shows all the markers.

2 In the Index, click the Markers tab.

The yellow marker added in the Edit page is displayed in the Index as a thumbnail.

3 At the top of the Index window, click the list view button to view the Index as a list.

4 Double-click the marker in the list.

The playhead jumps to the marker position and the name for this marker indicates that you need to set keyframes for the rest of the music track, so it dips under the narration.

You can set volume levels here much as you did in the Edit page.

5 Select the second sound effects clip on Audio 4.

6 In the upper-right corner of the interface toolbar, click the Inspector button to open the Inspector.

7 Drag the volume slider to the right to about 5.0 to increase the volume of this quiet prop plane.

Now let's continue the volume changes for the music that you started in the previous lesson.

You can use the other tab of the Index to open more room for a track.

8 At the top of the Index, click the Tracks tab.

In the Tracks tab, you can eliminate repetitive scrolling up and down within the timeline by hiding those tracks you are not currently working on.

9 Click the visibility control (the eye icon) for tracks 3 and 4 (SFX 01 and SFX 02).

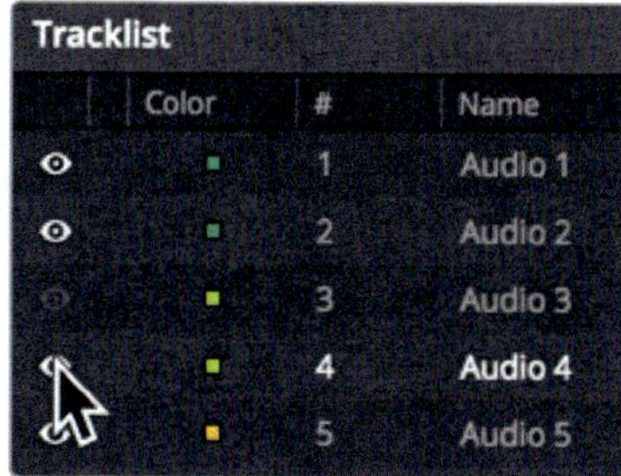

The SFX tracks are now hidden in the timeline (but not muted), thereby placing the dialog and music next to each other. Hiding tracks can be extremely helpful when working with tracks that are located far apart in large timelines.

10 Continue to dip the volume of the music under the narration by Option-clicking (Mac) or Alt-clicking (Windows) to add keyframes where needed.

11 When you are done, click the visibility control for the sound effect tracks, and close the Index.

Often a project will expand to include multiple dialogue and music tracks, Foley tracks, and sound effects tracks; so, it is critical to your creative and technical efforts to be able to show and hide tracks as well as maintain an up-to-date spotting list.

Organizing tracks into submixes

The audio signal from every track in the timeline flows from the track to the main output. When you monitor the main output, you hear all of the tracks combined or mixed together—hence the term, audio mixing. Once you have adjusted the levels of your clips and the corresponding tracks, you can simplify the mixing process with submixes. A submix combines the signals and effects from multiple tracks into a single channel strip on the mixer. It is common in post-production to create submixes for dialogue and SFX.

In this exercise, you'll combine the narration tracks into one submix, and the sound effects tracks into another. Doing so will make it easier to lower all the SFX by a small amount, or boost all narration clips to finesse the mix. The Fairlight page has a very simple but powerful way to create and assign tracks to busses (submixes). Now that you have set the relative levels of each clip, you can place all your narration into a submix and do the same for the SFX.

1 Choose Fairlight > Bus Format to open the Bus Format window.

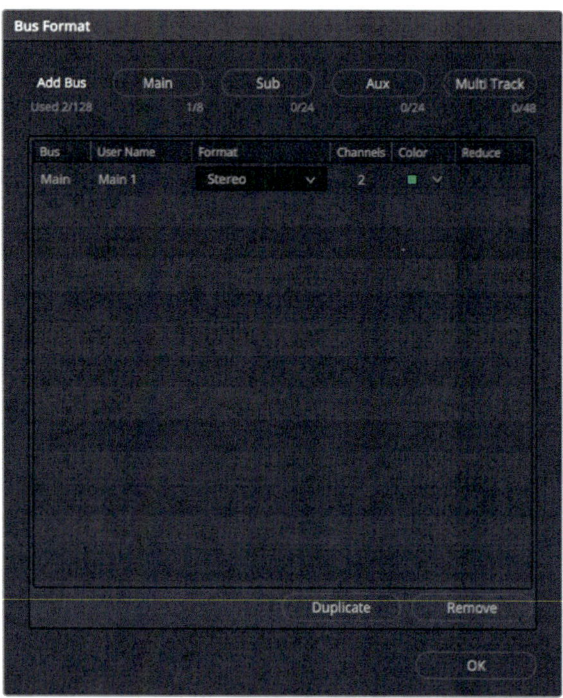

The Bus Format window is where you assign tracks to one of four types of busses. Submixes are just one of those types. This window currently contains the default main bus called M1, which is the stereo audio output for this project.

2 Click the Sub button to add a Submix bus to the bus list.

This new submix will be used for your stereo narration tracks. So, the first things you'll do is set it to stereo and name it more descriptively.

3 In the Format pop-up menu, choose Stereo.

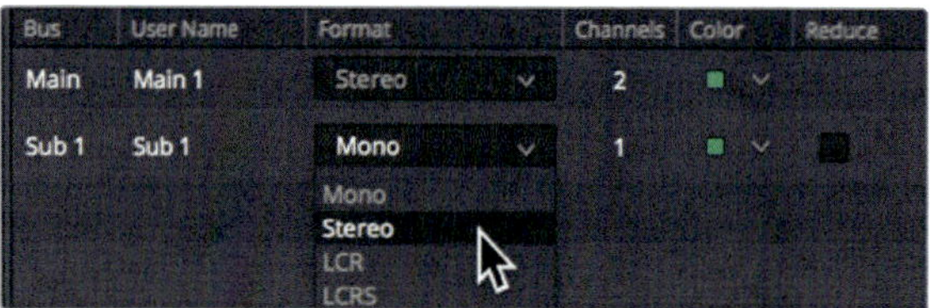

4 In the User Name column, double-click the Sub 1 name, and enter **Narration Sub**.

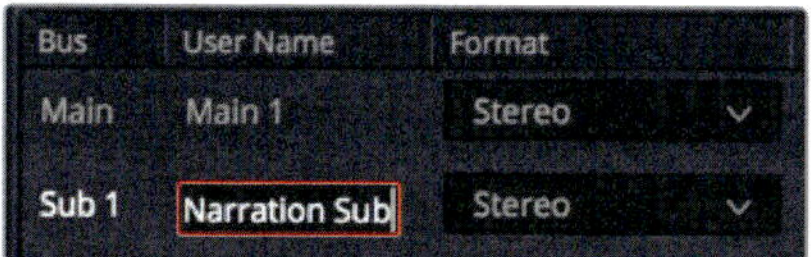

You'll also assign the same color to the submix as the individual narration track: Navy.

5 In the Narration submix's color pop-up menu, choose Navy.

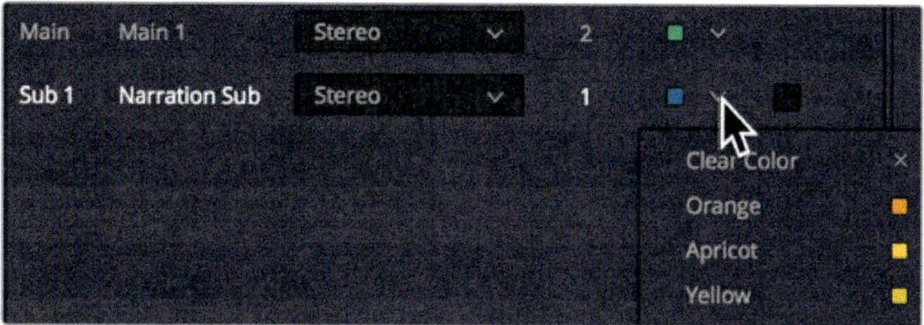

One bus is set up; now you can set up another for the sound effect tracks.

6 Click the Sub button to add another submix to the list.

7 In the Format pop-up menu, choose Stereo.

8 In the User Name column, double-click the Sub 2 name, and enter **SFX Sub**.

9 In the SFX submix's Color pop-up menu, choose Lime.

10 Click OK to close the Bus Format window.

You are all set up with submixes. The next step is to assign the tracks to go into those submixes.

Assigning tracks

The next window you will use is the Bus Assign window where you configure the tracks that go into your submix.

1 Choose Fairlight > Bus Assign to open the Bus Assign window.

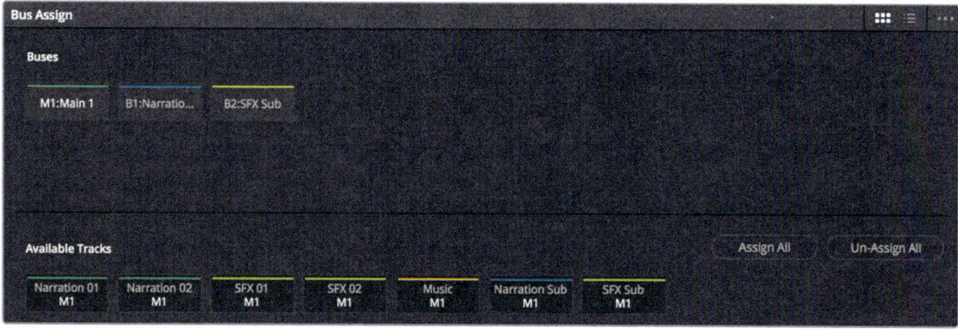

The Bus Assign window has buttons along the top representing the current Main and buses. The buttons along the bottom represent the possible tracks, submixes, and aux buses that can be assigned to the top busses.

2 In the buses along the top of the window, click the B1 Narration Sub button.

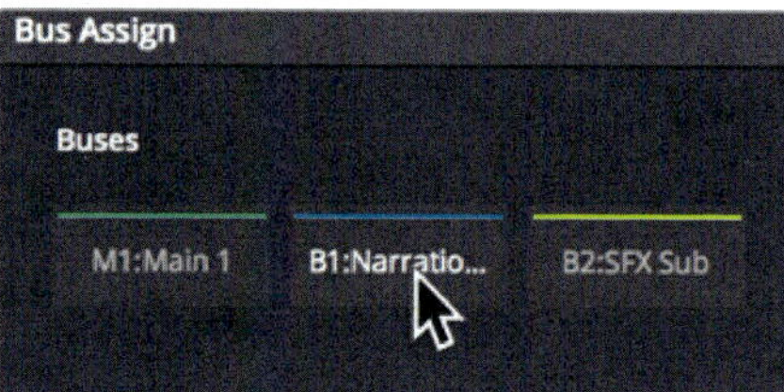

With the Narration Sub button selected, you can now choose the tracks from the bottom list that you want to include in this Narration Sub bus.

3 In the bottom list of tracks, click the two narration tracks, Narration 01 and Narration 02.

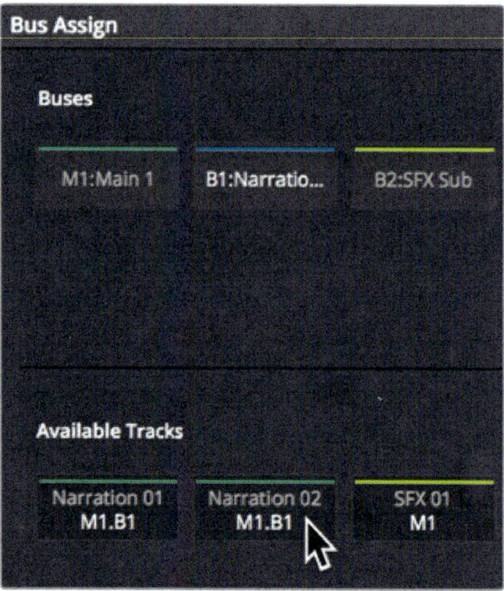

When the narration tracks are selected, the B1 label is added to the bottom of the button to indicate that they belong in that bus. Since you do not need the individual tracks to be in the main once they are assigned to the submix, you can remove them.

4 In the buses along the top, select m1:main.

5 In the available tracks area, click both narration tracks to remove them from the main.

Only B1 now remains in the narration 01 and narration 02 tracks to indicate that those two tracks are in the submix. You can now do the same for the SFX submix.

6 In the buses along the top of the window, click the B2 SFX Sub button.

7 In the bottom list of tracks, click the two SFX tracks, SFX 01 and SFX 02.

8 In the buses along the top, select m1:main.

9 In the available tracks area, click both SFX tracks to remove them from the main.

Only B2 now remains in the SFX 01 and SFX 02 tracks to indicate that those two tracks are in the submix.

10 Click Save to close the Bus Assign window and save the track assignments.

With the clip level adjustments done in the Edit page and the bus assignments completed, you can return to setting levels for each submix using the meters in the Fairlight page.

Measuring Loudness

Over the past 10 years or so, some parts of the film and television industries (as well as the music and radio industries) have moved from mastering programs using VU meters or RMS meters to loudness meters. Why do we need another meter for audio? When you watch television, you may have noticed the differences in audio levels between channels, programs and commercials that had you reaching for the volume control. Those differences are not easy to measure using a decibel scale with traditional RMS (Root Mean Squared) meters, or peak meters. And if you can't measure them, you can't normalize them. Hence, the need for new audio measurement tools.

1 In the upper-right interface toolbar, click the Meters button.

In DaVinci Resolve you can choose what audio meter standards are used so you can simply select the type of metering you feel most conformable with. The Fairlight page has standard RMS peak meters for each individual track on the left and meters for the master and submixes to the right. Your two submixes, labeled S1 and S2, can be seen with their color coding at the top.

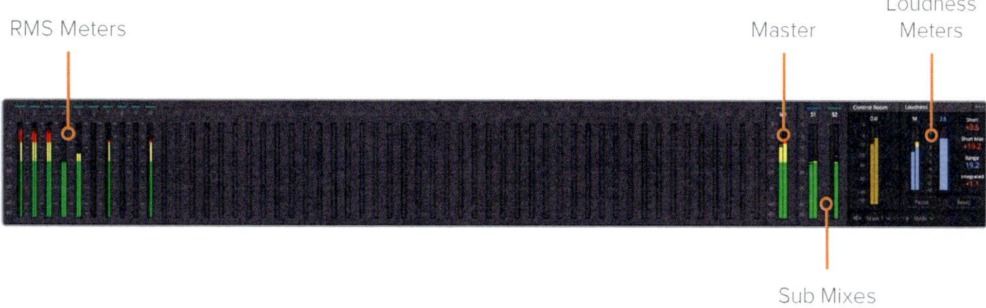

The last two meters to the right are the loudness meters. These two meters do not use decibels as a scale, instead they use loudness units.

The meters are part of a larger loudness standard (ITU BS.1770) that was introduced to measure the perceived loudness over the entirety of a program. The integrated sum of the loudness of a program is given a value in loudness units (LU). All broadcast programs from commercials to sitcoms to live events must meet the same loudness standards. Those standards use target levels for the entire program measured in loudness units full scale or LUFS. The targets for the entire program, called the integrated level, vary slightly depending on the region you are in and how your program is delivered.

In the North America and parts of Asia, an integrated target of -24 LUFS seems to be the norm for broadcast, whereas in Europe it is -23 LUFS. Film trailers, web videos, and other forms of distribution can have different targets.

On the Fairlight page, the meters and the side numeric display provide a variety of useful loudness measurements. The meter on the left shows the momentary loudness unit average, the meter on the right measures the short term (three seconds) perceived loudness of audio material.

However, generally the most important display to pay attention to when determining the loudness of your program is the integrated number on the right side of the meters.

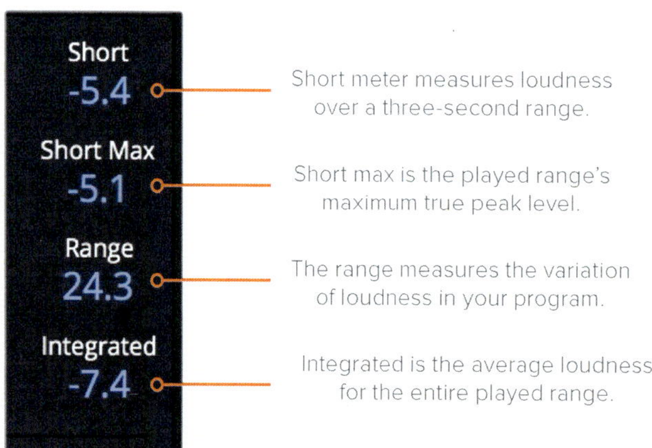

1. The integrated LUFS number is the average loudness over the played start to stop range. To set the start of your range, you need to click the Start button under the loudness meters just before playing. You then reset the resulting measurement before taking another one.
2. Move the playhead to the start of the timeline.
3. Click the Start button.

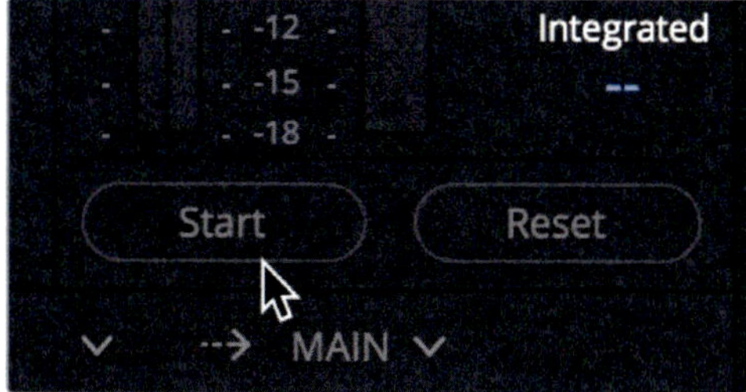

4 Press Spacebar to play the timeline, then stop playback at the end.

When monitoring the integrated level, you are looking to ensure that the numeric display hovers around 0 ±0.5 LU. In television broadcast, the target for an integrated LUFS meter is recommended at -23 or -24 LUFS ±0.5 LU. However, the loudness meters use a scale from +9 to -18 loudness units where -23 LUFS is equal to 0 on the scale.

TIP You can change the EBU scale for the meters as well as the -23 target in the general options category of the project settings.

Now, with an understanding of how to read the loudness in your program, you can use the mixer to adjust your submixes and main output to hit the target loudness.

Setting track levels

Audio balance is the key to the entire sound mixing process. When you achieve balance in your mix, you have successfully done your job. You've already performed volume adjustments at the clip level; but to achieve overall balance, you must adjust entire tracks relative to one another and monitor the overall program's loudness using the LUFS values.

1 In the upper-right corner of the Interface toolbar, click the Inspector button to close the Inspector.

2 Click the Mixer button to open the Mixer panel.

The Mixer panel includes a channel strip for each track in your timeline, as well as a master audio strip and two strips for your submixes. Each channel strip is color coded based on the track color that you assigned.

3 Position the playhead at the start of the timeline, and press the Spacebar to play the entire program.

   Overall, the sound effects seem too low. No problem. You can the adjust the volume level for the entire track by using the corresponding sub (S2) fader on the mixer.

4 Play the timeline again, and drag up the S2 fader between -10 and -5 dB.

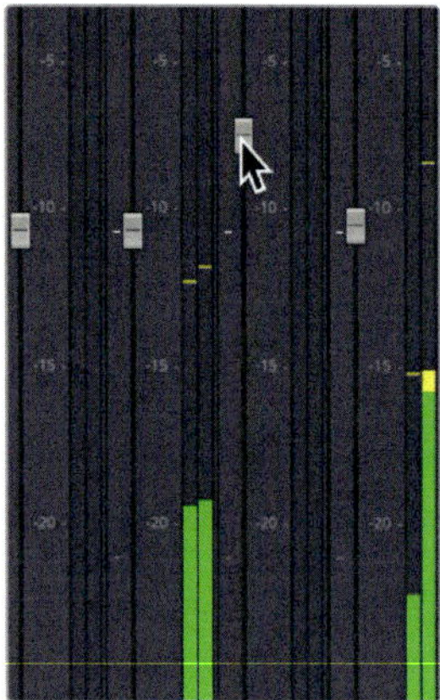

5 Adjust the S1 Narration submix and the Music on track A5. Monitor the loudness levels so they fall around -24 LUFS.

As you make level adjustments, it is important to keep an eye on your loudness levels, and also to monitor the audio results using quality speakers in a calibrated system. Just as you want to view your video content on a calibrated display, you must hear your audio content in a calibrated environment. So, always take off those headphones and mix your audio using professional speakers set to a calibrated loudness level.

Adjusting fade handles

The Fairlight page includes fade handles at the head and tail of each audio clip just like the Edit page. However, if you look closer, you'll see that the Fairlight fade handles include a curve control in the middle of the handle.

1 In the timeline, select the Music clip.

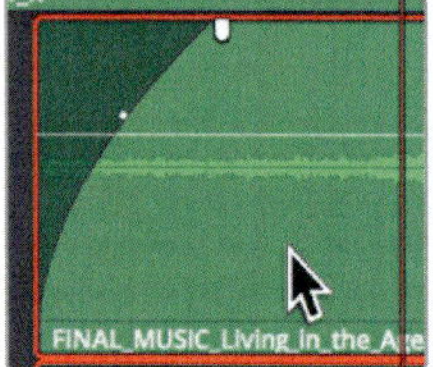

When a clip is selected, the fade handles appear as they do on the Edit page. You can shorten or lengthen a fade by dragging those handles.

2 At the start of the music clip, drag the handle to the right to extend the fade to double its current duration.

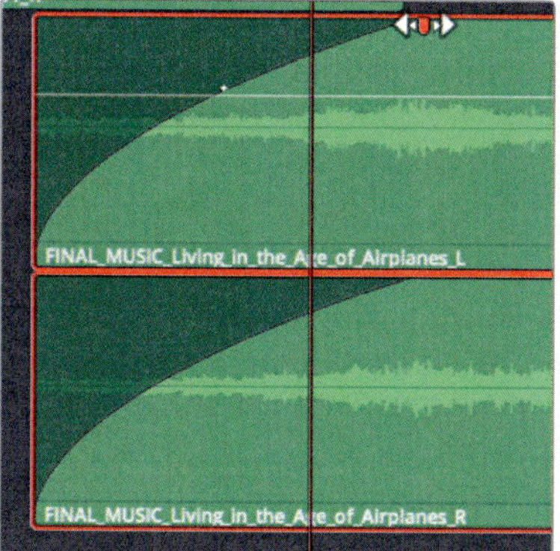

Adjusting the curve control lets you control the power and shape of the curve.

3 In the fade curve, drag the center handle to the right to accelerate the fade.

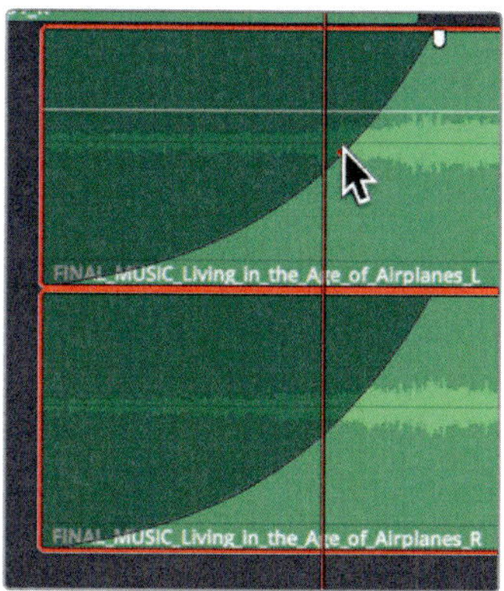

TIP You may need to expand the track height and width to see the handles.

The power of the fade will increase the closer it gets to completion.

When you have been around film and television industry for any period of time, you'll hear colleagues talk about how sound is half of the video experience. Not a filmmaker alive will dispute that, yet sound still seems to get the least attention when it comes to independent filmmaking. So, let's give attention to a lesser-known George Lucas quote: "Filmmakers should focus on making sure the soundtracks are really the best they can possibly be, because in terms of an investment, sound is where you get the most bang for your buck."

An Introduction to Color Correction

Before you get into the technical side of color correction and learn how DaVinci Resolve 14's powerful color correction tools work, it's important to take a moment to understand color correction and the creative medium it is.

Color correction is not something you can do by learning the controls of the color corrector, and it's not something you can do well just using the scopes. It's a highly creative skill in itself. Just as a good editor can tell a story and bring a dramatic flow to a program, the colorist evokes an emotion in a viewer via visual manipulation of the image. While it can take time to learn how to be a top-level colorist, like all creative skills, it never gets boring because you'll always have something new to learn and a new creative style to explore!

When using DaVinci Resolve, you have the advantage of over thirty years of color correction experience. DaVinci pioneered the development of color correction hardware and software specifically designed to artistically enhance visual images acquired from film, video, and digital sources. As a result, DaVinci Resolve possesses an incredibly deep, sophisticated, and efficient toolset for adjusting the look of the clips in your program, and managing these adjustments over an entire timeline.

Furthermore, DaVinci Resolve has continuously evolved thanks to feedback from countless professional colorists worldwide working at all levels of the film and broadcast industry. So, the DaVinci Resolve Color page has been developed to work the way colorists think. Still, for all its technological sophistication, it's important to remember that DaVinci Resolve is merely a tool that requires an artist to realize its full potential. But of course, that's the fun part!

The following lessons cover the basics you'll need to learn to begin harnessing the power of the Color page in your own projects--be they feature films, episodic television, web series, short subjects, spots, promos, or corporate videos. No matter what you work on, these formats employ the same fundamental grading techniques and the same basic tools; so, if you're new to the world of professional color grading, don't worry. All rock star colorists once had to learn these first steps for themselves, and you'll use the fundamentals you learn here for the rest of your career.

Gone are the days when high-quality color grading was unaffordable. Blackmagic Design has put the powerful color tools of DaVinci Resolve within reach of any editor that has a reasonably capable workstation or laptop. The polish you'll need to achieve world-class results is only a click away on the Color page.

However, before you start getting into the specifics of color, it's important to step back and consider, what are these tools really used for?

Why color correct your work?

It's a tempting question, and one that countless producers and directors have asked. "The program looks fine the way it was shot, why spend the time to grade it?" It's a good question in an industry where time is money; if the program you've cut in the Edit page looks fine, why bother grading it?

The answer is because your program won't look as good as it will after being graded.

The process of adjusting the contrast and color of every clip in a program is variously called color correction, color grading, or just grading. The difference in terminology is largely superficial, but most experienced colorists prefer "grading" because "correction" implies you only adjust things that are wrong, whereas "grading" implies that you're holding each clip in your program up to a higher artistic standard. A colorist doesn't ask, "Does this clip look good?" A colorist asks, "Could this clip look better?"

Setting the tone of the visuals

Much has been said about the emotional power of color to shape audience mood, and nobody would argue that a scene lit by cool blue lighting is going to have a very different vibe than one that's lit by warm orange lighting. The greenish tinge of fluorescent fixtures, and the salmon-hued wash of mercury vapor streetlights each paint the scenes of a show with different atmospheric feelings that, when done right, add to the narrative and how your audience perceives it.

Cooler Warmer

Of course, what these varied illuminants mean depends on the visual palette you develop. Warm lighting that denotes romance in one film may instead portray roiling, desert-bound discomfort in another. Their impact, depends on the associations that your grading makes between the visuals and the story. Should this scene seem later in the day? Should the colors be more subdued? Should the sky be an inescapable presence? You control these audience perceptions when you exercise subtle control over the picture via color grading.

The important takeaway is that the Color page gives you the tools to mold these associations to suit your needs--intensifying, attenuating, or completely counteracting their effect, as necessary, to strike the right tone for each and every scene.

Portraying the world subjectively

Narrative cinematography is rarely concerned with capturing objectively lit renditions of locations with perfectly accurate, neutral color and tonality. Instead, truckloads of lighting instruments and careful art direction manipulate the light and color of the location to make it look somber, magical, frightening, or sultry. These efforts extend to the grading suite, where your job is not to portray the world as it is, but the world that the cinematographer and director want the audience to see.

What the camera saw (left) What you want audiences to see (right)

Documentary photography may very often be concerned with presenting a supposedly unvarnished (yet gloriously rendered) look at the world. And yet, even this "realistic" look at the world is a fabrication, as every adjustment you make to improve the visibility of a subject, enhance the glory of nature, clean up some archival footage, or push the surroundings of the frame to recede artfully into the background, can be as carefully thought out and manipulated as any music video grade.

The point is, whether you're making a horror movie, an architectural documentary, a sales video, or an automotive spot, you're using the tools and techniques of color correction to create a subjective representation of the imagery. The more control you can exercise over this representation, the larger palette of emotional response you'll have to draw from.

Evolving to do high-end work

If you want to learn and stay competitive, and especially if you intend to work on client projects rather than your own, it's good to make yourself aware of current styles and trends. You've no doubt heard that if you want to write, you should read as much as you can, and the same holds true for color grading. Watch movies, television, music videos, and web shorts. And if you're watching television, make yourself watch the ads. Once you've had a chance to learn the grading controls that DaVinci Resolve offers, you'll start to see how different looks correspond to adjustments you can make in your own projects.

And finally, get out into the world and look at other visuals. Flip through fashion magazines, go to art galleries, take a hike in the woods, and observe. Fill your mind with diverse images, and analyze them to see what inspires you. The more aware you are of other visual disciplines, the more ideas you'll bring to your own work.

A last issue to consider is the effect affordable color grading has had on the television industry. In most current episodic television production, the visual style is now as good as in a feature film. This dramatic change in quality has made television programming better than ever.

An unintended benefit of this change is that top-level feature film actors now move into television work and back to film with amazing freedom because television no longer looks like an inferior medium. Also, high-level film crews and facilities are able to do a wider range of both television shows and feature films, while still retaining their premium status. It's an exciting time when you consider the additional increase in the number of distribution platforms for high-quality work, such as streaming services. The industry is growing more quickly than ever, which means talented editors and colorists are more in demand than ever!

The goals of color grading

Color correction can be considered the process of choosing which parts of the raw image data to display to create a pleasing image for the viewer.

Developing the image

The latest generation of digital cinema cameras are almost all capable of either shooting raw color space image data, or at the very least, recording RGB image data with a log-encoded exposure. Doing so preserves the maximum amount of image data for manipulation during the color correction process. While this is great for flexibility in workflow and for making high-quality adjustments, acquiring media in this way forces you to take the extra step of transforming it into a viewable image for editing and finishing (in much the same way that film negative required development and printing to yield a viewable image.

DaVinci Resolve simplifies this task with built-in camera raw controls, DaVinci Resolve color management, and LUT support so you can quickly get your media to a sound starting point upon which to build the rest of your grade.

Log encoded source (left) The same source normalized and corrected (right)

Making every clip look its best

While the job of the cinematographer is to light and expose the image with an artistic intent, your job as an editor and colorist is to realize this intent by adjusting the color and contrast of the image of each clip so the final result is as close to the director's and cinematographer's intentions as possible. In the process, you can overcome inconsistencies with exposure and color balance that were otherwise unavoidable. Furthermore, you can subtly adjust warmth and contrast to realize looks that were not achievable during the shoot, but that the director and cinematographer would have liked.

An underexposed image (left) The corrected image for the audience (right)

Of course, in some situations, you may find it necessary to fix media that has more substantial problems in color and exposure. In these cases, the tools exist to make far more involved changes to the image; however, the quality of your results will depend heavily on the quality and "latitude" of your source media. For example, Blackmagic URSA Mini cameras record quite a bit of image data within raw or minimally compressed media formats, allowing you to make extreme corrections that would be impossible on consumer cameras. Happily, in either case, the Color page provides the tools to process images in many different ways to adjust the image to achieve a better look.

Quality control

While you're doing all this, it's important to keep in mind that for all the creative possibilities that DaVinci Resolve affords, it's still important that the deliverables you give to your client have appropriate signal levels relative to their distribution requirements. In particular, programs destined for cinema, broadcast, or streaming usually have very specific outer boundaries of luma, chroma, and gamut that you must not exceed, or you'll risk having a show kicked back to you for quality control violations.

DaVinci Resolve provides tools specifically designed to help you keep an eye on how the image data is affected, and to fine tune the image. In particular, the scopes display the standard Waveform, Parade, Vectorscope, and Histogram graphs that you can use to objectively analyze image data. These scopes let you see the boundaries of what's possible, and make it easy to spot subtle problems, and compare the characteristics of one image to another.

Balancing scenes

It's rare for uncorrected shots to match one another seamlessly. Even the most carefully exposed angles of coverage can have small variances to be evened out. For example, run-and-gun programs using available light often result in edited scenes with huge changes in lighting and color as one shot cuts into the next.

Small or large, variations between shots can call undue attention to the editing, and jar the audience in ways that throw them out of the program. Balancing these differences is another fundamental task of the colorist. You know you're finished when every shot in a scene looks like the same time and the same place, and the color and contrast adjustments you've made flow unnoticeably from one clip to the next.

Adding style or custom "looks"

Of course, it's not all about subtlety and correction. It's often appropriate, when grading music videos and commercials, for instance, to bring some radical visual style to a piece. Here, too, DaVinci Resolve provides an abundance of features for manipulating unexpected aspects of the image. For example, you can use custom curves to create an illusion of chemical cross-processing.

Grading an image (left) with curves to create a cross-processing effect (right)

The tool Hollywood uses

If all that isn't enough incentive to plunge forward into the next few lessons and exercises, keep in mind that DaVinci Resolve has become the tool of choice for some of the largest post-production facilities in the industry, worldwide. And yet, thanks to its accessibility, within the last several years, DaVinci Resolve has also become the go-to tool for a wide variety of smaller boutique post companies and individual artists. Considering only projects completed recently, DaVinci Resolve was used to grade blockbusters like "Independence Day: Resurgence" and "X-Men: Apocalypse," along with indie productions such as "The Big Sick" and "A Ghost Story;" not to mention television shows including HBO's "Westworld," AMC's "The Walking Dead," and the CW's "Arrow."

Whether you're looking to build a foundation of skills to enter the post-production industry as a contributing artist, or you want to develop the ability to finish your personal creative work in your own way, the following exercises will usher you into a much larger world of image manipulation and artistic expression than has ever been available in the average non-linear editing application.

Lastly, color grading is just fun! The feeling of resting your hands on the trackballs and holding the emotion of your images in your hands is exhilarating. It's like no other feeling in the world; you can make adjustments in real time, instantly see the results, and feel the emotional impact in your heart. We believe that color correction is one of those tasks that is more creative than cerebral. It's also one of those jobs that surprises you every day and has an emotional connection that reminds us why we fell in love with the television industry in the first place!

Enjoy the journey!

Lesson 9

Quickstart: Color Correction

Like editing, color correction is an art form that takes time to learn and master. Color is an incredibly powerful creative tool that can define the style and convey the mood of your film. If you give yourself the time to practice and learn, you'll be able to master this exciting skill and create images that look amazing!

This lesson will teach you the fundamentals of color correction by providing an overview of the most important tools. You'll learn about the primary corrector, secondary adjustments, nodes, tracking, and even applying DaVinci Resolve FX for special effects. Experience is key, and with so many controls at your fingertips, this lesson will give you the start you need toward learning this creative skill.

Time

This lesson takes approximately 60 minutes to complete.

Goals

Remember, DaVinci Resolve was originally developed for high-end color correction and finishing on feature films. That means you're about to work with the same tools that Hollywood's top colorists use to correct and finish the biggest blockbuster films, episodic television shows, and commercials

Learning the color page layout

The technical and creative process of color correction takes place within the Color page in DaVinci Resolve 14. Let's start by examining the Color page layout.

1. Open DaVinci Resolve to the project manager window.

   To begin exploring the Color page, let's switch projects. The project you will use was archived using DaVinci Resolve. An archive is a self-contained project that includes all of its media. All you have to do is restore the archive and the project will be available with all the media already linked. You'll learn more about archives later in the book.

2. Right-click the window, and choose Restore.
3. Navigate to R14 lessons > Lesson 09, select the Citizen Chain Cyclery.dra folder, and click Open.
4. In the project manager, open the restored Citizen Chain Cyclery project, and from the Rough Cuts bin, open Color Grading Quick Tour into the timeline.

   One of the best aspects of DaVinci Resolve is that editing and color grading are completely integrated into a single application, so you can easily move between the two with a single click.

5. At the bottom of the DaVinci Resolve window, click the Color button to go to the Color page.

The Color page is divided into seven main areas.

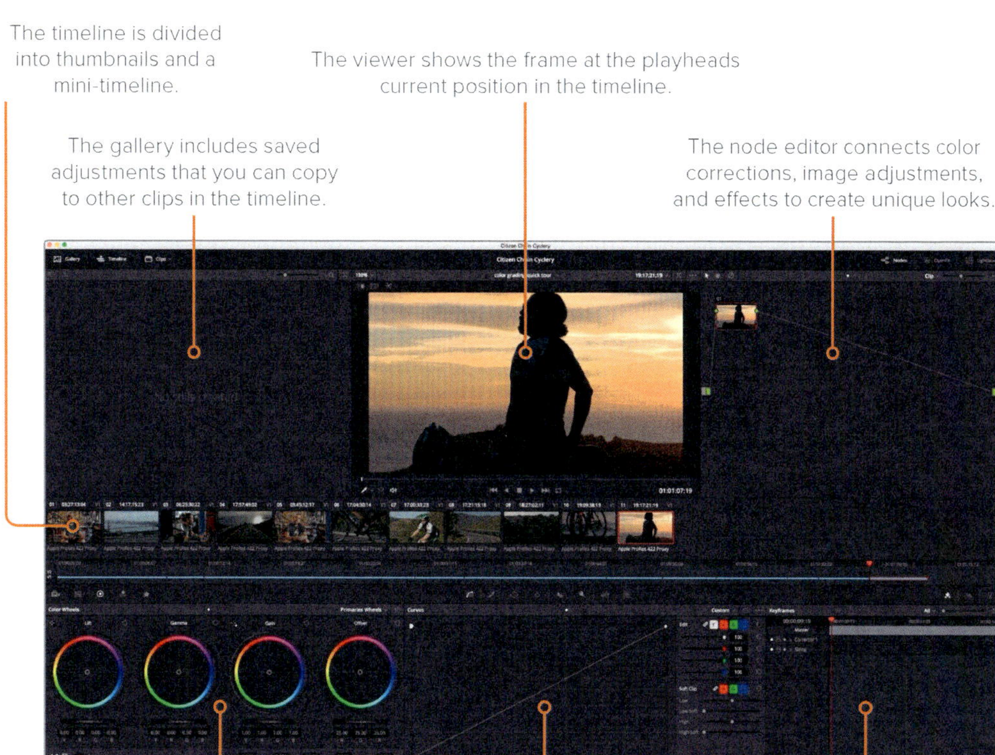

The left palettes contain primary adjustments for color, contrast, and RAW image processing.

The center palettes provide access to color curves, power windows, tracking, keying, and stereoscopic controls.

The lower-right area may display the keyframe editor, color/luminance scopes, or a histogram display.

NOTE When you're using DaVinci Resolve on a computer display with a resolution lower than 1920 x 1080, some panels and buttons will be consolidated and not look exactly like the images in this lesson.

When you switch to the Color page, whatever was loaded into the editing timeline is displayed in the Color page timeline. The Color page does not change or alter any cuts or transitions. It just provides a way of looking at your timeline that is more appropriate for color correction.

6 Click the center of the last thumbnail to select it.

An orange outline appears around the selected thumbnail, and the playhead jumps to the first frame of that clip.

7 Below the thumbnail, double-click the Apple ProRes 4 22 Proxy name to switch to viewing a display name for each clip.

The Mini-timeline below the thumbnails displays thin bars to represent each clip. A bar's width is proportional to a clip's duration.

8 Drag the Mini-timeline playhead to the left to scrub through the timeline until you reach the first clip.

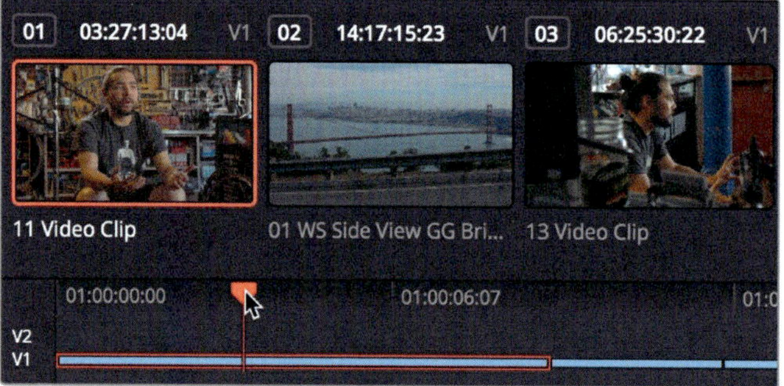

TIP If a track is disabled in the Edit page, it will be dimmed in the mini-timeline.

As you scrub through the timeline, the clip under the playhead highlights in orange to show that it is selected, a behavior similar to the thumbnail display outline. The transport controls under the viewer, as well as all the playback keyboard shortcuts, are the same as you used on the Edit page.

Now that you have a basic understanding of the Color page layout, you're ready to make some adjustments.

Modifying lift, gamma, and gain

The most popular controls for creating styles and balancing your shots are found in the primaries corrector. Because DaVinci Resolve includes many controls in the primary corrector, you will find that you spend much of the time using the primary corrector while you are on the Color page.

Using the primary corrector, you can achieve a refined result because you can divide the image into tonal regions.

The lift, gamma, and gain controls broadly correspond to the dark, midrange, and bright regions in the image. Within each region, you can adjust hue and luma values to create a wide variety of styles. For instance, if you wanted to adjust color or brightness in the dark areas, you would move the lift region.

Let's make a few adjustments to get the feel of some of the primary corrector's controls. In this exercise we'll use the color wheels controls.

1 In the timeline, click thumbnail 11 to relocate the playhead to that clip.

2 In the Lift color balance control, drag the color balance indicator slightly toward blue to add blue to the darker areas.

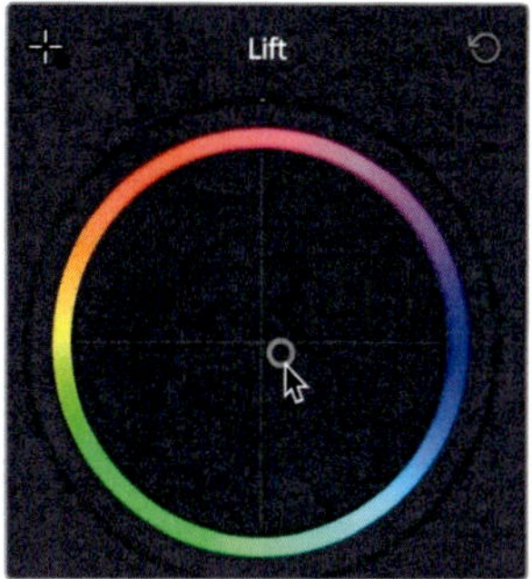

TIP Adjustments made in the color balance controls are subtle. In most cases, the edge of the control will remain very close to or still touch the crosshair.

The master wheel controls, below the color balance controls, adjust brightness in the lift, gamma, and gain regions. Adjusting the Lift wheel will increase or decrease brightness in the darker areas.

3 Drag the Lift wheel to the right until the Y luminance value reads 0.05.

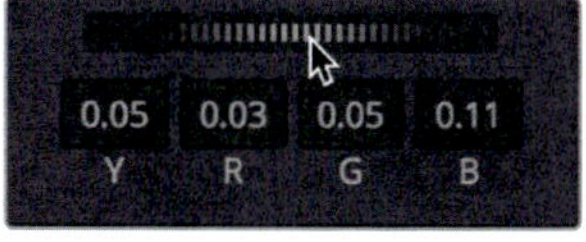

The master wheel under the Lift color balance control adjusts the black point for the image. When dragging it to the right, the darker areas in the image become brighter.

Let's look at the other end of the spectrum by adjusting the Gain control.

4 In the Gain color balance control, drag the color balance indicator slightly toward orange to add orange to the brighter areas.

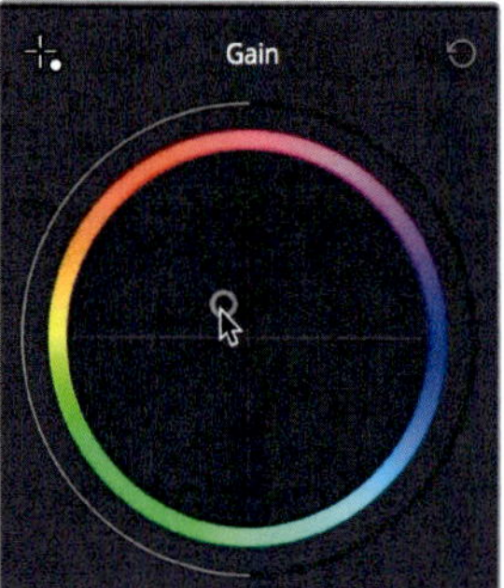

The Gain color balance control tints the brighter areas of your image.

The master wheel under the Gain color balance control adjusts the white point for the image. When dragging it to the right, the brightest areas in the image become brighter.

5 Drag the Gain master wheel to the right until the Y luminance value displays 1.3.

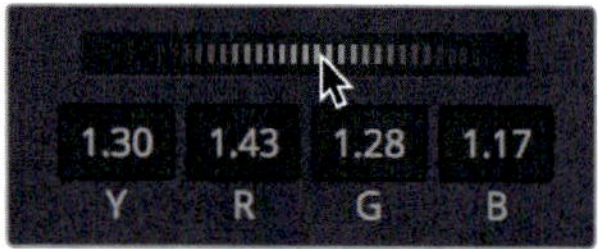

By adjusting the Lift master wheel and the Gain master wheel, you have effectively adjusted contrast in the clip. Instead of using a simple contrast control, you have exercised greater control over the black point and white point using the Lift and Gain master wheels.

Now let's look at gamma.

6 In the Gamma color balance control, drag the control slightly toward magenta.

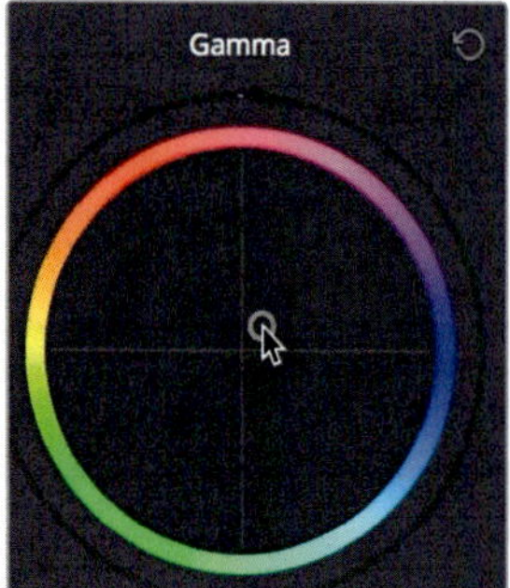

The Gamma color balance control tints the midrange of your image.

7 Drag the Gamma master wheel to the left until the Y luminance value displays –0.03.

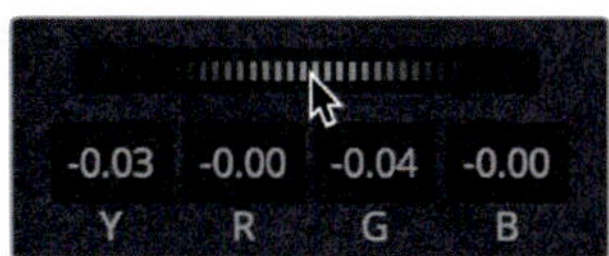

The master wheel under the Gamma color balance control adjusts the overall brightness while maintaining the black and white points that you set previously. When dragging this control to the left, the overall image becomes darker. Let's compare the corrected image you've made to the original image.

8 Choose View >Bypass All Grades, or press Shift-D, to see the original image. Then choose View > Bypass All Grades, or press Shift-D again, to view the corrected image.

These adjustments were made only to give you a feel for the controls and what they do. They clearly didn't produce anything worth keeping. You can reset each control or the entire primary corrector using the reset buttons.

NOTE In the Color page, each clip has its own undo/redo history. That is, choosing Edit > Undo will undo previous changes depending on which clip is the current clip. The Edit page also has its own undo history, which is separate from the Color page. The Edit and Fairlight page Undo/Redo commands operate on the entire timeline, not on each clip.

9 In the upper-right corner of the Gamma color balance control, click the Reset button.

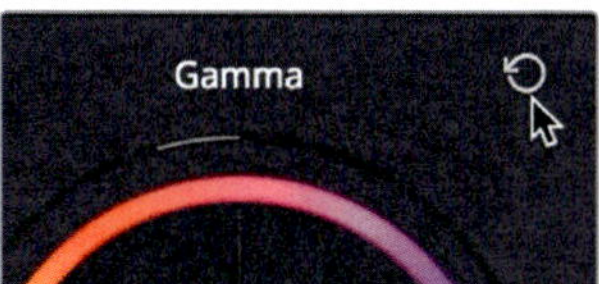

You can also reset the entire primary corrector using the panel reset button.

10 In the upper-right corner of the primary corrector panel, click the reset button.

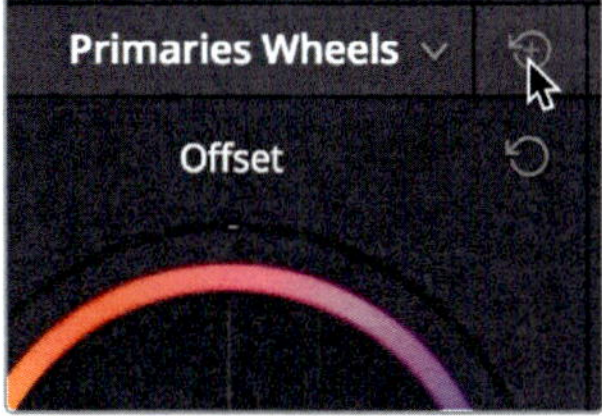

The Lift, Gamma, and Gain controls are not isolated adjustments that change only the dark, midrange, and bright areas. In fact, their ranges overlap by a considerable amount. When you adjust the lift, for example, most of the adjustment affects the darker areas, but the midrange and even some bright areas also may be affected.

Similarly, when you adjust gain, most of the adjustment operates within the brightest areas of the image, but you'll also see a fall-off impact within the midrange and a small amount of dark areas. This overlap helps you make more natural, smoother-looking adjustments, but it also means that you'll need to move among the three color balance controls to achieve best results because making an adjustment to one control visibly impacts the others.

Using other primary corrector controls

The primary corrector controls that you have used are adjusted mainly to achieve an overall look by controlling lift, gamma, and gain within the image. You can also make a few primary corrector adjustments that you may be more familiar with from other video or photo applications. The shared adjustment controls are located across the bottom of the primary corrector panel. These global adjustments affect the entire image, not just the lift, gamma, or gain.

1 Select thumbnail 03.

   The color in this shot appears a bit low in contrast and under saturated. Although you can modify saturation and contrast in DaVinci Resolve, the primary corrector includes some very quick and easy options.

2 In the adjustment controls, position the pointer over the Contrast value field.

   TIP Depending on your screen resolution, the names of each adjustment control may not be visible. Use the icons to identify each parameter, or hover your mouse pointer over an icon to view a tool tip of its name.

   Dragging the Contrast value to the right increases the contrast.

   TIP As with all value fields in DaVinci Resolve, when the pointer is over the value field or the parameter's name, you can drag to modify the value.

3 With your pointer in the Contrast value field, drag to the right until the value reaches 1.2.

The adjustment controls have a second page of parameters for globally improving your shot.

4 Click the 2 button to display the second page of the adjustment controls.

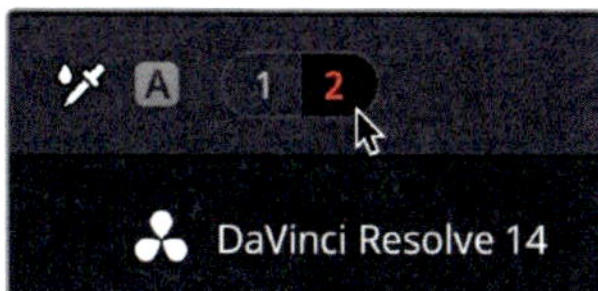

Boosting neutral colors and detail

On page 2 of the adjustments, color boost and midtone detail are two good adjustments that can add a nice visual punch to your shots. Color boost allows you to increase the saturation of lower saturated colors without too much increase in high saturated colors.

1 With your pointer in the Color Boost value field, drag to the right until the value reaches 30.

TIP Double-clicking on any name in the adjustment controls will reset the value.

Now his skin has more color in it and the already saturated blue pole hasn't increased as much.

There are also adjustments for controlling shadows and highlights, similar to the master wheels.

2 With your pointer in the Shadows value field, drag to the right until the shadows in this image are not as dark and crushed.

The other adjustment that can add a good pop to your images is the Midtone Detail control.

The Midtone Detail control really is the best of all sharpen filters because it does a brilliant job of sharpening edges without increasing noise.

3 With your pointer in the Midtone Detail value field, drag to the right to increase the local contrast around edges in the image.

In both the Color Boost and Midtone Detail controls, a little goes a long way. A typical way to adjust these is to first add more correction than you want and then slowly back off until you find an acceptable result.

Making white balance adjustments

You can perform straightforward white balance correction using the temperature and tint controls. Let's locate a shot that needs a white balance correction.

1 Click thumbnail 02 to view the shot of the Golden Gate Bridge

This shot clearly has too much blue. The clouds, which should be purely white, are the most obvious signs. Using the temperature control, you can shift the color in an image either warmer toward yellow, or cooler toward blue.

> **TIP** For off-axis color casts that come from irregular light sources, you can use the tint control to shift an image away from magenta or green.

2 Drag the Temperature value field to the right, until it is between 500 and 600, to increase the warmer tones.

As always, when you make a significant adjustment, you should compare it to the original image.

3 Choose View > Bypass All Grades, or press Shift-D, to see the original image. Choose View > Bypass All Grades, or press Shift-D again, to return to the corrected image.

TIP You can use either the DaVinci Resolve Micro or Mini Panel with dedicated primary corrector hardware controls for quicker access to all of the most common tools.

With the adjustment you have made, this image now has a much better white balance.

In most color correction situations, you would likely bounce between the two images a few times while refining your adjustments. Rarely do you set a control once, compare it to the original, and move on. Color correction is an iterative learning process. It takes time; but the more you explore your options, the more you find the adjustments that work best for you.

Understanding nodes

Instead of stacking color corrections and effects as layers, you can add as many color correctors and filter effects as you like using nodes. Each node is a full DaVinci Resolve color corrector and you can place as many of them as you like.

You can view the nodes as a color correction flow chart for each individual clip. The clip, or the input, starts at the left, flows through nodes, and ends on the right side of the screen with the corrected image output.

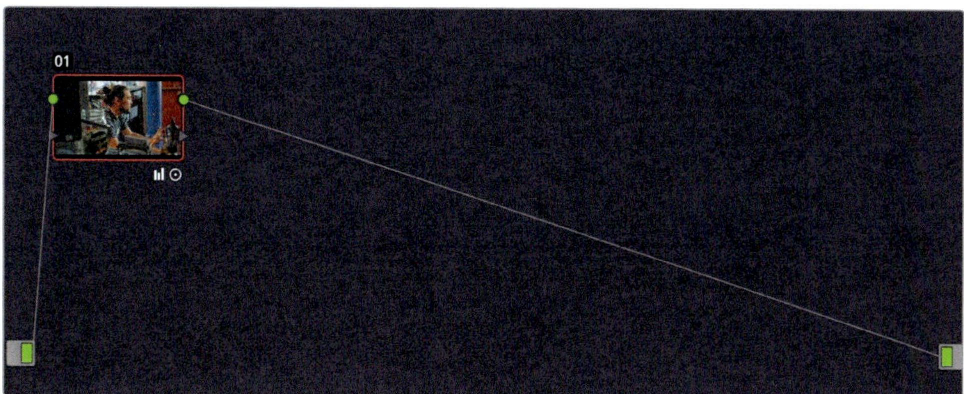

The adjustments you made in the preceding exercise were performed using the first node, which automatically is provided for you in the node editor. As you create more sophisticated corrections, you can add more nodes that target different parts of the image or add effects.

Making secondary color corrections

So far, you've made adjustments to an entire image, so-called primary color corrections. Secondary color corrections isolate only parts of an image to make very specific changes to a particular image area. A classic but extreme example is turning an entire image to black and white except for one object. Let's apply secondary color correction to a much more common task that you'll be able to use on many of your productions.

1 Click the second interview clip, thumbnail 03.

2 Press the Spacebar to play the clip.

3 After you have viewed the clip a few times, press the Spacebar to stop playback.

   This clip looks perfectly fine except for the distractingly large bright blue pole in the middle of the room. You can't remove the pole altogether, but you can dim the color to make it less intrusive in the shot.

Using nodes and curves

The first part of making a secondary color correction on one object is to isolate the adjustment to its own node.

Using the node editor (located in the upper-right quadrant of the DaVinci Resolve window), you can combine one or more individual nodes to create more intricate corrections.

By default, every clip has one node in the node editor that contains the initial adjustments you made. The node shown in the current project contains all the adjustments you previously applied to create the neutral correction.

However, you also have the option of creating additional nodes in which each node contains one or more corrections that affect the image. Using multiple nodes, each containing separate adjustments, you can exercise more precise control over the order of those adjustments and more easily track and modify them.

Let's add a second node to the node editor to separate the adjustment that you will apply to the blue pole from the color boost correction you made earlier.

1 In the Nodes menu, choose Add Serial Node, or press Option-S (Mac) or Alt-S (Windows).

The node is added after the initial node in the node editor. The new node is given the number 02 and has a red outline to indicate that it is the currently selected node. You can rename the node to be more descriptive.

2 Right-click the node, and from the contextual menu, choose Change Label.

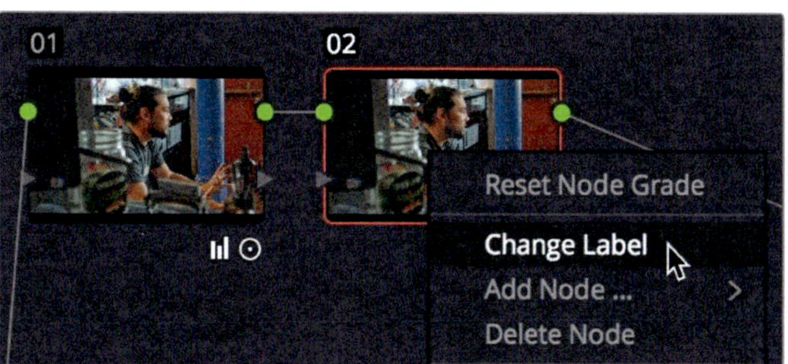

3 Type Blue Pole, and press Enter to rename the node.

With a new node in place, you can now use the curve controls to modify the color of the blue pole. The default curve controls are custom curves that you may be familiar using in Photoshop or other image-editing applications. However, DaVinci Resolve includes several curve controls for isolating various hues, as well as luminance and saturation levels. To lower the blue of the pole, you'll use a hue vs. saturation curve.

4 In the curves pop-up menu, choose Hue Vs Sat to display the hue vs. saturation curve.

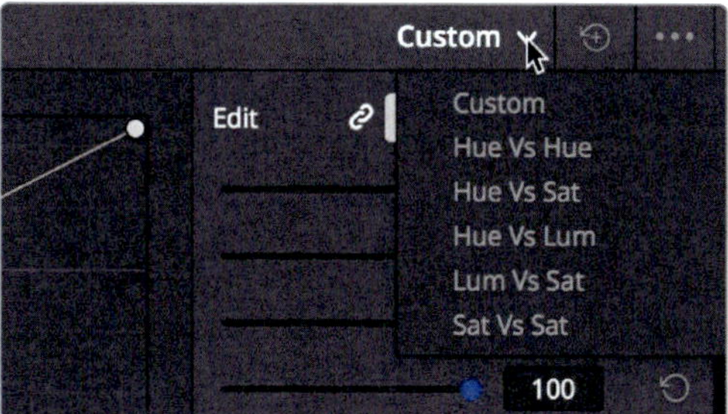

The hue vs. saturation color spectrum is displayed under the timeline. These controls let you isolate a specific hue in the image just by clicking the image.

5 In the viewer, hover the mouse pointer over the blue pole.

The pointer changes to an eyedropper to indicate that you can sample an area of the image you want to isolate.

6 Click anywhere on the blue pole to sample its hue value.

Once you click the viewer, three points are added to the line in the hue vs. sat palette. The center point is the actual hue you selected. The other two points constrain the range of hues you will be adjusting as you drag the center point.

7 In the hue vs. sat palette, drag the center point down to lower the saturation of the blue hue.

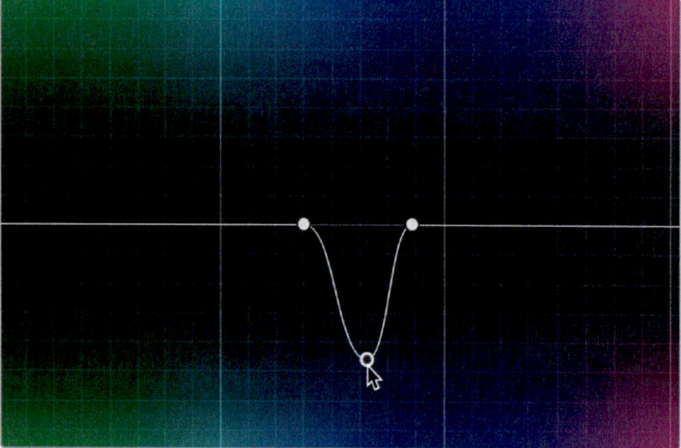

Only items with that specific hue are affected by the decreased saturation. These hue, luminance, and saturation curves simplify a very complex process, so you do not have to worry about getting a perfect key or manually rotoscoping.

As always, after you make a significant change to an image, it is a good idea to compare your change with the original image. In this case, you want to compare only the changes to the pole. Previously, you were comparing changes to the original image. In this case, because you made the adjustment on a new node, you can temporarily disable the node to compare the change.

8 Choose Nodes >Enable/Disable Current Node, or press Cmd-D (Mac) or Ctrl-D (Windows).

Now you are viewing the neutral correction without the toned-down saturation adjustment to the blue pole.

9 Choose Nodes > Enable/Disable Current Node, or press Cmd-D (Mac) or Ctrl-D (Windows).

You may want to quickly compare before and after images a few times to review the change. You can also tweak the saturation level further if you think the image requires it.

Applying DaVinci Resolve FX

DaVinci Resolve includes many high-quality filter effects called DaVinci Resolve FX that include blurs, glows, film grain, and lens flares. You can apply these effects to an entire clip or you can combine them with spline shapes to isolate an effect to one area of the frame.

1 Click thumbnail 07.

2 Press the Spacebar to play the clip.

This clip already has basic color correction on its first node, so you'll place a second node to add an effect.

3 Choose Nodes > Add Serial Node, or press Option-S (Mac) or Alt-S (Windows).

The effect to add is a common one: blur out the logo on the cyclist's shirt. This is something you frequently must do to avoid violating copyright laws. The first step is to isolate the area you want to blur.

4 Under the viewer, drag the playhead to the start of the clip.

5 In the middle of the toolbar, click the power window button.

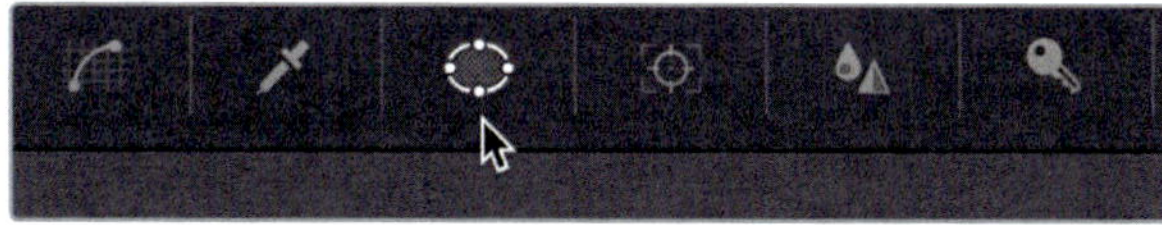

You can use power windows to isolate a part of the frame. Unlike the hue vs. sat curve, you use spline-based shapes instead of a hue selection.

6 Click the circular power window shape to add it to the clip.

7 Drag one of the corners of the circular power window so that the inner circle is roughly the size of the logo on the cyclist's arm.

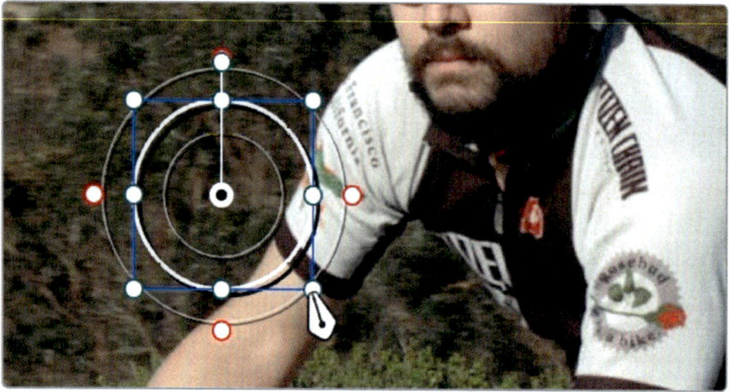

8 Drag the center of the circular power window shape to cover the logo on the cyclist's arm.

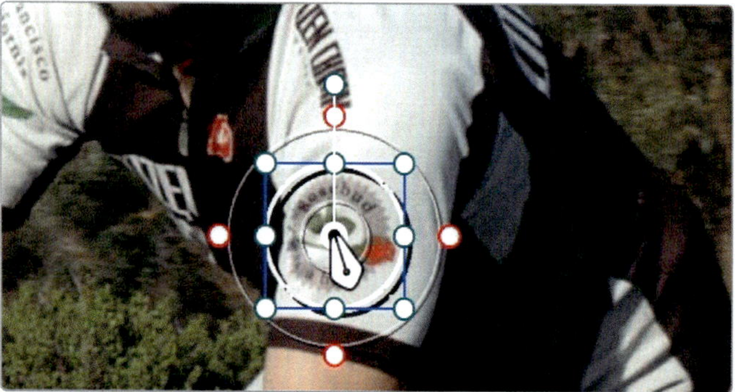

You can use a DaVinci Resolve FX to "fill" the power window and obscure the logo.

9 Above the node editor, click the OpenFX button.

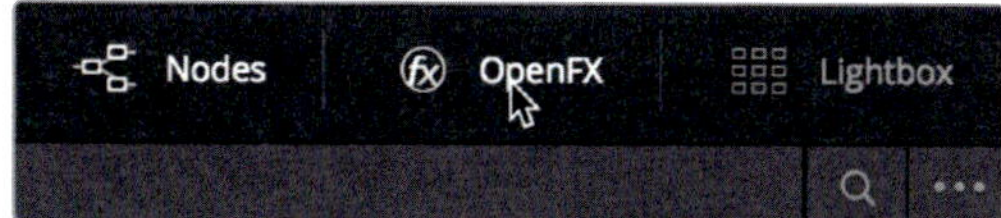

In the OpenFX panel, you'll typically find a list of the third party filter effect plugins that you added to DaVinci Resolve. However, it also contains a list of DaVinci Resolve FX that are included with the application.

10 Scroll down the OpenFX panel to find the Mosaic Blur effect. Drag the effect onto the second node that contains the power window.

When the mosaic blur is added, it fills in the power window to obscure the logo. You can modify the mosaic in the settings panel that automatically appears.

11 In the OpenFX settings panel, raise the Pixel Frequency to around 100 to increase the number of mosaic squares used in the power window.

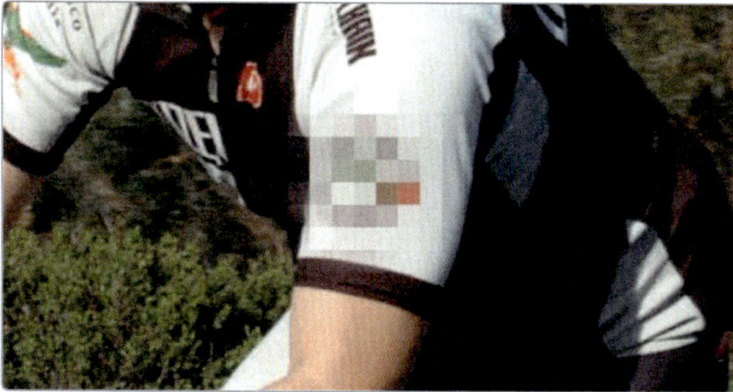

The mosaic is placed for this one frame only. Because the cyclist moves throughout the shot, you will need to track the power window to ensure that it follows the logo around the screen.

Tracking power windows

DaVinci Resolve features an incredible 3D perspective tracker that not only follows objects, but can also determine if that object changes rotation or perspective. As a result, you can track the most challenging objects to apply DaVinci Resolve FX or color correction.

1 In the toolbar, click the tracker button.

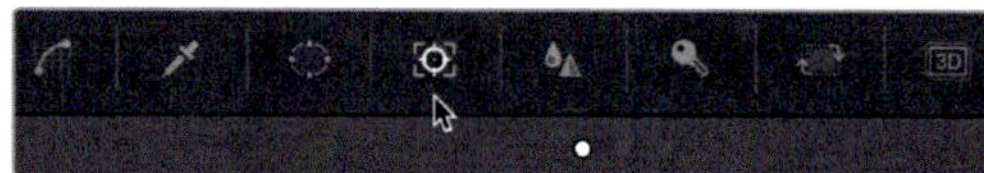

2 Make sure the playhead in the Tracker panel is at the start of the clip.

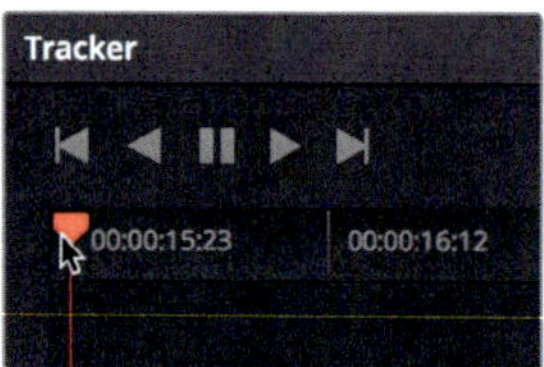

The tracker can track various aspects of objects within a clip. You can enable and disable the transform types using the checkboxes at the top of the panel. Disabling parameters can sometimes help improve tracking, so long as you needn't track those disabled aspects of the object. For instance, in this clip, you do not care if the logo zooms or rotates; however, you do care about tilting and perspective changes.

3 Click the Zoom and Rotate checkboxes to deselect those transforms.

4 Begin tracking by clicking the track forward button.

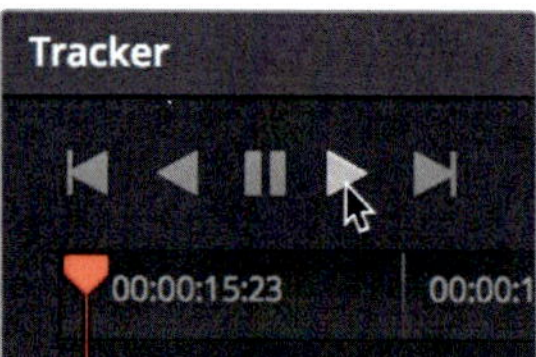

5 Once the track is completed, scrub or play through the clip to see the results.

The tracker causes the power window to perfectly follow the logo on the cyclist's sleeve.

This quickstart lesson was only a brief example of how to use the controls in the Color page. You'll dive deeper into the Color page in the next three lessons where you'll learn more about controls in the primary corrector, additional secondary tools, and how to copy and compare the creative looks you make.

Lesson 10

Performing Primary Corrections

With so many tools at your fingertips, the creative possibilities in DaVinci Resolve 14 are endless! In addition to the creative judgment you'll develop as a colorist, you'll also want to master some fundamental techniques. In addition to stylizing the look and being creative, you're also responsible for ensuring that images look right.

As a colorist, you'll need to fix image exposure and color in bad shots. The ability to balance shots is also important because in a single scene the shots from one camera must match other cameras' shots and no shot should have an unwanted color cast. The colorist is also responsible for making sure that images are "legal" and fall within the established technical parameters based on whether you are finishing for television broadcast, film, or web distribution.

This lesson teaches you how to create a base neutral grade and to balance shots. You'll use the primary corrector for most of this lesson and rely on built-in scopes to check your work. You'll save and copy grades, so you can quickly apply them and match up multiple shots.

Time

This lesson takes approximately 45 minutes to complete.

Goals

Balancing color and brightness with the Color Wheels

Because unforeseen or unavoidable situations occur during production, clips can be recorded with an incorrect color tint, or have highlights that are dull or shadows that are too bright. Therefore, you always start your color correction process by balancing shots. Doing so prepares your shots for the creative look you may later apply to them.

1. Open DaVinci Resolve to the project manager window.

   In the next three lessons, you'll learn a color grading workflow using a real project: a documentary called Gnarly in Pink about a girls' skateboarding club. Once you restore the archived project, you will begin working just as if you had edited the project in DaVinci Resolve and were ready to color grade it.

2. Right-click in the project manager window, and in the pop-up menu, choose Restore.
3. Navigate to R14 lessons > Lesson 10. Select the Gnarly in Pink.dra folder, and click Open.
4. In the project manager, open the Gnarly in Pink project. In the Master bin, open the Gnarly in Pink edit into the timeline.

   The major time saver here is that this timeline was edited in DaVinci Resolve and you do not have to conform it from another editing application. You'll learn more about the conform process in Lesson 13.

5. At the bottom of the DaVinci Resolve window, click the color button to go to the Color page.

   You are now set to make some adjustments. The standard approach is to tackle luminance first, and hue and saturation second.

6. Click thumbnail 08 to select it.

You can start evaluating this clip just by watching it in the viewer.

It is a low-contrast shot in which the shadows around the cabinet appear muddy and gray, whereas the white walls are dull.

To improve contrast, you can use the primary correction tools you used in the previous lesson.

7 Drag the Lift master wheel to the left until the Y luminance value is near -0.10.

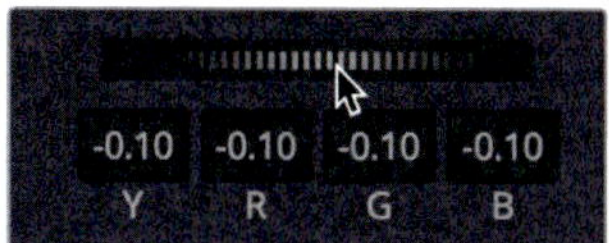

As you drag the Lift master wheel to the left, the darkest areas in the image become darker.

8 To increase the highlights, drag the Gain master wheel until the Y luminance value displays 1.30.

Now that you have maximized the contrast, you can compare those changes with the original image by temporarily disabling the adjustments.

9 Choose View > Bypass All Grades, or press Shift-D, to see the original image; then choose View > Bypass All Grades again to view the corrected image.

With just two simple adjustments you have created a better-looking shot in which the image shadows appear rich and dark while the highlights are bright.

Checking adjustments on scopes

As an objective way to evaluate adjustments made while color correcting, DaVinci Resolve includes four video signal scopes. You can use the Waveform, Parade, Vectorscope and Histogram to check the luminance, exposure, hue, and saturation of a clip.

1 Click the video scope button to the far right of the toolbar.

2 Click the disclosure arrow to open the video scope pop-up menu.

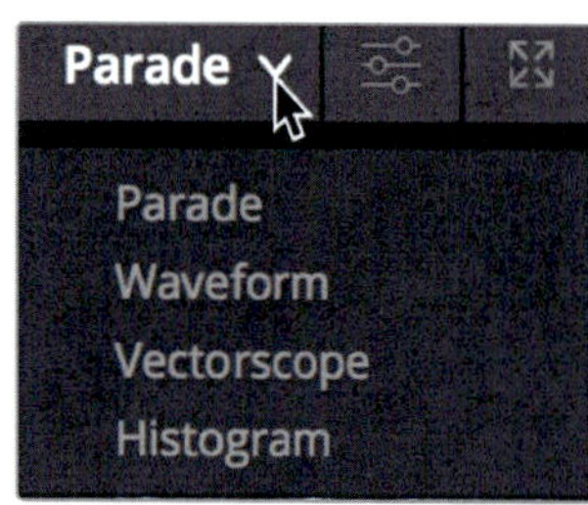

TIP You can view two or four scopes at once by choosing Workspace > Videoscopes > On.

Using the menu, you can switch between the four different scopes depending on what you want to monitor.

3 Choose Parade.

The Parade scope graphs each color channel individually. The graph is read from bottom to top with absolute black at line 0 and absolute white at line 1023. When balancing shots, the image in the waveform, called the trace, should not go below 0 or above 1023, otherwise the image will clip—cut off parts of the waveform--and lose detail.

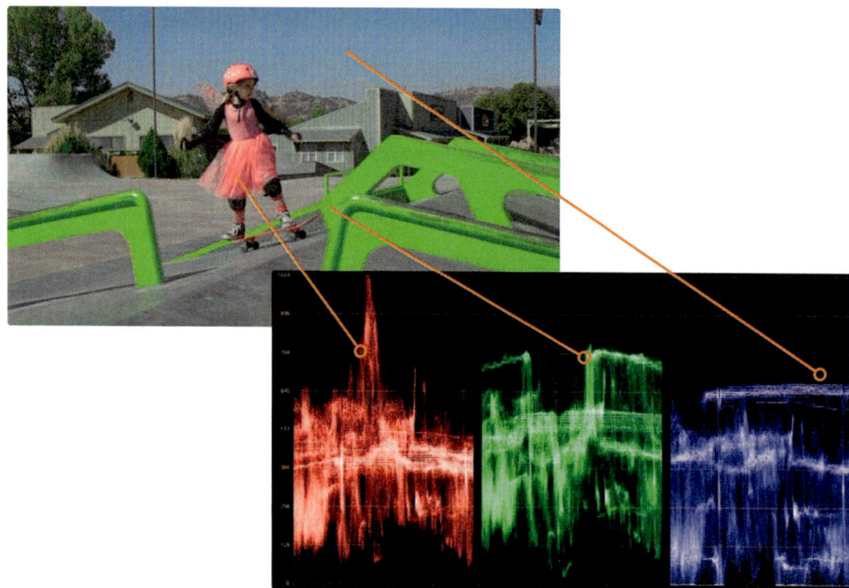

Reading the Parade from left to right, each channel corresponds to the image displayed in the viewer. For instance, the left part of the red, green, and blue trace corresponds to the left part of the image. This layout makes it easy to look at the scope and know exactly which area you are evaluating.

TIP You can switch back to the keyframe editor and hide the video scopes to free up your graphics card's processor and improve playback performance.

In general, the bottom of the trace should fall somewhere between 0 and 128 on the graph. If some elements of the shot are absolutely black, then the trace should fall closer to 0. If the darkest part of your image is more of a dark gray, then it might fall closer line 128 on the graph. For this shot, you can fine tune the Lift and Gain master wheels so your trace stretches between 0 and 1023.

4 Drag the Lift master wheel until the bottom of the trace falls somewhere between the 0 and 128 lines.

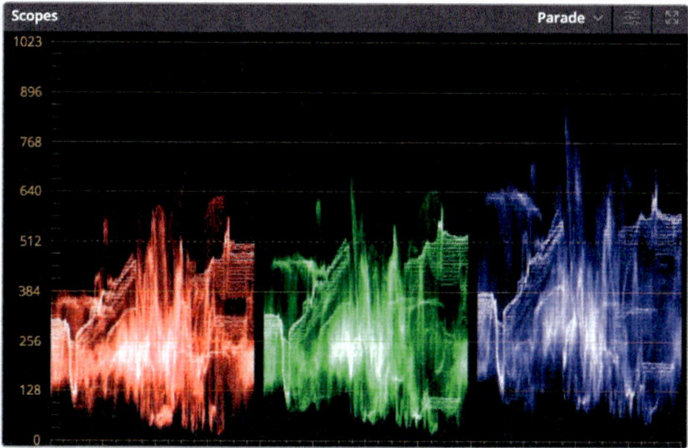

5 Drag the Gain master wheel to the right until the trace in the waveform is between the 896 and 1023 lines.

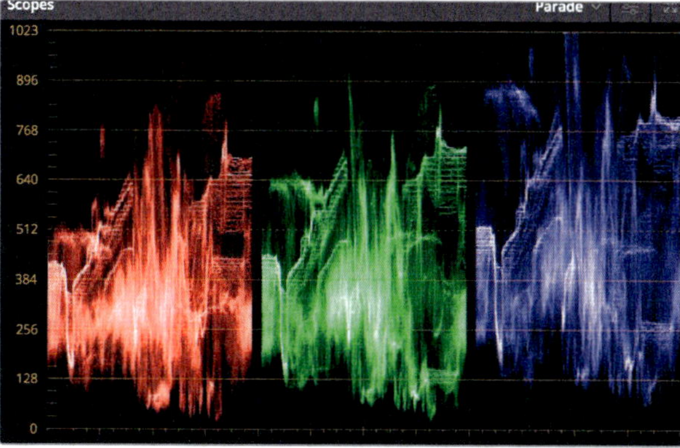

Next, you'll fine tune the hue and saturation. However, let's discuss a little additive color theory before you evaluate the color. To create pure white using additive colors, you mix red, green, and blue at full intensity. On a parade scope, a white image would have the red, green, and blue traces completely level along the top of the graph. That being the case, you'll use the white walls in your shot to make color balancing easier.

It is clear this image appears to have a blue tint. While the tint is easily seen with your eyes, you can confirm it in the Parade.

The blue trace is shifted higher than red or green which is an indication that a blue tint exists in the highlights. Having a distinct blue or orange tint, or cast, in an image is a common white-balance issue. In the previous lesson, you used the temperature control to correct white balance, but you can just as easily use the primary corrector's Color Wheels to deal with this common issue.

6 In the Offset color wheel, drag the color indicator toward yellow/orange until the walls appear white; and the red, green, and blue trace in the Parade are level along the top.

TIP Unless you have a completely white frame, a scope will always display some uneven areas. Use the trace as a guide but use your eyes to make the final decision when performing manual white balance adjustments.

You've created a balanced, neutral color correction using the primary corrector on one shot. You repeat this for every shot in your timeline when you begin the color grading process. Doing so builds a consistent neutral starting point for your shots.

Using the viewers for color grading

DaVinci Resolve's viewers are previews that are not intended to be color-critical displays, especially for projects intended for television broadcast or digital cinema. For those purposes, you can use a Blackmagic Design UltraStudio or DeckLink card to connect to a broadcast or digital cinema calibrated display.

Making a neutral color grade with the primaries bars

A lot of the power in DaVinci Resolve comes from the flexibility of its toolset which provides many ways to do the same thing. To learn more about the primary corrector toolset, you'll explore another method for creating a balanced correction on a new shot.

1 In the timeline, click thumbnail 04.

This is one in a series of opening shots that are extreme close-ups. As in the previous shot, this shot requires a color and brightness balance; but instead of using the Color Wheels, you'll use another set of tools in the primary corrector called the Primaries Bars.

2 In the upper-right corner of the primary corrector palette, in the pop-up menu, choose Primaries Bars.

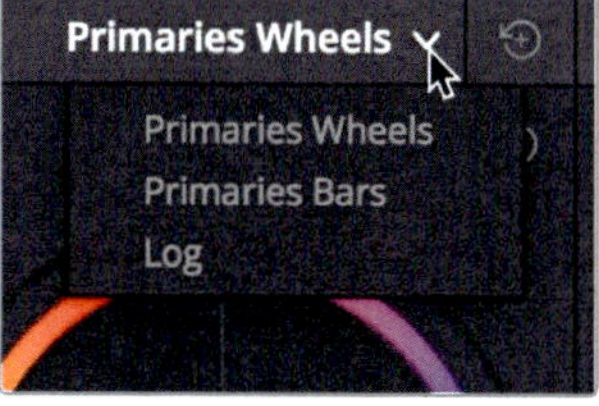

The Primaries Bars allow you to make color and luminance adjustments similar to the Color Wheels, but the bars provide explicit red, green, and blue adjustments in the lift, gamma, and gain regions. That being the case, some may find them more effective tools for balancing specific color channels in different regions of a shot. For tonal adjustments, the master wheels are directly below the bars, just as they were with the Color Wheels.

3 Drag the Lift master wheel to the left until the Y luminance value is near -0.05.

Once again, the darkest areas in the image become darker. Unfortunately, the highlights in this image are trickier to adjust because, unlike the previous shot, this image has no white anywhere in the frame. However, using your eyes, you can see that you need to add additional brightness to the highlights.

4 To increase the highlights, drag the Gain master wheel to the right until the highlights in the image are bright and the top of the trace is in the upper quarter of the scope.

Not having white in any part of the image makes it more challenging to accurately correct for any white-balance issues, However, you do have a large area of pavement that provides a neutral gray tone. Just like white, a neutral gray consists of equal parts red, green, and blue. Because it is very difficult to see neutral gray on a scope, you'll need another method for measuring the red, green, and blue values. In this case, you can use an eyedropper to display RGB values for a region of pixels in the frame.

5 Hover the mouse pointer over the viewer until it becomes an eyedropper.

6 Right-click in the viewer, and choose “Show picker RGB value”.

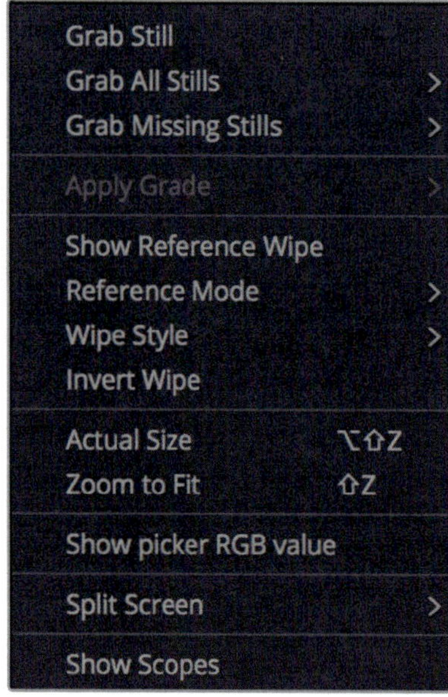

The 8-bit RGB values (0-255) are displayed as a tooltip next to the eyedropper. These values show you if a pixel has more red, green, or blue in it.

7 Hover the eyedropper over the gray background on the right side of the shot.

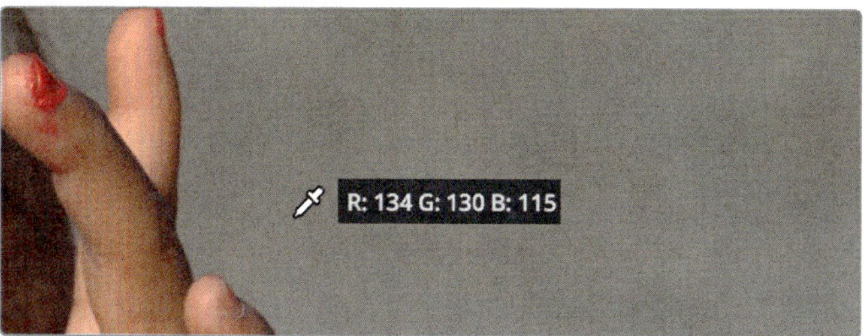

The RGB values show a much lower level of blue and a higher level of red in the midtone gray area of the cement. If the cement should be a neutral gray color (which you are assuming it should be), then you need to balance these RGB values. This type of channel adjustment is easiest using the Primaries Bars.

> **TIP** You can display 10-bit numeric values (0-1023) with the RGB picker by choosing View > Show RGB Picker Values In > 10 bit.

8 To increase the amount of blue in the midtones—where the tone of the cement lives—drag the blue Gamma bar slightly up and the red Gamma down.

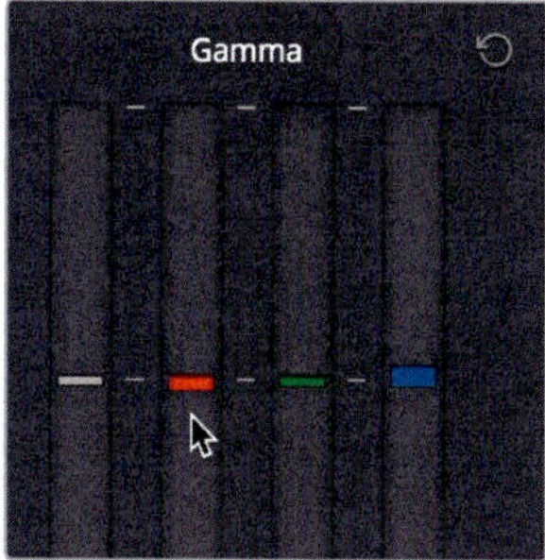

Once you have evened out the midtones, you can check that the darkest shadows do not have any color tint.

9 Hover the eyedropper over the right side of the girl's face in the shadow of her hair.

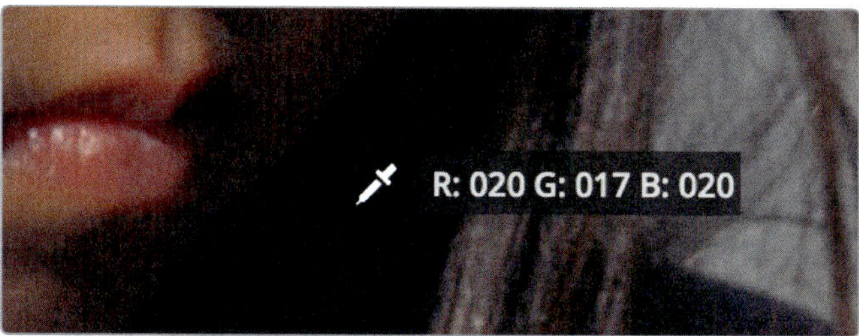

10 Keep making small adjustments to the red, green, and blue Lift bars until the RGB values are nearly even.

After making critical adjustments, it's good to compare your changes to the original shot.

11 Choose View > Bypass All Grades, or press Shift-D, to see the original image; and then press Shift-D again to view your corrected clip.

With these adjustments, the image now has a much better neutral balance and contrast. You did that quickly. Rarely do you set a control once, compare it to the original, and move on. Color correction is an iterative process—adjust, compare, adjust, compare. It takes time; but just like cooking or gardening, the more you explore, the more you find what works best.

Creating a style using the main primary controls

Beyond neutral grades, the primary corrector includes the main set of tools for the more creative process of giving your shot a style. Now that you have a balanced shot, you can give it more of a stylized look. When beginning to create a look, the most important task is to figure out your goal. What are you trying to convey with the look? In this case, your documentary is about a little girls' skateboarding club, but the opening extreme close-ups are of the girls polishing nails and painting with glitter. Let's give this action a warmer, softer, sunset magic hour style to juxtapose it with the livelier skateboarding scenes.

1 In the node graph, right-click node 01, and choose Add Node > Add Serial.

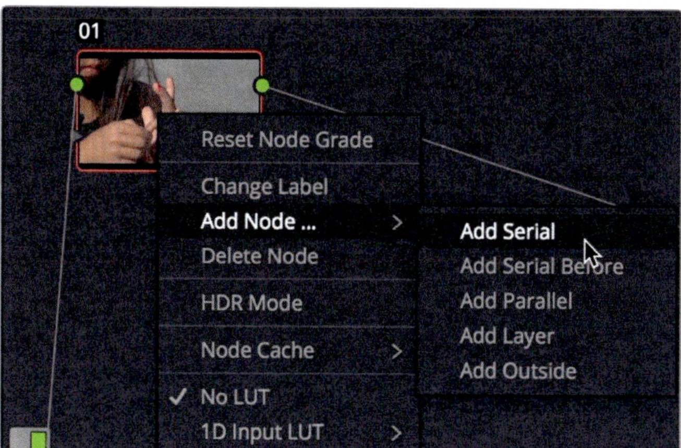

Adding another node gives you a way to separate the color balance from the stylized look. Separating a realistic balance from a stylized look not only makes it easier to compare their difference, it makes it much easier to find certain corrections as the number of nodes in your graph increase over time. You can make it even easier to identify corrections by naming them.

2 Right-click node 01, and choose Change Label.
3 Type balanced as the name of the node.
4 Right-click node 02, and choose Change Label.
5 Type style as the name of the node.

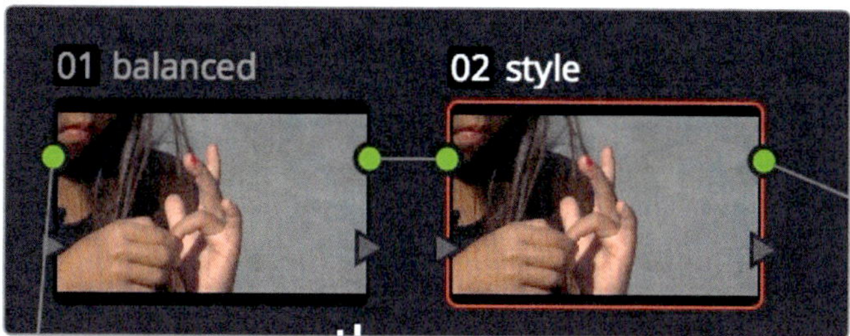

Now with a bit of organization in place, you can start creating the magic hour style you want for this shot.

6 Make sure that in node 02, the style node is selected in the node graph.
7 In the primary corrector panel, in the pop-up menu, choose Primaries Wheels to return to the Color Wheels.

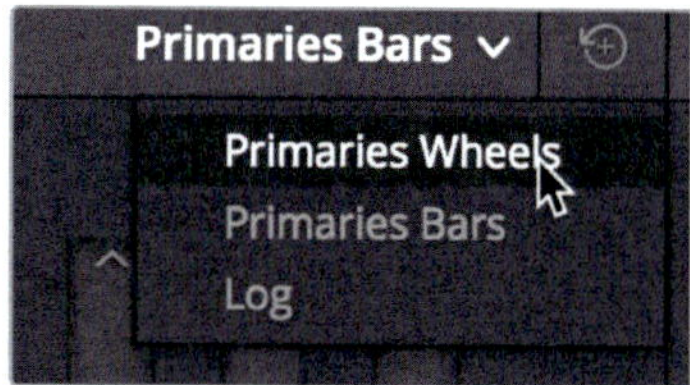

While the Primaries Bars are great for adjusting a specific color channel, the Color Wheels give you more freedom in mixing the exact color you want for a look. This is especially true when you are using a DaVinci Resolve Micro, Mini or Advanced control panel because you can use both hands to dial in the color using the track balls.

8 In the Gain color wheel, drag the color indicator toward yellow/orange to warm up the highlights.

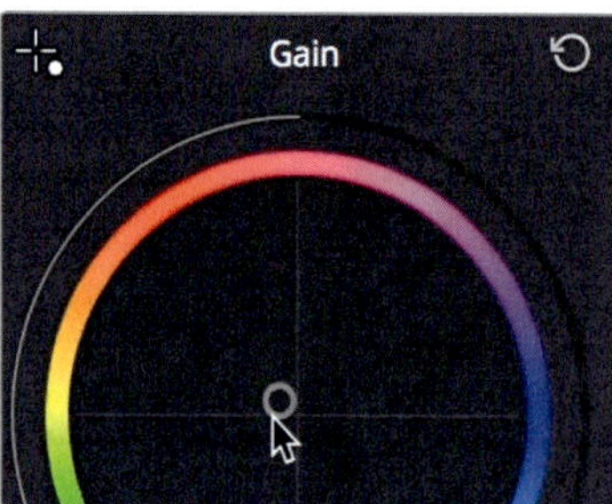

Don't get too wrapped up in wondering which toolset to use from the primary corrector. Which you choose is mostly a matter of comfort. The more you work with them, the sooner you will start to gravitate toward a set of tools that you feel best suite your working style.

Enhancing styles with the additional primary controls

The additional primary adjustments located under the master wheels can also be added to subtly modify brightness, color, or sharpness. To warm up the skin tones a bit more, you can again use the Color Boost adjustment.

1 In the adjustments strip, click 2 to show the second page of adjustments.

2 Drag the Color Boost value to the right to slightly increase the warmer skin tones.

To add a soft glow to this shot, you'll alter the Midtone Detail value. You used this in the previous lesson to sharpen edges a bit. In this case, let's apply a negative value.

3 Drag the Midtone Detail value to the left to soften the image and add a slight glow.

Applying this type of softness is a subtle adjustment. If you go too far, human skin can appear plastic.

4 To compare the style version with the balanced image, click 02 above the style node to disable it, or press Cmd-D (Mac) or Ctrl-D (Windows).

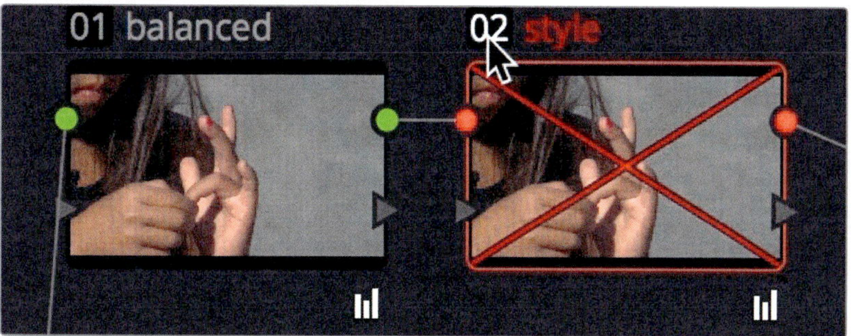

This ability to compare is one of the benefits of separating corrections into different nodes.

5 To return to the stylized look, click 02 above the style node, or press Cmd-D (Mac) or Ctrl-D (Windows).

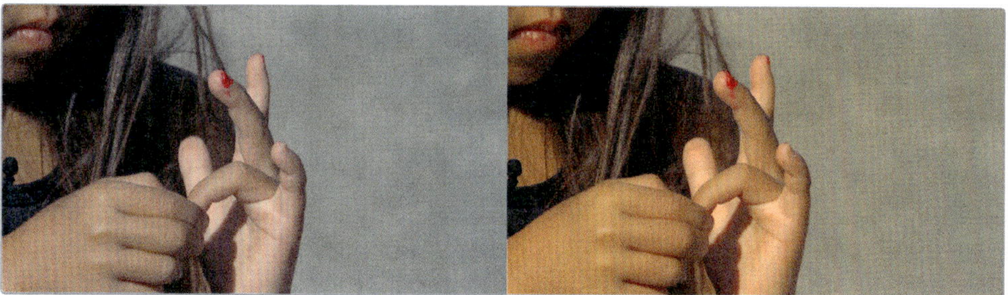

Before After

This shot now has a great look that you can apply to other similar shots in the same scene. Later in this lesson, you'll learn how to copy corrections from one clip to another; but right now, you have one more primary correction to do, and this one gets a bit trickier.

Using curves for primary color corrections

In this exercise, you'll look at another shot that is more complicated in its problems, and correcting it requires more sophisticated adjustments. You will begin again by creating a neutral grade; but instead of using the primary corrector, this time you'll make the adjustments using custom curves.

1 Select thumbnail 33.

2 Click the play button to review the clip.

In this case, it might be better to color correct this shot using a frame from the middle of the clip where the girl is in the frame.

3 Drag the viewer's playhead into the middle of the clip until you see the girl in the air.

You will approach this shot in the same way you did the previous shot: first, you will correct luminance, and then tackle the color imbalance. The "elephant in the room" is the mixed light sources in this shot. Some of the ceiling lights are florescent and give off a green light. You'll correct for them last.

Looking at your image, you can see a couple of issues that need to be fixed. The contrast is again low. A parade scope will confirm this because the trace is bunched up in the middle. The blue is weak and the green from the fluorescent lights is very strong. The blacks also look grey and they need to be deeper. Those are the issues you need to fix.

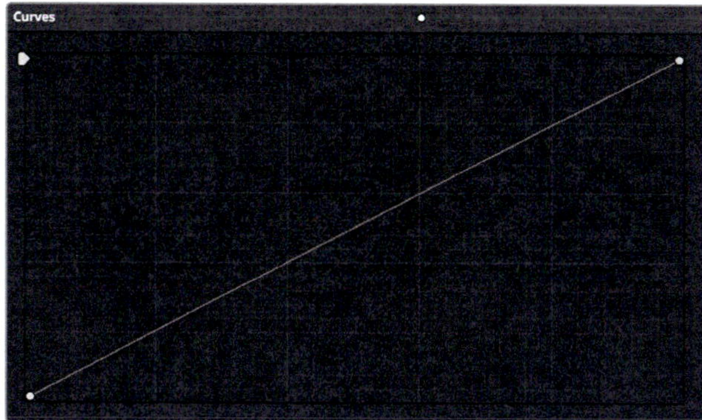

NOTE On computers with lower screen resolution, you may have to click the curves button to display the Curves palette.

The custom curve graph is a plot graph in which you can perform incredibly flexible adjustments on specific tonal ranges of images. The X axis represents the image's tonal values going from the darkest shadows on the left to the brightest highlights on the right. Along the Y axis are the output, or offset, values with darker adjustments placed lower in the graph and brighter adjustments placed higher.

TIP When you hear the terms tonal values or the tonal range, they're talking about brightness values.

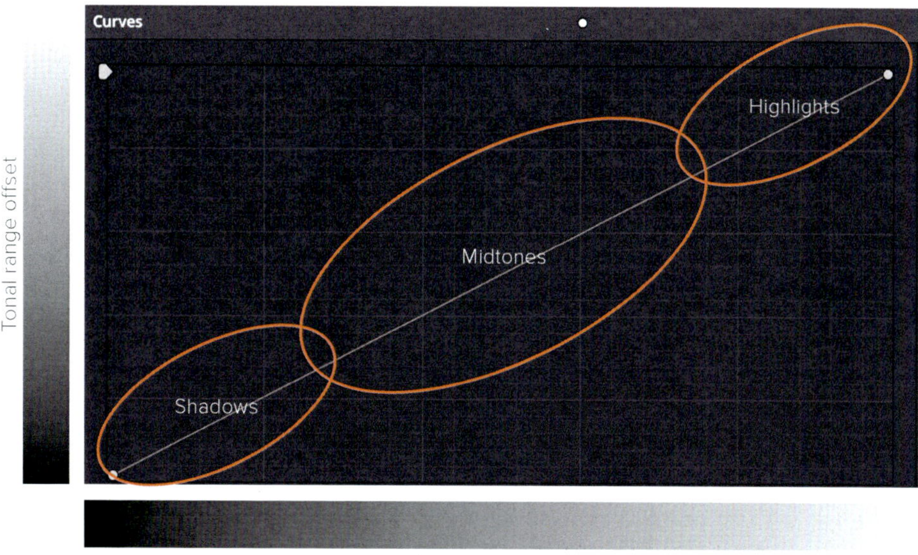

TIP In the Color page, each clip has its own undo/redo history. That is choosing Edit > Undo will undo different steps depending on which clip is currently selected.

To first tackle the low contrast issue, you can drag the end points in the custom curve graph to adjust the black and white points.

4 Position the mouse pointer over the control point located in the lower-left corner of the custom curve graph.

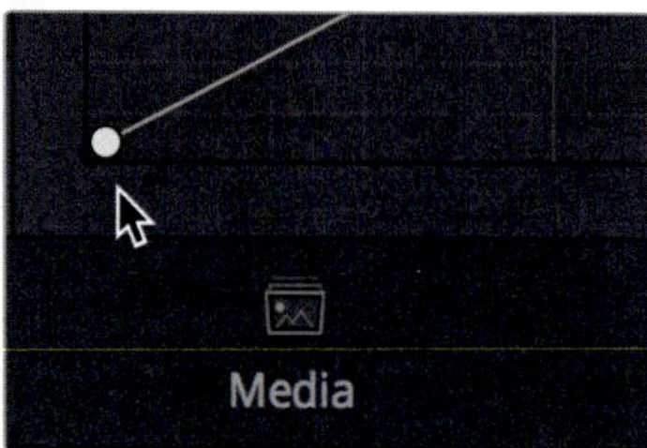

This point is the black point control. Similar to the Lift master wheel, adjusting this point raises or lowers the black point in a clip.

5 Drag the point to the right until the foam pit is darker, but still contains some detail. This should place the trace in the Parade near the bottom of the graph.

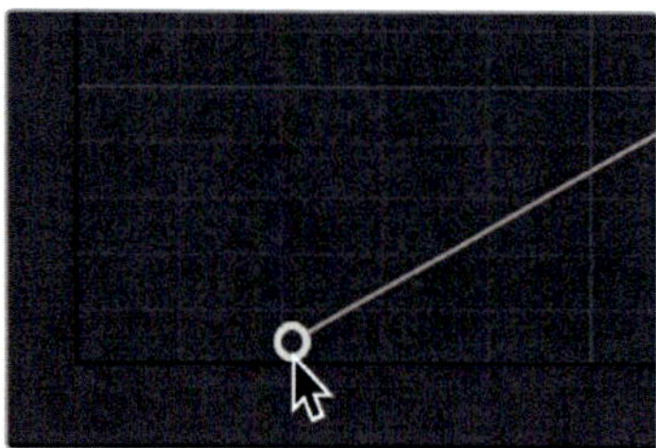

Moving the black point to the right darkens the black point, the darkest part of the image.

6 Position the mouse pointer over the control point located in the upper-right corner of the custom curve graph.

This point is the white point control. Similar to the Gain master wheel, adjusting this point raises or lowers the white point in a clip.

7 Drag the point to the left until the bright lights on the ceiling are bright but not blown out. This should place the trace in the Parade near the top of the graph.

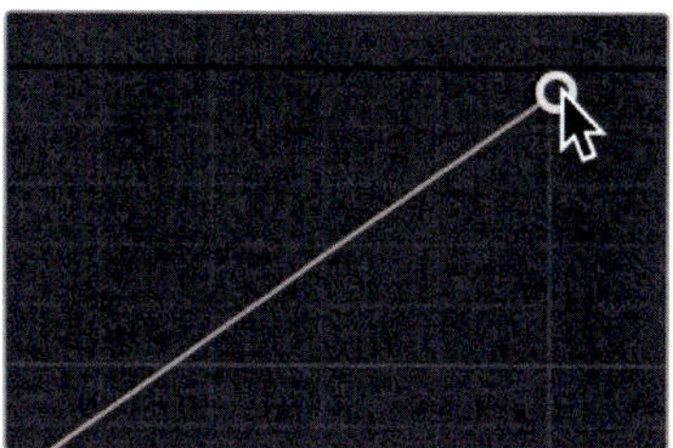

Dragging the control point to the left brightens the white point, the brightest part of the image.

Moving those two points as you did is the equivalent of increasing contrast using the Contrast slider, or adjusting the Gain and Lift master wheels.

8 Choose View > Bypass All Grades, or press Shift-D, to see the original image, and then press Shift-D again to compare your corrected clip to the original.

   Now you can work on color balance. Although the custom curves combine all the color channels and the luminance by default, you can independently adjust any of them by unlinking the controls.

9 In the curve controls area, click the link button to unlink the controls.

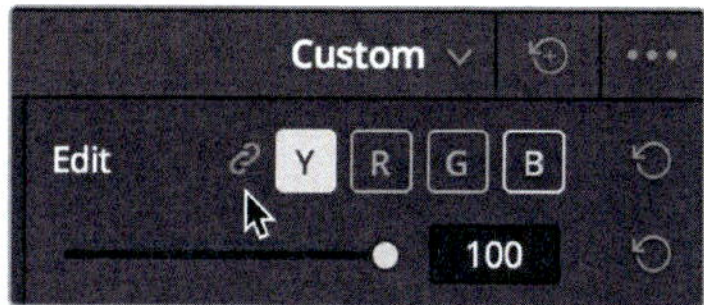

   Looking at the ceiling, it does appear brighter. But it should look white, whereas it appears more beige. This appearance indicates an excess of red and green in the color channels. You can also see that the blue channel in the Parade is much lower in the highlights than the red or green channels. Using the unlinked controls, you can boost the blue channel so that the ceiling is better balanced.

10 In the curve controls area, click the B (blue channel) button to display the blue custom curve in the curve editor.

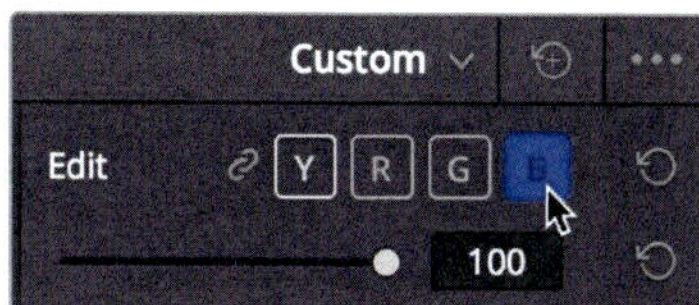

11 Drag the blue control point to the left until the ceiling appears white. The blue trace in the Parade scope will align at the top of the graph with the red and green traces.

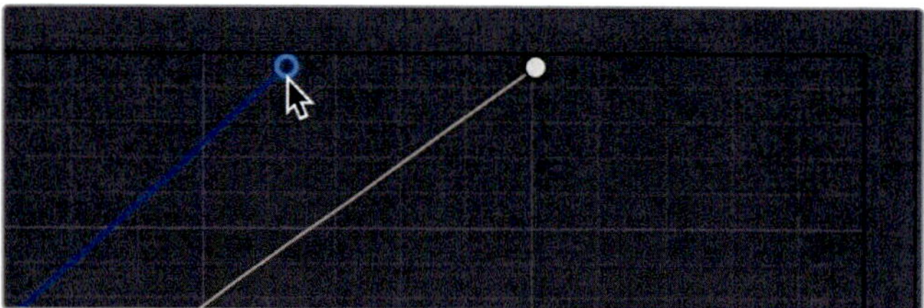

13 Choose View > Bypass All Grades, or press Shift-D, a few times to compare the original with your corrected clip.

Before After

Fixing difficult lighting using secondary curves

When a clip in a timeline is balanced, your next task is to correct any small areas of inconsistent lighting. These might be caused by someone's shiny forehead; the exaggerated darkness of a pair of eyes or, as in this case, the florescent lights that cast a green glow on the ceiling. Instead of bringing the entire green channel down to fix one small area, corrections such as these are best performed by isolating the problem area. You'll start making that isolated adjustment by adding a new node in the node graph.

1 In the node graph, right-click node 01, and in the menu, choose Add Node > Add Serial.

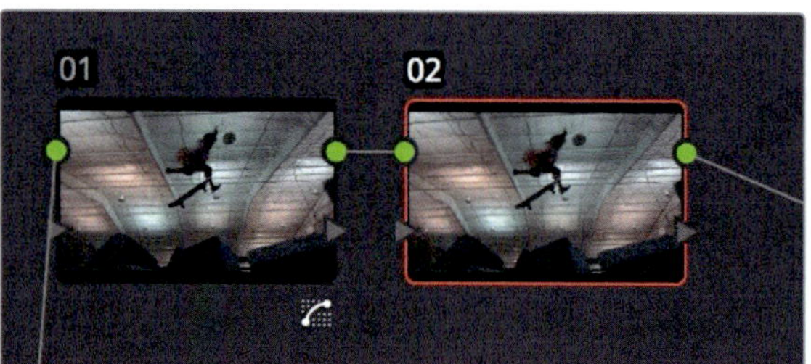

You can think of a node as a transparent layer stacked on top of your clip. When you make any color adjustment, those adjustments are applied to the node, and not to the clip. As a result, every adjustment in DaVinci Resolve is non-destructive because you can enable and disable a node at any time.

TIP When should you add a new node, and when should you keep working on the current node? There is no hard and fast rule. Think of nodes as a trail of bread crumbs. When you want to make sure you can get back to a previous color state of your clip, add a node.

You'll use the second node that you just added to fix the green glow.

2. In the node graph, double-click node 02 to ensure that it is selected.
3. To hide the timeline thumbnails, Gallery, and node graph, and enlarge the viewer, press Option- F (Mac) or Alt-F (Windows), or choose Workspace > Viewer Mode > Enhanced Viewer.

   Just as in Lesson 9 when you wanted to adjust the blue hue of the pole, you can adjust the saturation of a specific hue, such as the green glow, using the hue vs sat curve.
4. Above the curve controls, click the curves pop-up menu.

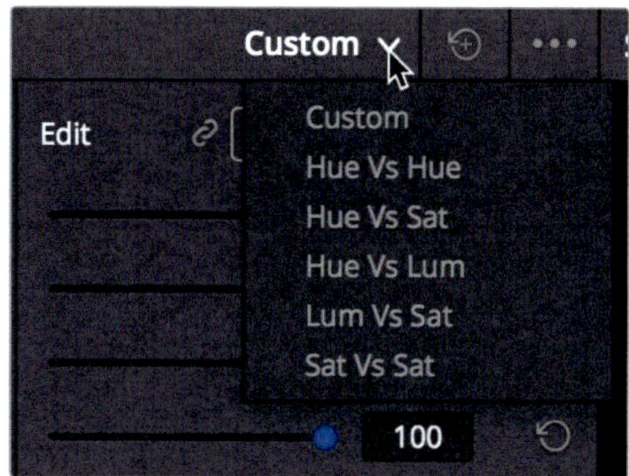

5. In the pop-up menu, choose Hue Vs Sat.

   Using this curve, you can select the green hue and then desaturate it, so it matches the white ceiling. When dealing with subtle colors like the green you have here, it can help to temporarily increase the saturation to select the hue.
6. In the adjustments below the primary corrector, click 1, and increase the Sat all the way to the right.
7. Now, with the green more prominent, in the viewer, click two or three areas where the green glow is visible on the ceiling.

Each time you click in the viewer, a point is added along the curve to identify the selected hue.

8 To lower the saturation of the selected green hues, drag the center points down to the bottom of the graph, leaving the outer two points in place.

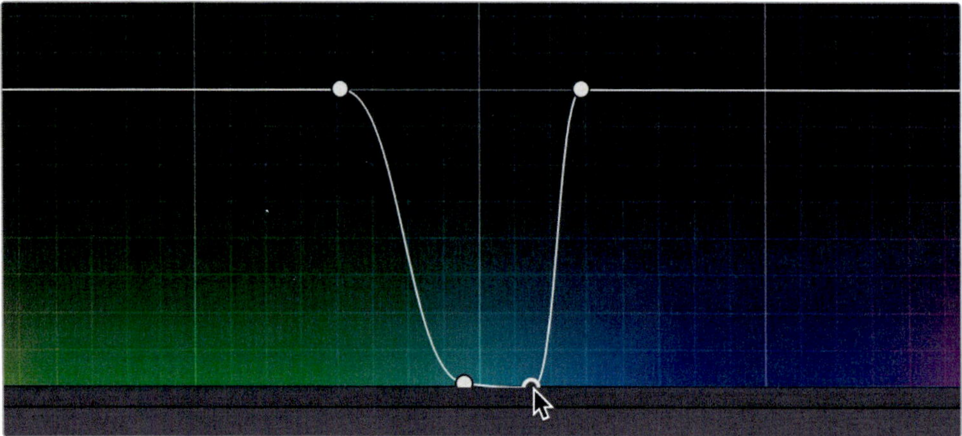

Dragging the points lower in the graph decreases the saturation of their hues. The outer two points define the range of hues that will have decreased saturation. If you are not removing all the green with this adjustment, you can modify the range by widening the distance between the outer two control points to include more hues.

9 Drag the leftmost control point to the left so it includes more green hues.

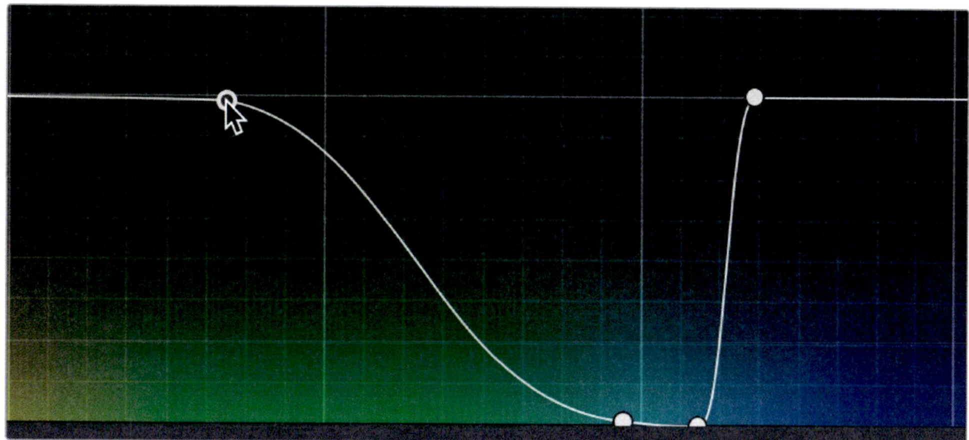

10 In the adjustments below the primary corrector, double-click the Sat label to reset the saturation back to normal.

To compare the changes you just made, you can disable node 02, the node in which the changes were made.

11 To show the timeline thumbnails, Gallery, and node graph again, press Option-F (Mac) or Alt-F (Windows), or choose Workspace > Viewer Mode > Enhanced Viewer.

12 With node 02 selected in the node graph, click 02 above the node, or press Cmd-D (Mac) or Ctrl-D (Windows), to see the image without the hue curves adjustment. Press Cmd-D (Mac) or Ctrl-D (Windows) again to view the corrected clip.

Before After

Nodes allow you to organize your color adjustments in flexible ways. On some simple grades, you may use only the default node. For complete grades, you may add a dozen nodes. Using nodes, you can quickly navigate to the exact adjustment you are looking for and combine color adjustments in multiple ways.

Copying corrections between similar shots

Color correcting moving images gets very challenging the moment you edit multiple angles together into a single timeline. Not only does each shot need to look correct on its own, it needs to look correct when compared to the shots immediately before and after it. This process of blending the look and feel of two or more shots is called shot matching.

The point of shot matching, or scene balancing, is simple: in real life, as you glance around, perhaps while having a conversation, you'll notice a consistency to what you see. Color saturation, skin tone hues, and brightness levels don't dramatically alter from glance-to-glance unless some external event forces the lighting to change. (For example, a light is turned on or clouds cover the sun.) If you want to maintain the illusion that your stories are real, then you need to mimic this consistency by matching shots and balancing your scenes.

In this exercise, you'll explore a few of the simplest tools that DaVinci Resolve provides for shot matching. You'll use these tools to create a seamless flow of sequential shots.

1 In the thumbnail timeline, select thumbnail 08.

This is the first shot of the girl that you color corrected earlier in this lesson. One of the simplest shot-matching situations is matching shots from the same camera that were recorded at the same time. Thumbnails 06, 07, and 08 are jump cuts, seemingly from the same recording session. One of the easiest ways of copying a correction from one clip to another is to simply use copy and paste.

2 In the node graph, select node 01, and press Cmd-C (Mac) or Ctrl-C (Windows) to copy the node's settings.

3 Select thumbnail 07. In the node graph, select node 01, and press Cmd-V (Mac) or Ctrl-V (Windows) to paste the settings.

The color correction in clip 08 is copied and pasted onto clip 07. Let's try a similar operation using the middle mouse button.

4 Select thumbnail 06.

5 Using a three-button mouse, middle-click the thumbnail of clip 08.

You copied clip 08's correction to clip 06.

6 Play through this three-shot sequence.

The shots all feel like one continuous sequence. Brightness, hue, and saturation look consistent. There's nothing distracting your audience, even subtly, and they can enjoyably watch this scene.

Copying corrections from clip to clip is one of the simplest and most common shot matching tasks you'll perform. But DaVinci Resolve has additional tools to assist in the task of matching shots.

Saving and applying stills

Instead of copying and pasting nodes from one clip to another right on the spot, you can save your grades into the Gallery and recall them whenever you need them.

1 Select thumbnail 04, which shows an extreme close-up of the girl you graded earlier.

You can save the grade you created on this shot by saving a still into the gallery

2 Right-click in the viewer, and choose Grab Still.

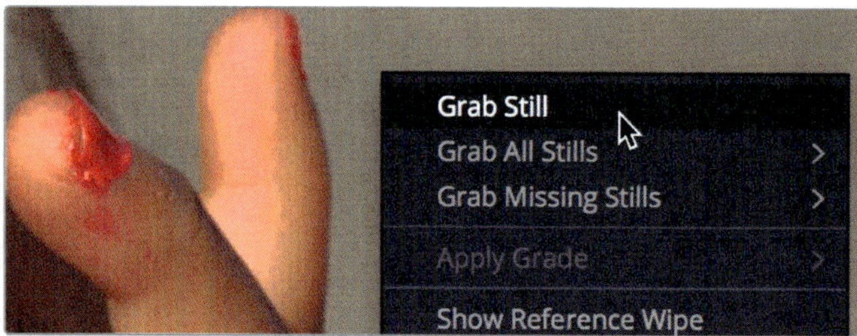

A reference still image is saved into the Gallery. The still image also contains all the instructions to rebuild the correction for that shot. It's helpful to name these stills so you will later know exactly what they do.

3 Right-click the still, and choose Change Label.

4 Name the still magic hour.

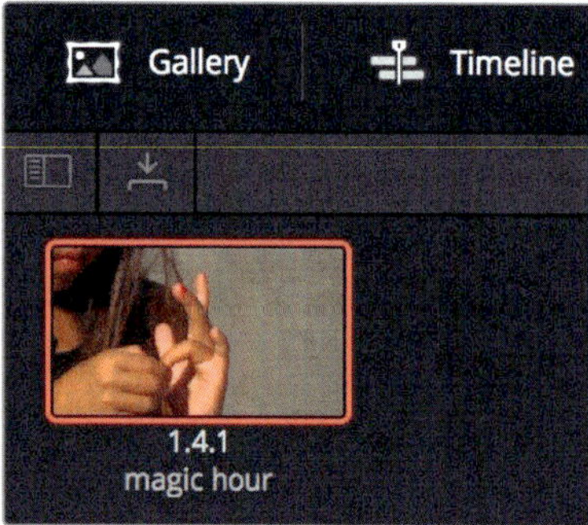

Gallery stills make it easy to apply the same correction to one or more clips. Unlike copy and paste which only takes one node at a time, the gallery still contains all the nodes that make up that look.

5 Click thumbnail 03.

You can select multiple thumbnails in the timeline and apply the grade from the Gallery.

6 Cmd-click (Mac) or Ctrl-click (Windows) thumbnail 02 to select it.

7 In the Gallery, right-click the still, and choose Apply Grade.

Both clips now have the same grade applied to them. You can play over all three clips and see that they have a consistent appearance. Next, you'll learn how to put the image part of that still to use.

Comparing shots

One of the keys to success when shot matching is to pick a few reference shots as you work, save them as representative stills to the Gallery, and match your other shots to those reference stills. By comparing the stills in your Gallery to other shots in your scene, you'll improve the color continuity across your entire project.

1 Select thumbnail 05, a wide shot of the girls painting their nails in the skate park.

This wide shot has audio of the girls continuing to speak about their nail polish, but it establishes the skateboarding setting. That being the case, it should have a similar tone and warmth that you gave the extreme close-ups. However, it is more than likely this shot was captured at a different time and possibly a different skate park than the extreme close ups. So, let's start just by comparing the two shots.

2 To compare thumbnail 05 to the color correction on thumbnail 04, double-click the still in the Gallery to open it into the viewer.

By default, a horizontal split appears in the viewer. The shot from the timeline (thumbnail 05) is on the left. The selected still in the Gallery is on the right.

3 Choose Workspace > Viewer Mode > Enhanced Viewer, or press Option-F (Mac) or Alt-F (Windows), to expand the viewer.

4 In the viewer, drag to the left, and then drag to the right to wipe between the still and the current timeline clip.

When comparing the cement of the wide shot with the cement in the still, the timeline clip appears a bit cooler, as you would expect. So, let's apply your warm magic hour grade you saved in the Gallery.

5 Choose Workspace > Viewer Mode > Enhanced Viewer, or press Option-F (Mac) or Alt-F (Windows), to display the Gallery.

6 In the Gallery, right-click the still, and choose Apply Grade.

The timeline clip now has the color correction applied from the still. A glance over at the node graph shows that the two nodes from the magic hour look reconstructed in that panel. However, as is often the case with clips shot at different times and places, the look is slightly different when applied to this clip. It is a bit warmer.

7 In the viewer, drag to the left, and then drag right to compare the warmth of the two shots in different areas.

Because the timeline clip is warmer than the reference still, it is a good idea to cool it off a bit using the Color Wheels.

8 In the node graph, select node 02.

9 In the Offset color wheel, drag the color indicator toward blue until the cement from the timeline clip closely matches the cement in the reference still.

10 To disable the image wipe in the viewer, click the image wipe button in the upper-left corner of the viewer.

Shot matching is made easier when you start using the Gallery and reference stills to help your analysis and inform your color-correction moves. You should also use the scopes to minimize any visual quirks because your visual perception naturally tends to force the shots to match. The combination of reference stills and scopes will make the shot matching process more accurate, giving your entire project perfect color continuity.

Lesson 11

Making Secondary Adjustments

Primary adjustments let you work on the entire image, whereas secondary adjustments let you isolate and work on specific parts of an image.

For example, you might want to change the color of a car from blue to red without affecting the rest of the shot, or add warmth and saturation to an actor's skin, or create a vignette to help focus the viewer's attention in a specific area of the scene. DaVinci Resolve features many powerful tools to do this.

In this lesson, you will use qualifiers to isolate elements based on their color, as well as apply basic geometric shapes to isolate areas of your frame. Then, you will use the tracker to follow a moving element and grade it as it moves throughout the scene.

Time

This lesson takes approximately 45 minutes to complete.

Goals

Using qualifiers

The qualifier palette allows you to select a color from a shot with the intention of grading it separately from the rest of the footage. This palette also includes several ways to select that color using hue, luminance, and saturation. This detailed level of control enables you to get a clean isolation, or **key**, of objects even when other elements in the shot are of a similar color.

TIP Qualifiers are used only as a method of creating a key. They are not color-grading tools in themselves, and you only start to see their effects when you begin adjustments in the color-grading palettes.

1. Open DaVinci Resolve 14, if necessary, and then open the Gnarly in Pink project you have been working on.
2. Open the Gnarly in Pink timeline to load it into the viewer.
3. Click the Color page, and then click thumbnail 12.

For this shot, you'll create a much-requested stylized image in which one color will be preserved, while the rest of the image is black-and-white. You've seen many movies with this effect and maybe even tried creating it in other applications. The critical task is being able to cleanly isolate the color you want to keep.

4. In the central palettes, click the qualifier icon.

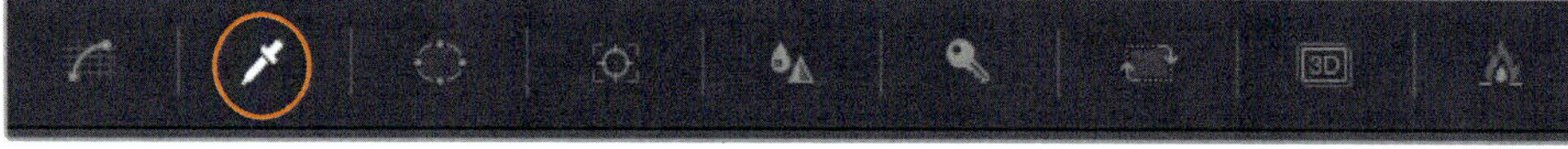

In this clip, you will want to preserve all the pink helmets and clothes.

When the Qualifier palette is selected, hovering your mouse pointer over the viewer will display an eyedropper that is used to select the color you want to preserve.

5 Click the back of the middle girl's helmet.

TIP The eyedropper is not only used with the Qualifier palette. It will affect whichever palette you have open at the bottom of the Color page. If custom curves are open when you clicked your image, control points are added to the curve.

The clip in the viewer will not change, but you should see some activity in the Qualifiers palette. The settings now reflect the information you have fed them by clicking the image. Your next step will be to refine and clean up your selection. To do so, you first must change the viewer's output to show your selection.

6 In the upper-left corner of the viewer, click the highlight wand icon.

You are now seeing a representation of your selection against a gray background. As you can see, the selection is not very clean. There's a variety of ways to improve it. The first will be to add additional samples of the color into the Qualifier palette.

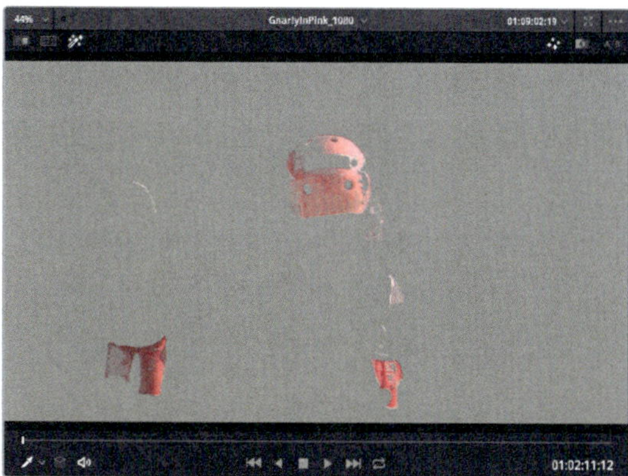

TIP You can change how your selection is presented using the three icons in the upper-right corner of the viewer. Next to the standard grayscale highlight icon, you also have the highlight B/W icon that outputs your image as a black-and-white matte. This might be familiar to users who have previously used compositing programs and have experience with keying.

7 Click the add color range button located in the Selection Range of the palette.

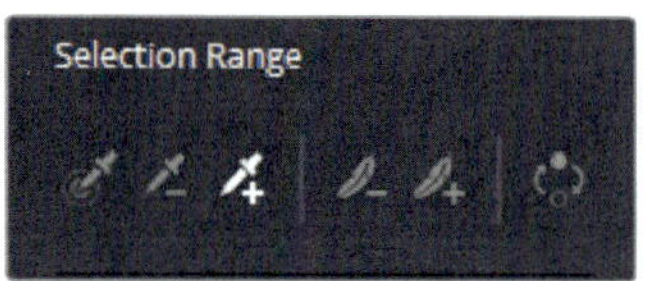

The add color range button allows you to append additional colors to the currently selected range.

8 In the viewer, click a few more times around the helmet to pick up more of the pink color. It might be easier at this stage if you temporarily turned off the highlight view to avoid selecting the wrong colors.

TIP You can press Cmd-Z (mac) or Ctrl-Z (Windows) to undo an incorrect selection.

You should now see a fuller selection that still has imperfections.

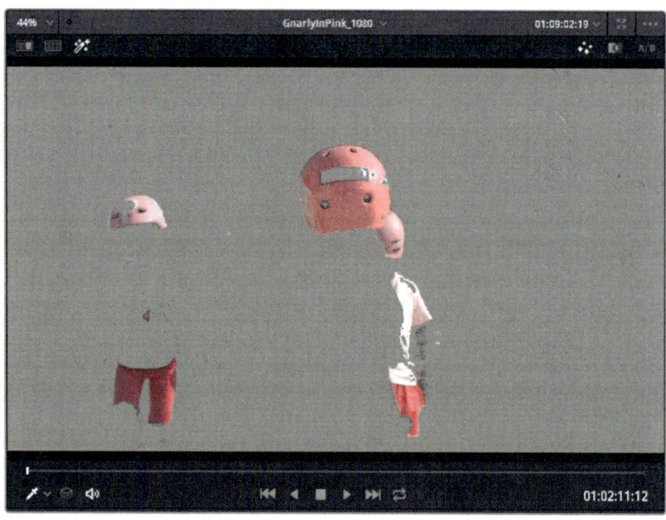

When the range is fully selected, you can fine tune your selection by adjusting the Hue, Saturation, and Luminance values in the Qualifier palette. The qualifier's controls are very sophisticated and powerful. It can take some time to master as you figure out how different shots react to different parameters. The best way to learn is to just start making changes.

9 Adjust the Hue Width and the Low and High values under each Saturation and Luminance control bar to refine the range of the selection.

> **TIP** It can be helpful as you experiment to turn some of the sliders on and off by clicking the red dot to the left of the name. Doing so allows you to check if that change would improve the selection.

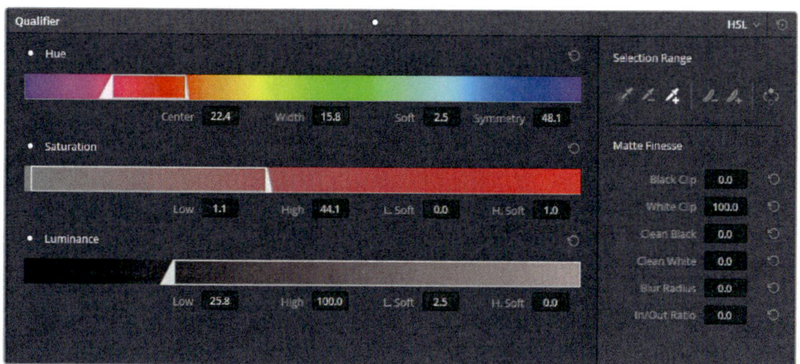

The last form of clean up would be to adjust the image matte. This technique does not rely on color selection as much as it manipulates the selection you have already made.

10 In the upper-right corner of the viewer, click the highlight B/W icon.

11 Adjust the Matte Finesse controls Black Clip and White Clip to remove smaller, unwanted white and black specks that appear in the matte.

The goal when adjusting the Matte Finesse controls is to produce a clean black-and-white image. Gray areas are seen primarily in transparent areas of which there are none in this shot.

12 If the selection starts to become too harsh or pixellated, adjust the Blur Radius to soften the edges of the matte

Now you're ready to start grading!

13 Turn off the Highlight view by clicking the Highlight Wand icon.

You can test to see if your selection includes all the colors you want by shifting the color of the selected area. A good method is to drag the Hue values at the bottom of the Color Wheels palette.

14 In the adjustments strip below the Color Wheels, drag the Hue value in either direction to shift the pink color in the viewer to another color.

15 In the adjustment strip, double-click the Hue label to reset the value.

If you notice some issues with the selection, return to the Qualifier tool, and use one of the three methods described previously to clean up your selection.

Using outside nodes

Occasionally, you will want to focus your attention on the area that is the exact opposite of your selection. For example, in this shot, you might be happy to leave the helmets pink, and manipulate everything that isn't pink. With the pink selection already made, doing so will be easy. You will add an outside node, and use your qualifier key to create an inverted version of the original node.

1 With node 01 selected in the node editor, choose Nodes > Add Outside Node, or press Option-O (Mac) or Alt-O (Windows).

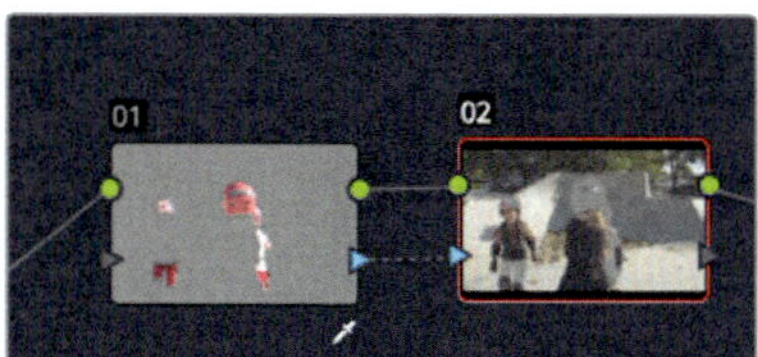

A second node is now present in the node graph. As with previous nodes you've created, it will share a connection with the preceding node via the green RGB input/output circles. However, this time you'll see a new connection: the key input/output blue triangles. The key is the portion of the image that you isolated using the qualifier tool. When you create an outside node, it receives the key from the previous node and automatically inverts it.

2 With node 02 selected, in the adjustment controls strip, drag the Sat field to a value of 0.

You can increase the drama in this shot by increasing the contrast in the black-and-white portion of the image.

3 With node 02 selected, in the adjustment controls strip, drag the Contrast slider to 1.2.

The ability to reuse key data is a beneficial component of node-based workflows. It speeds up the grading process by requiring you to perform only a single key, and requires less computer processing power than a layer-based systems.

Combining qualifiers and power windows

Like qualifiers, power windows allow you to specify which area of the image you want to alter. Unlike using a qualifier, power windows do not rely on a selection of hues, but rather a spline shape within the frame that identifies what you want to mask out.

You can refine an area of the frame you want to modify even more precisely by using these two tools together. Often, an image will have several instances of a hue that you are trying to manipulate. Instead of focusing your efforts on cleaning up the selection in the Qualifier palette (and likely compromising the quality of the key), sometimes the best option is to use a power window to further indicate which part of the frame in which you want to focus the selection.

1 In the timeline, click clip 08.

   In this exercise, you will change the color of the pink box to match the color scheme of the rest of the room.

2 With node 01 selected in the node graph, choose Nodes > Add Serial Node, or press Option-S (Mac) or Alt-S (Windows).

3 In the toolbar, click the qualifier button, and then click eyedropper in the Selection Range of the Qualifier.

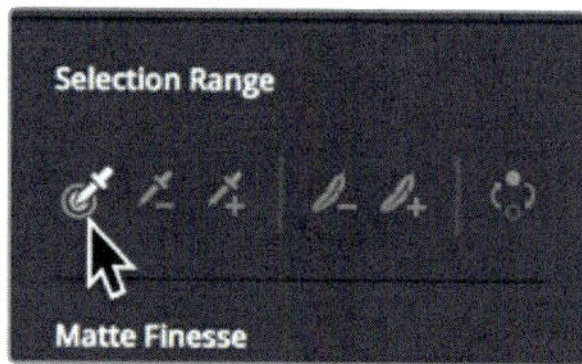

4 In the viewer, using the eyedropper, click the the box next to the girl.

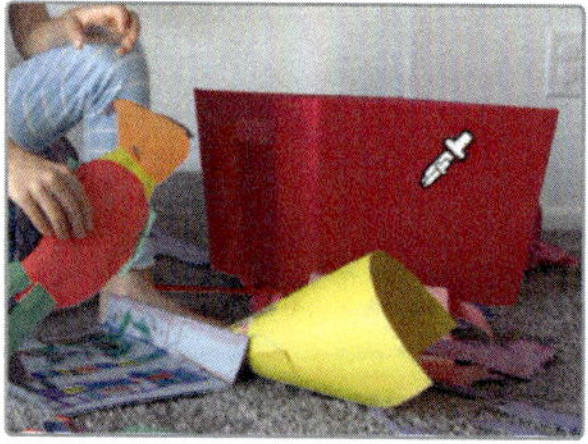

TIP If your qualifier is not working, it may be because you still have the "Highlight +" tool activated in the Selection Range of the Qualifier palette. Switch back to the regular qualifier selection tool to make a new selection.

5 Enable the highlight mode in the viewer to see the result of your qualifier selection.

6 In the HSL qualifier settings, increase the Hue Center and Width to clean up the selection.

7 Further adjust the Saturation/Luminance High and Low controls in the main section of the palette, and then use Matte Finesse controls to realize the final result.

   Again, the qualifier takes trial and error (and some experience) to use most effectively. Still, almost immediately, you will find it impossible to completely isolate the box from the other pink items in the room (most notably the clothes hanging in the back.)

   In cases where your efforts are complicated by similar colors, your focus should be on getting the cleanest selection for your subject (the box, in this case), while ignoring everything else.

8 In the toolbar, in the central palettes, click the window palette icon.

9 In the list of power windows, click the curve button to activate it.

10 Double-click next to the curve button, and label the window as **Box**.

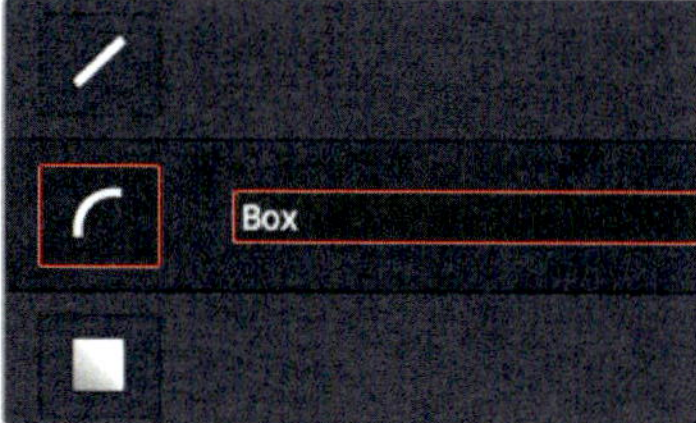

11 Move your mouse pointer into the viewer.

   When a window is active in the palette, hovering your mouse over the viewer shows a pen tool, indicating that you are ready to draw a custom window shape.

12 Draw a shape by clicking around the box, taking care to exclude any areas that you do not want to include in your final selection.

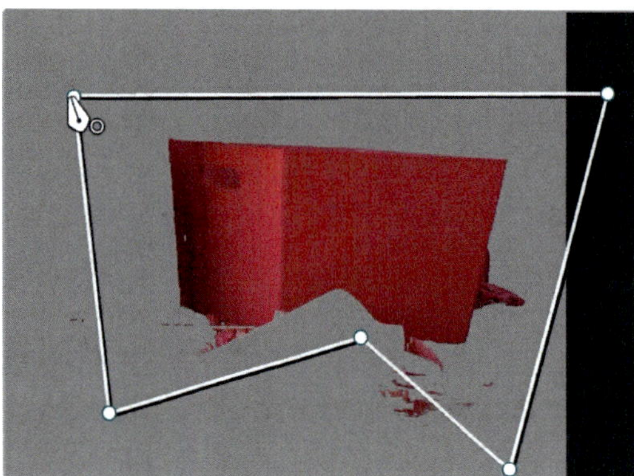

13 To complete the custom shape, click the first point in the shape to close it.

When the shape has been created, the power window will exclude the rest of the image, keeping only the selection inside the box.

14 In the viewer, turn off highlight mode.

15 Below the Color Wheels palette, in the adjustment controls strip, drag the Hue control until you reach a blue color similar to the tones seen in the rest of the shot.

Combining the qualifier with a power window has allowed you to quickly make a clean selection based on the hue of the element, while just as quickly excluding interfering elements from your selection. Using both tools meant that you didn't have to draw an overly precise power window around the box. Combining the two tools has saved time and resulted in the cleanest key.

Tracking power windows

As convenient as the power window is as a selection tool, its use is limited to either frame-wide applications (such as vignettes) or still elements in locked-off camera shots. If you use a power window to target a person or an element in a roaming/panning shot, the selection will be lost as soon as any movement occurs. While your target drifts off in the frame, your power window will remain static in the viewer.

Thankfully, you have another powerful set of tools at your disposal in the Tracker palette. The tracker is able to read the visual information in a frame and detect its recurrence in subsequent frames. So, it can build a motion path that you can attach to the power window. With the power window moving directly with your target, you now have the freedom to apply grades (and qualifiers) to very precisely targeted areas of your footage.

1. In the timeline, click clip 19, and move the playhead to the last frame of the shot.

   Your goal in this exercise is to change the pony's eye color.

2. With node 01 selected in the node graph, choose Nodes > Add Serial Node, or press Option-S (Mac) or Alt-S (Windows).
3. In the toolbar, click the qualifier icon, and select the blue part of the pony's eye.

4. Just as in the previous two exercises that used the qualifier, tweak your selection until you get a completely clean selection of the eye.

   Because other blue elements are present in the shot, they will also be included in the final qualifier selection. However, you now know that the best way to fix this is to draw a power window around the area you want to retain.

5 In the toolbar, click the window button, and click the circle window to add a circle window to the viewer.

6 Rename the circle window to **Iris**.

7 In the viewer, drag one of the corner control points of the circle to rescale it to the size of the iris.

8 Drag the center of the circle to place it directly over the eye.

9 In the Softness controls, reduce the Soft1 value to 0 because you are not trying to blend the selection into its surroundings.

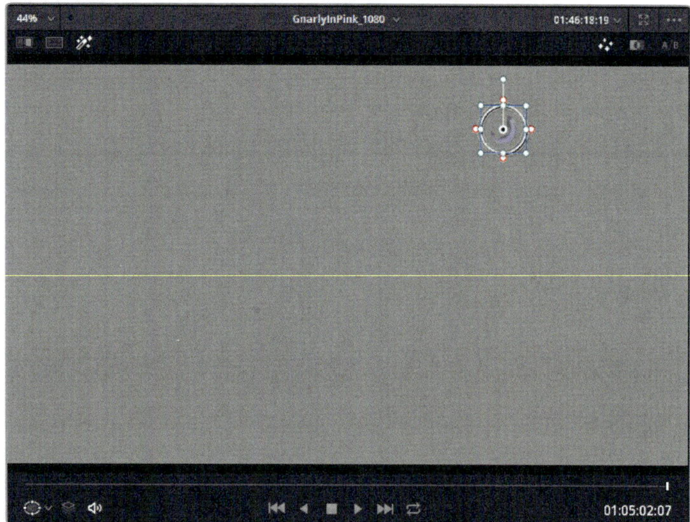

TIP Your power window doesn't have to be in its final form when tracking. In fact, sometimes it makes more sense to create a larger/smaller power window to produce a clean track, and later to reshape it to fit the content after the tracking is successful.

If you scroll back and forth through your clip right now, you'll see that the circle window remains static whereas the pony moves forward. Go back to the last frame to ensure that the window is still directly over the eye.

10 In the toolbar, click the tracker button.

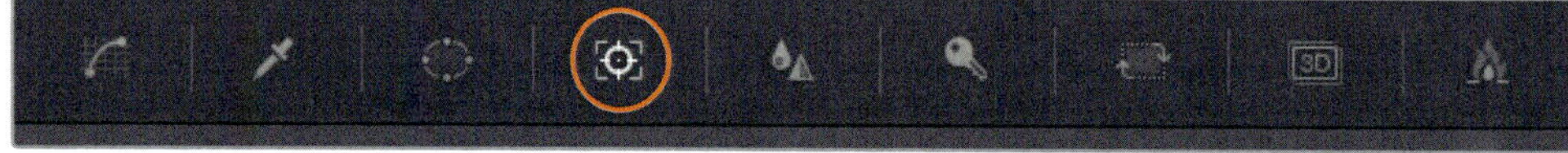

The Tracker palette is primarily a large empty box with some controls that allow you to indicate which aspects of an element you want to target.

TIP In general, you should leave these default controls (Pan, Tilt, Zoom, Rotate, Perspective 3D) unchanged. However, you may come across some projects in which a track would benefit from having one or more of the criteria removed from the analysis.

A series of buttons in the upper-left corner of the palette are responsible for performing the tracking. These do not control the playhead and should not be confused with transport buttons!

The tracking buttons allow you to analyze your footage forward and backward, either continuously or one frame at a time (sometimes a necessity for tricky tracks.) A pause button is available to stop the analysis. Because your playhead is at the end of the clip, you will need to continuously analyze backward.

11 Click the track reverse button, the second button from the left.

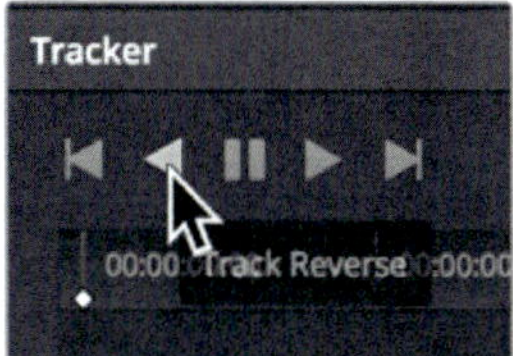

In the viewer, the clip starts playing backward, and the circle window follows the eye along as the pony moves closer to the right edge of the screen. Simultaneously, the Tracker palette graph is populated with tracking data.

12 When the track is completed, drag the orange playhead in the Tracker palette back and forth to verify that the track is successful.

TIP You can use the scroll bars at the bottom and right side to zoom in horizontally and vertically to inspect the track data.

If the track was successful, you are now free to proceed with the grade.

13 Drag the Offset color wheel in the direction of the color red.

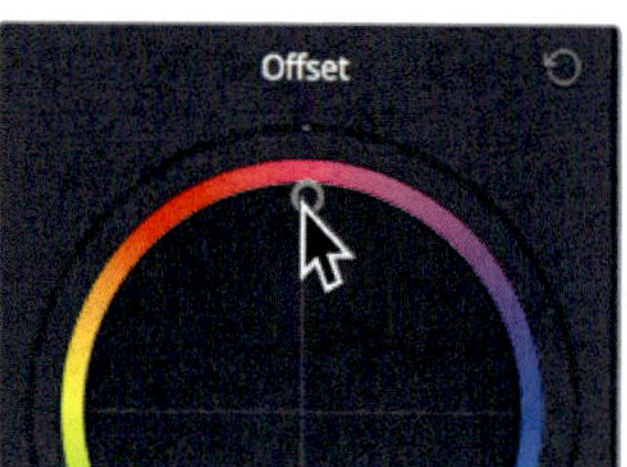

You now have a secondary grade in which you have cleanly altered a single moving element, while not altering the surrounding visual data that shared its color properties. Understanding and utilizing tracking can open a whole new dimension in secondary color grading which can now include moving subjects and cameras.

Tracking the Pink Helmets

In the first exercise, you intended to keep one helmet pink, but turn the rest of the image black and white. But you didn't really do that because the shirts and pants were left pink. Now that you know more about secondary color correction techniques, you can track the helmet of the girl in the center of the shot.

1 In the timeline, click clip 12.

2 Double-click node 01, the node that contains the qualifier.

3 In the toolbar, click the window button.

4 Click the circle window to add it to the viewer.

5 Rename the circle window to **Helmet**.

6 In the viewer, drag the center of the circle to place it directly over the helmet on the girl in the center.

7 Drag one of the corner points of the circle to roughly rescale it to her size.

8 Reduce the Softness to 1.

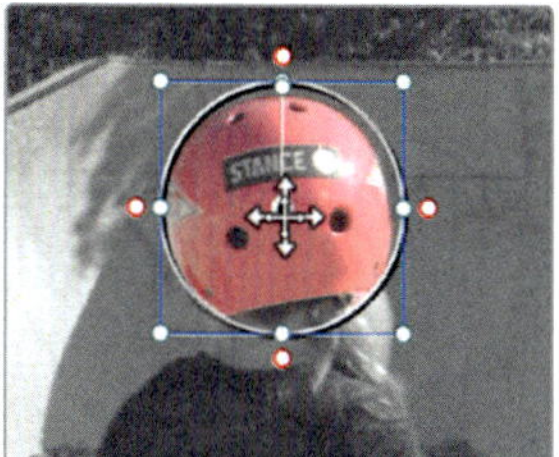

9 In the toolbar, click the tracker button and then click the track forward button.

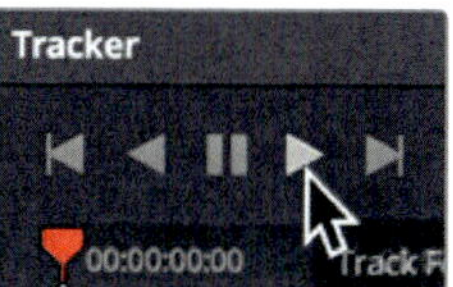

10 Play the clip to ensure that the tracking is correct throughout.

Understanding how to use tools such as the qualifiers, power windows, and the tracker palette enables you to perform secondary color grading with substantial control over the final look of your image. However, it is only when you combine these tools that their true potential is unlocked.

Lesson 12

Designing Creative Looks

Now that you're familiar with the primary and secondary tools, you should begin exploring the creative implications of grading. The colors within a scene can influence how your audience should feel. Cold tones denote a brooding character and warm colors tend to signify that everything is going well.

You can also use color to quickly communicate place and time by creating distinct looks for each location and/or time period. Finally, your grading can stylize your project and give it a unique, memorable look.

In this lesson, you will gain an understanding of how to employ these practices as you create, save, and compare three distinct looks.

Time

This lesson takes approximately 45 minutes to complete.

Goals

Mixing a black-and-white shot

When working with the Color Wheels palette in previous lessons, you saw that one of the adjustment controls was Sat, or **Saturation**. This simple setting sets the strength of your colors, allowing you to push them beyond their originally recorded values or to decrease them completely, thereby turning the image to black and white.

In the Color page, you have several methods for turning a color image to black and white. Setting your Sat adjustment control to 0 is the obvious way, but it happens to be the least flexible way.

In the RGB Mixer, you have the option to fine tune the strength of the individual red, green, and blue channels even for a black-and-white image, which can produce stark and interesting results. You'll experiment with this look by creating a new version for it.

1 Open DaVinci Resolve 14, if necessary, and open the Gnarly in Pink project you have been working on.

2 Open the Gnarly in Pink timeline to load it into the viewer.

3 In the timeline, select thumbnail 29.

4 In the toolbar, click the RGB Mixer button to open the RGB Mixer palette.

5 At the bottom of the palette, select the Monochrome checkbox.

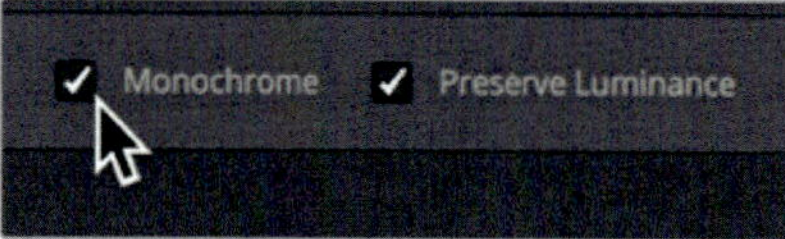

Doing so will turn your image black and white. Many of the controls under the individual red, green, and blue output bars will be dimmed and unavailable. When an image is set to monochrome, you can control only the degree to which red, green, and blue contribute to the mix.

6 In the RGB Mixer, on the green output bar, drag the grey bar up to the top, and then drag it down to the bottom.

Notice that as you drag, only some parts of the black-and-white image are affected.

You can repeat this step a few times while keeping your eye on the viewer to see the change.

7 Experiment by dragging the red and blue output bars to see how they affect the image.

Notice how strongly some elements are affected when they directly correspond to the channel color. By adjusting the red output, you can make the pink helmets change from dark gray to white. The blue channel is often tweaked in exterior shots to make a sky look more dramatic.

8 Set the red output around -0.45, the green output to -1.25, and the blue output to -0.50.

As you can see, there is no such thing as a single, definitive black-and-white version of your image. Even when the colors are completely desaturated, you can still control the prominence of individual RGB channels. This technique can result in some carefully-crafted, highly dynamic monochrome images.

Creating new versions

So far you have created only one black-and-white look. As you continue experimenting with new looks, you will want to save those looks, and later compare them to see which you prefer. Versions allow you to try out new looks on a clip without deleting your previous work. Even more conveniently, using versions, you can quickly load older versions of grades for comparison and creative decision making.

1 In the timeline, right-click thumbnail 29, and in the menu, choose Version1 > Rename. In the text field that appears, type **B&W**. Click OK.

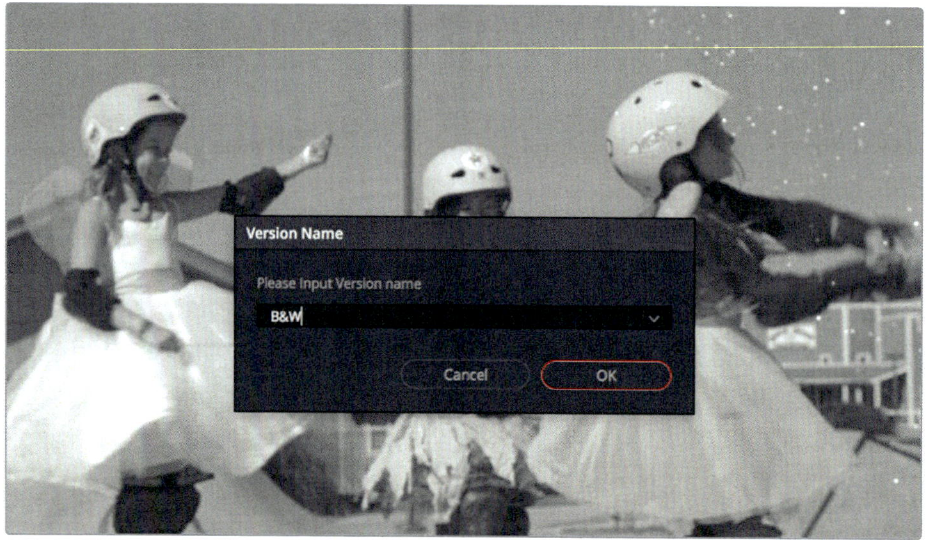

TIP DaVinci Resolve can use both local and remote grades. Local grades, which you are using here, are saved for the selected clip in the current timeline. Remote grades are shared across timelines into which that clip is inserted.

The first version is automatically created based on the current state of the clip. However, you can create a new version at any time to save changes that you make.

Think of a version as a new document. You first create the document, and then add the content. In this case, you first create the version, and then create the grade that it will contain.

2 Right-click thumbnail 29, and in the menu, choose Local versions > Create New Version. Name the version **LUT look**, and click OK.

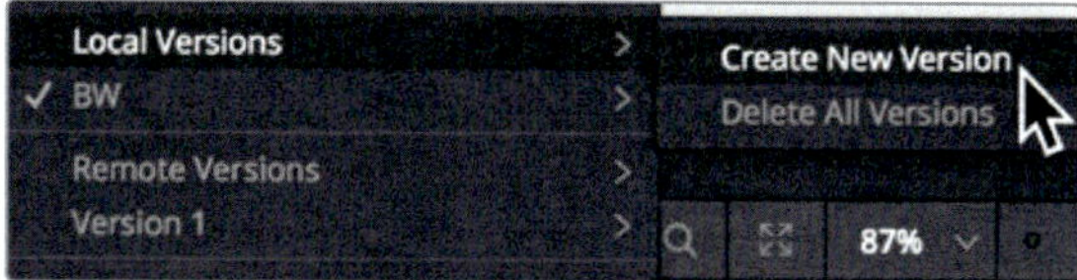

From now on, all changes you apply to your clip, from the simplest preset grades to the most advanced node graphs, will be saved in the LUT Look version of that clip.

Using a look up table for quick looks

With a new version created and named, you can create a new look using a **look up table**. At first glance, look up tables, or **LUTs**, appear very similar to presets in that they affect the color and luminance of your image with a click of a button. However, look up tables have many uses: they can assist with gamut conversions, monitor calibration, and creative looks.

Still, they are different in the way they affect pixel data. Whereas presets can be described as a series of calculations that will change a pixel based on its hue and luminance, LUTs are a mathematically precise way of modifying specific RGB values in an image into new RGB values by changing the hue, saturation, and brightness values of the source image.

Loading LUTs

DaVinci Resolve comes with a variety of LUTs that you can start using right away. However, one of the strengths of DaVinci Resolve and its LUT workflows is the ability to create custom LUTs and share them with other colorists, as well as adding LUTs created by third-party developers.

You will begin by installing a third-party LUT. To begin, let's assume you downloaded a LUT from one of the popular LUT-creation websites.

1 Choose File > Project Settings, and in the project settings, click the Color Management category.

2 Click the Open LUT Folder button.

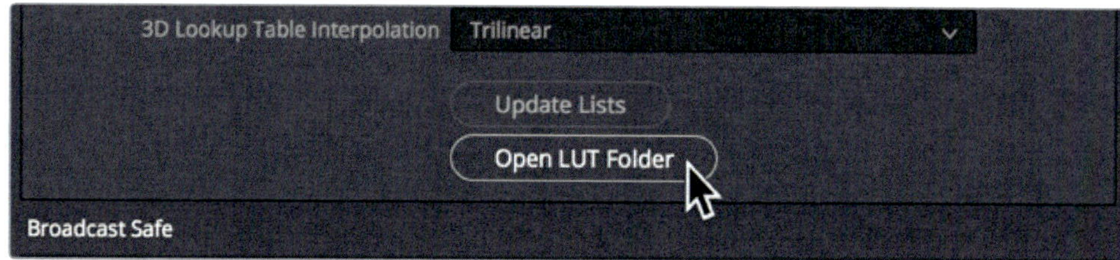

A window opens that contains a list of LUTs and folders of LUT categories.

3 In the window, create a new folder called **Training LUTs**.

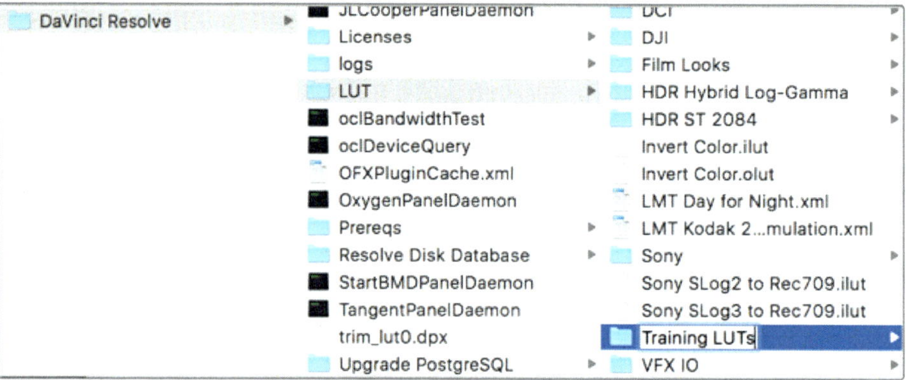

4 Open a second Finder (Mac) or Explorer (Windows) window, and navigate to R14 lessons > Lesson 12.

5 From the Lesson 12 folder, drag the Lesson_12_LUT.cube file into the Training LUTs folder.

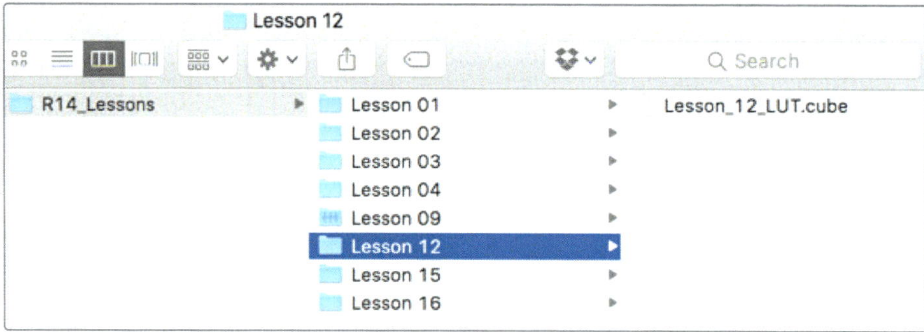

TIP Resolve 14 creates and uses LUTs in the DaVinci Resolve .cube format. This is an open LUT format and can be reviewed technically in a simple text editor..

6 Click in DaVinci Resolve, and in the color management project setting window, click Update Lists.

7 In the project settings window, click Save.

You now have a LUT loaded into DaVinci Resolve and you will be able to access it in the node graph.

Be cautious when downloading LUTs from third-party sources. Their implementation could result in the footage looking very different from your originally intended hues. This is why LUTs are popular for in-house use at production studios. Post-production companies are able to generate LUTs that are precisely calibrated to their environments.

LUTs for monitoring

LUTs are commonly used for monitoring. They allow you to change how a video signal displays when content is captured on camera. When the signal is captured using a filmic non-linear gamma curve, an HD monitor will display a very flat image with low contrast and saturation. If you apply a LUT to the display, it can convert the signal to appear as HD (Rec 709)—the contrast will become much more pronounced, and the colors more saturated. If your project has a specific look, you can even create a custom LUT, save it, and upload it onto a Blackmagic Design camera, where you can shoot the scene while seeing an approximation of how the final footage will look.

The applied in-camera, monitoring LUTs will not alter the footage being captured. It will affect only how the image appears on the camera viewer. When the footage is transferred to DaVinci Resolve, it will still have full grading capacity.

Applying a LUT

Look up tables allow you to quickly recalibrate the way your color pixel data is displayed, essentially providing another form of grading. Conveniently, you can easily apply LUTs once they are installed in DaVinci Resolve.

1. Select Color > Reset All Grades and Nodes to give you a blank slate to work with.
2. In the timeline, select thumbnail 29.
3. In the node graph, right-click node 01, and in the menu, choose 3D LUT > Training LUTs > Lesson_12_LUT.

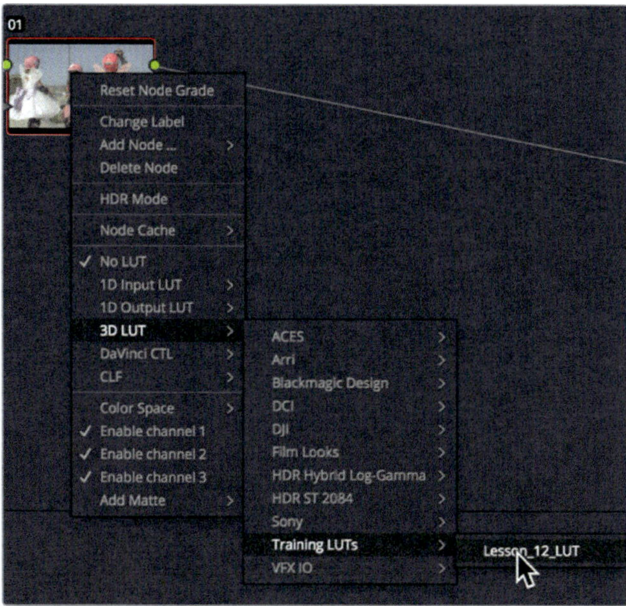

You have now applied the look up table you installed previously. The LUT has changed the look of the clip. As with the Gallery presets, you can continue working on the grade of your clip, using the LUT merely as a starting point.

TIP LUTs will change the appearance of your clip within a single node. They do not have node structures and do not employ secondary grading practices as do keys and mattes.

4 Right-click node 01, and in the menu, choose Add Node > Add Serial, or press Option-S (Mac) or Alt-S (Windows).

5 In the adjustment controls, drag the Contrast field to increase the contrast slightly.

6 In the Color Wheels palette, drag the Offset wheel toward yellow/orange to add warmth.

The effectiveness of a LUT is largely determined by its compatibility with your footage and monitor. For a LUT to look exactly as intended, it must be applied to the same type of footage generated by the same model camera, and displayed on the same calibration monitor. If any of these factors are different, the LUT will often look unlike its intended look; and in some cases, that difference may be dramatic.

Saving LUTs

Once you've customized some LUTs, you'll want to save them for future use.

1 Right-click thumbnail 29, and choose Generate 3D LUT (CUBE).

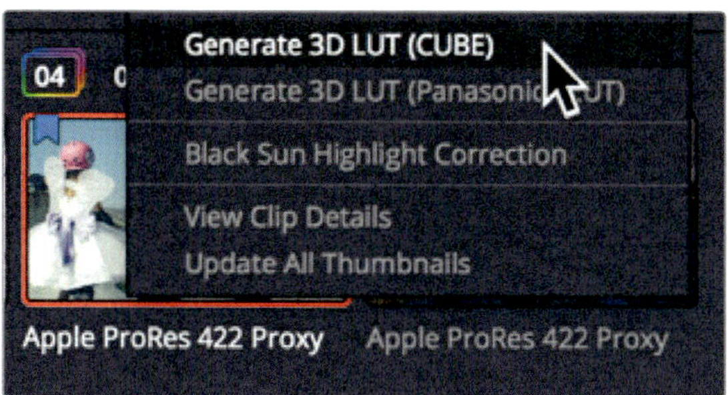

2 In the "Save LUT as" dialog, type the name **Cold Air LUT**.

3 Click Save.

   You have now generated a LUT based on the configuration of your pixel color data. To verify that you have been successful, you should be able to access the LUT in the node graph.

4 In the node graph, right-click a node to open the 3D LUT pop-up menu.

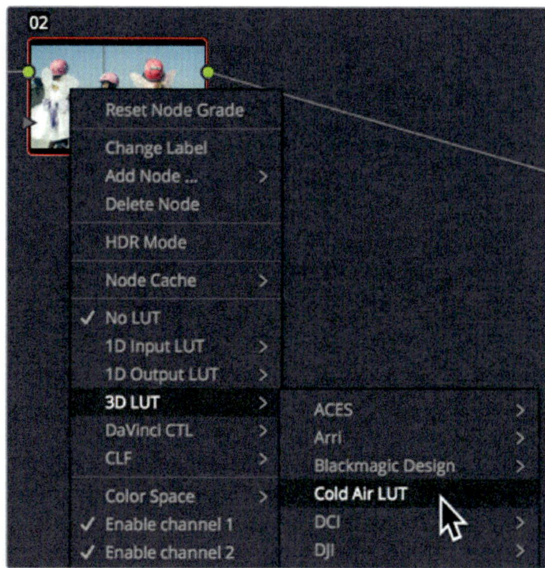

   The LUT you created appears in the 3D LUT list.

In this exercise, you used look up tables to create a look for a clip. However, in DaVinci Resolve 14 you can use LUTs in many parts of your image-processing workflow. You can add them to every clip, to the output of every timeline, or just to your viewer or video monitor.

Creating a bleach bypass

In this exercise, you'll create a bleach bypass look for the same shot, and then compare it to your other looks to choose your preferred look.

The bleach bypass process, sometimes called a **silver retention** or **ENR** process, basically layers a black-and-white copy of the image over a color copy to achieve a low-saturation, high-contrast look. It stems from a film development process in which the bleaching stage was, well, bypassed. It can be seen in many television shows and films including **Reds**, **Saving Private Ryan**, and **Seven**. Let's set it up, first by naming and saving a version.

1. Right-click thumbnail 29, and in the menu, choose Local versions > Create New Version.
2. Enter the label as **Bleach Bypass Look**.
3. Choose Color > Reset All Grades and Nodes.

   The bleach bypass look uses a low color saturation. In this first node, you'll decrease the saturation using the shared controls under the master wheels.
4. In the shared adjustments controls of the Color Wheels palette, lower the Sat value to 25.

   Up to this point, you have used serial nodes that are strung into your project, one after the other. They pass along color adjustments as they progress. Now, you'll use a Layer node that works more like Photoshop and other layer-based applications in that it composites images together, one on top of the other.
5. Choose Nodes > Add Layer Node, or press Option-L (Mac) or Alt-L (Windows).

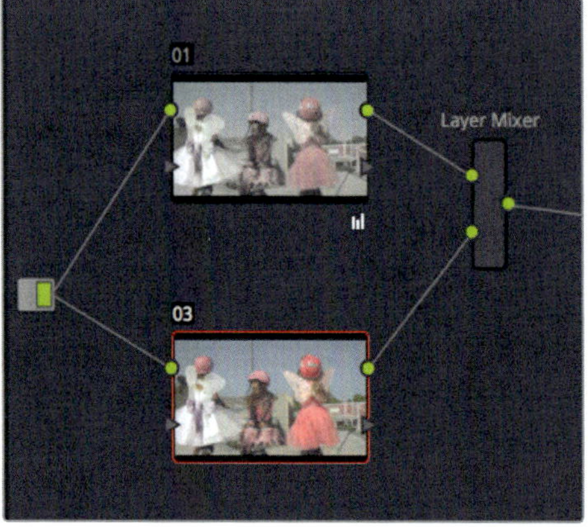

Adding a layer node automatically connects the selected node and adds a second node.

The node with the low saturation (node 01) is automatically connected to the layer node into the upper connection. A second node (node 03) is automatically added and connected into the layer node below node 01. Nodes connected below another node are actually composited over the upper connected node.

6 In the node graph, double-click node 03.

7 In the toolbar, click the RGB mixer button.

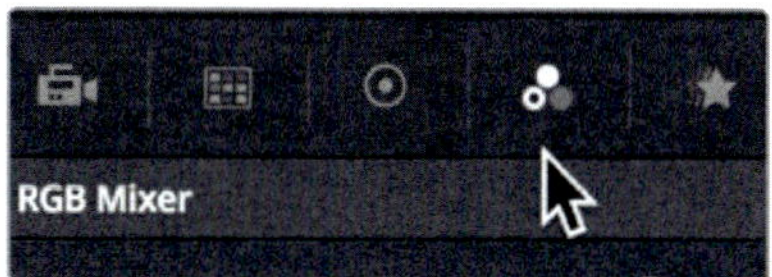

8 Select the Monochrome checkbox.

9 In the RGB Mixer, increase the red output to 0.8, bring the green output down to -0.40, and drag the blue output all the way to the bottom of the bar. Doing so produces a very nice high-contrast image.

TIP The bars can be very touchy to adjust. It is sometimes easier to use the scroll wheel on the mouse to adjust the bar your mouse pointer is over.

These two nodes will form the basic bleach bypass look, but they still need to be blended. Right-clicking the small layer node will allow you to choose how the nodes blend on top of one another using composite modes.

Composite modes perform mathematical operations to blend the color and luminance of nodes. Using them is mostly trial-and-error because their results can be hard to predict. Still, you can speculate at a high level as to what they may do when compositing images.

You can categorize composite modes based on the way they impact the layers.

- Darken Colors: Darken and Multiply
- Lighten Colors: Add, Lighten, and Screen
- Increase Color Contrast: Hardlight, Overlay, and Softlight
- Cancel Out Colors: Subtract (simple version of Difference)
- Invert Colors: Difference (similar to Subtract but with negative results)

The most commonly used composite modes tend to be Multiply for darkening, Screen for lightening, and Overlay or Softlight for increasing contrast.

Let's use the Softlight composite mode to blend the nodes, and thereby increase the contrast.

10 Right-click the layer node, and in the menu, choose Composite Mode > Softlight.

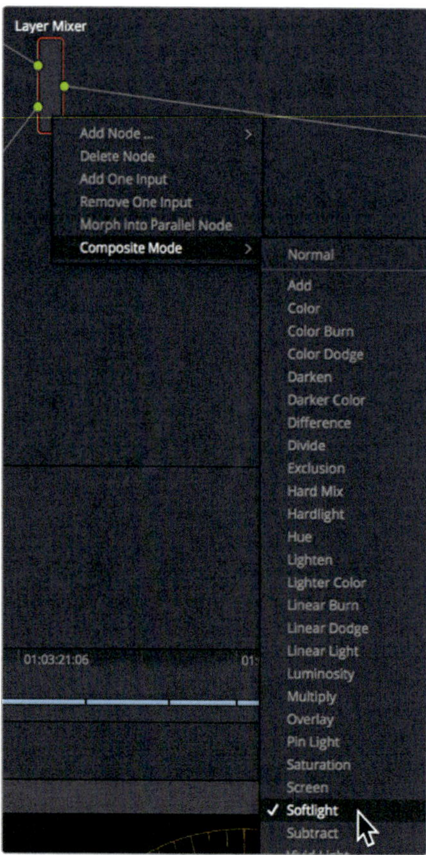

TIP You can add more inputs to connect new nodes to the layer node by right-clicking the layer node, and choosing Add One Input.

Let's add another node to make some minor adjustments to the overall composite. You can add nodes after the layer node to modify the entire composite.

11 Double-click the layer node, and press Option-S (Mac) or Alt-S (Windows), or in the contextual menu, choose Nodes > Add Serial Node.

The node is added after the layer node, just like the serial nodes you've added previously. Because the bleach bypass look can typically crush shadows, you can make final corrections to the tonal range using this last node.

12 In the toolbar, click the curves button.

13 Click the Y channel.

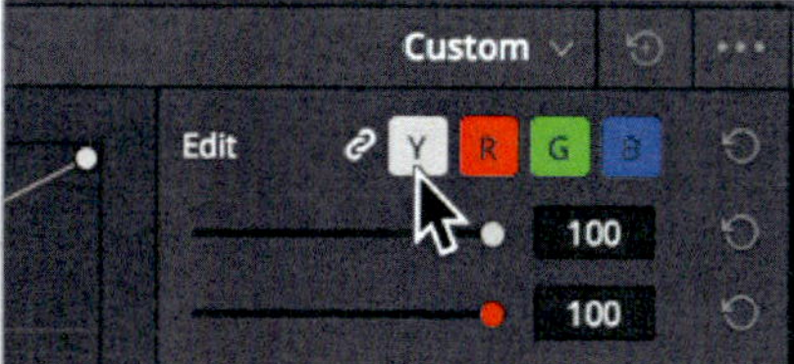

You will add anchor points to this curve in the options menu.

14 In the curve options menu, choose Add Default Anchors.

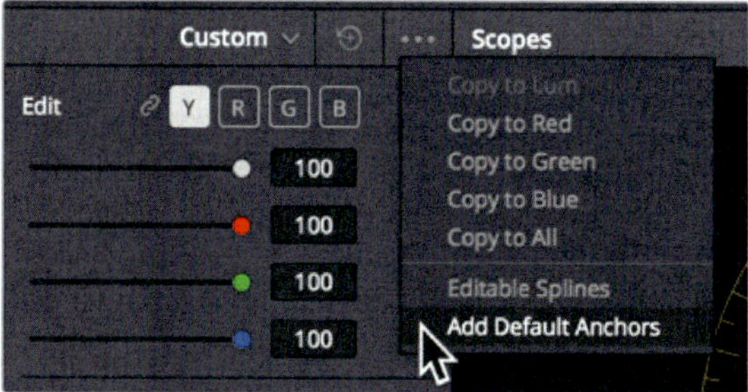

15 Drag down on the lower midtones anchor (third anchor from the bottom) to darken the sky a bit.

Now you have a classic bleach bypass look. Once the basic look is in place, you can decide to return to any node and further adjust parameters based on the entire look.

Understanding additive color

What color do you get when you mix all the colors of rainbow?

Some people will think of how light refracts from a prism and say that every color combined produces white. Others pause and think about what happens when you mix all the colors in your paint palette, resulting in a colorless gray sludge. However, the true answer depends on whether you are treating colors in an additive or subtractive sense. Subtractive colors are used in painting and print mediums. Additive color is used in light-based mediums such as sunlight, stage lights, or computer screens.

Because grading is performed using a computer screen, it uses an additive color system; however, when grading, you think about color design in a subtractive system.

For instance, in a subtractive color space, complementary colors produce aesthetically pleasing combinations, so you tend to create looks using these complementary colors. But when it comes to grading, complementary colors combined on an additive color wheel will neutralize one another. Adding blue to yellow will produce white (or some variety of gray). This is a vital principle to understand for color correction workflows.

Comparing versions

You previously saw how you could use the Viewer to compare active clips with the stills in your gallery. You can use a similar technique to compare multiple versions of the same clip.

1. In the timeline, select thumbnail 29.
2. Right-click in the viewer, and in the menu, choose Split Screen > Versions and Original to display all versions of your clip's grades in the viewer.

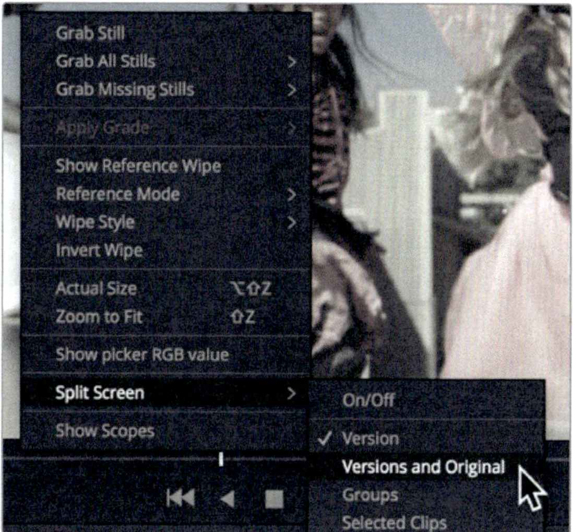

3. Click between the versions in your viewer to see their respective node graphs and the settings of the grading tools at the bottom.

You'll notice that selecting versions in the viewer will also change which version is applied to the clip in the timeline.

4 Double-click the version you like the most to apply it to the clip.

5 Right-click the viewer and choose Split Screen > On/Off to disable the split screen.

> **TIP** To expand the viewer for a larger preview, choose Workspace > Viewer Mode > Cinema Viewer, or press Cmd-F (Mac) or Ctrl-F (Windows).

Versions are most effective when used consistently throughout the grading process. Every time you make a new attempt at a look, you should change the version and reset the image to give you the option of switching back to an earlier version. This is a particularly convenient best practice in the early color-grading stages of a project when the grade has not yet been locked down. Without the director's immediate feedback, a colorist can attempt multiple variants of a look and hang onto the versions to be assessed and approved by the director at a later date. Once a look has been picked, the colorist might proceed to apply the same grade across the rest of the scene/timeline.

Saving grades across projects

You will often want to access your previous grades when working on new projects. A common reason could be that your work is episodic and requires a similar stylistic look from one episode to the next. Or, you might just like a specific grade you've created and want to use it again and again.

Previously, you used the stills gallery to save instances of your clips, and the expanded Gallery to access larger collections of stills and presets. You will now return to the features of the Gallery to understand how you can share grades across projects, and how you can quickly grab a still of every clip in your timeline as a method of cataloguing or archiving your project grade.

1 Right-click the viewer, and choose Grab Still to save the grade into the Gallery.

2 In the upper-left corner of the Gallery, click the still albums icon.

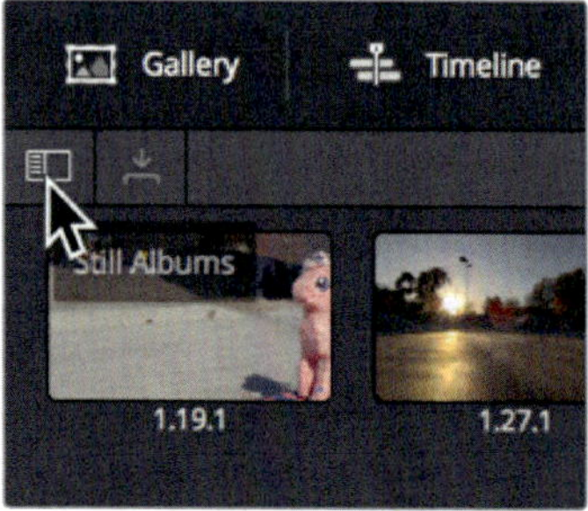

TIP If the resulting stills sidebar list is too narrow to read, drag the divider to expand the sidebar to the right.

In the list panel, you'll find a folder called PowerGrade 1. This folder can store files like a regular folder, but the PowerGrade folder is shared across all projects that are generated by the same user.

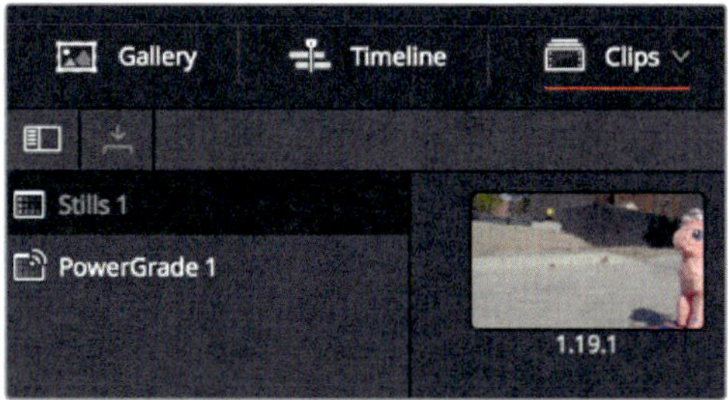

3 Drag a still from your stills album into the PowerGrade album beneath it.

When you open your next project, the Color page will already have this grade present in it.

With an understanding of color theory, you could quickly assemble a variety of looks using presets as a starting point for your grades. You could also employ the use of LUTs, or export your own to share with fellow collaborators. By saving versions of your grades as you progress, you will be able to quickly switch to earlier versions of your clips for reviews and approvals.

Lesson 13

Conforming from XML

DaVinci Resolve 14's editorial toolset is rich and powerful so preparing for color correction can be as simple as clicking the Color page button. But you can also color correct a timeline created in other non-linear editors.

In this lesson, you'll import a timeline created elsewhere and ensure that your imported timeline is 100 percent accurate. You will also prepare DaVinci Resolve to jumpstart your color correction process using the DaVinci Resolve Color Management workflow.

Time

This lesson takes approximately 60 minutes to complete.

Goals

Conforming a timeline using XML

Editors, finishing artists, and colorists often talk about the need to conform a timeline when they import a timeline from other non-linear editing software. What do they mean by **conform**?

The conform consists of the following steps, which you are executing in the first few parts of this Lesson:

- **Import a timeline:** You can import three timeline formats to DaVinci Resolve from other editing software: XML, EDL, or AAF. In this lesson, you are importing an XML file from Adobe Premiere Pro.
- **Link the timeline to the media on your hard drive:** Usually, the editor is on a completely different system from yours, so you must point DaVinci Resolve to the media's new file path.
- **Ensure your imported timeline is 100 percent accurate:** If your imported timeline isn't frame-for-frame accurate, then the conform process is unfinished (and the editor won't be happy!)

Let's begin this three-step conforming process by conforming the timeline for the documentary short film, Gnarly in Pink.

Importing the XML

You'll import a new timeline into a new project using an XML file exported from another editing system. You'll use two conforming techniques when importing an XML file. The first is to pre-populate your Media Pool with the clips used in the timeline. Because you already know how to add clips to the Media Pool, let's try the second method: letting the XML file import the media.

1. With DaVinci Resolve open, choose File > Project Manager, and create a new project called Gnarly in Pink conform.

   This is an empty project that you'll use to import the timeline and media. You'll use the same Gnarly in Pink media that you used in the previous color correction lesson.
2. Choose Workspace > Reset UI Layout, and click the Edit page button.
3. Choose File > Import AAF, EDL, XML.
4. In the file browser that opens, navigate to R14 lessons > Lesson 13.

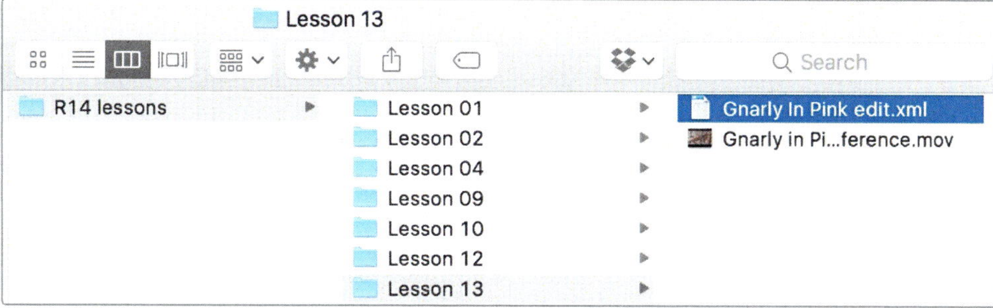

5 Select the XML file Gnarly in Pink edit.xml, and choose Open.

The Load XML window opens with multiple options. For this timeline, the default values are appropriate.

6 In the Timeline Name field, double-click to highlight the text, and type **Gnarly In Pink from xml** to rename the timeline.

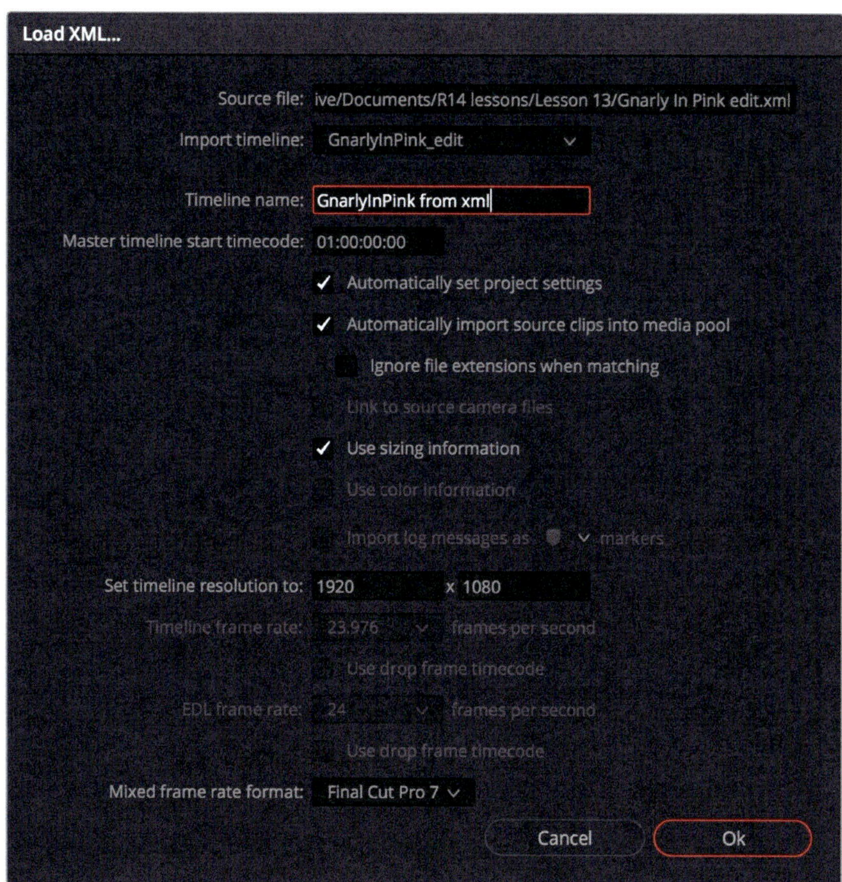

Before you move on, take another look at the checkbox options. The XML is not only setting your project settings, but also importing the media into the Media Pool.

7 Choose OK to close the window.

A new dialog appears to explain that the clips were not imported. XML files contain the file path to the original source media in the imported timeline. But the name of your hard drive and the file path to the media is more than likely different from the computer on which this timeline was created. This dialog gives you the opportunity to locate the lesson footage and link to it.

8 Click Yes to search for the clips.

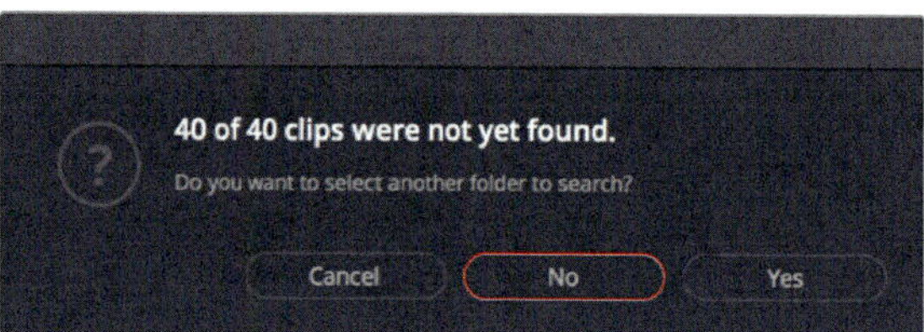

A new dialog opens in which you can point to the location of your media. This Gnarly in Pink media exists in the Lesson 10 folder, which is where you imported it the first time.

9 Double-click the name of your hard drive, and navigate to the R14 lessons > Lesson 10 folder. The media for Gnarly in Pink is located in the Lesson 10 folder, in the Gnarly in Pink archive.

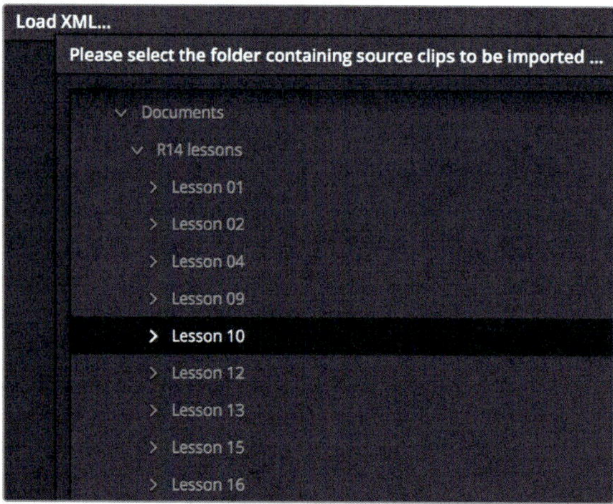

This next dialog informs you that 34 clips cannot be linked. You can choose an additional folder in which to search, but the folder you just selected does indeed contain all the media you need, so let's move on and use a different method of linking your clips.

10 Click No.

The Log window opens, detailing which files are missing. Often, if you closely read the Import Log, you can find patterns in the filenames of the missing clips.

11 Scroll down through the log. Pay attention to the file extensions.

The one thing all these clips have in common is that they have the DNG file extension, which is an extension used by Blackmagic Design cameras. So you need to make sure your clips have the same file extension.

12 In the Log window, choose Close.

Although you haven't yet linked the media successfully, you have completed the first step in the conform process: importing the timeline. In the next exercise, you'll do a quick bit of detective work to remedy the problem.

Importing the missing media

Conforming a timeline is often filled with many troubleshooting steps. Rarely does a timeline import perfectly. When you have problems, you need to put on your troubleshooting hat to figure out what's going on.

The question in front of you now: does the media folder you downloaded for this book contain the media for this timeline or not?

1 At the bottom of the DaVinci Resolve interface, click the Media page button.
2 In the media storage browser, navigate to R14 lessons > lesson 10 > **Gnarly in pink.dra folder**.

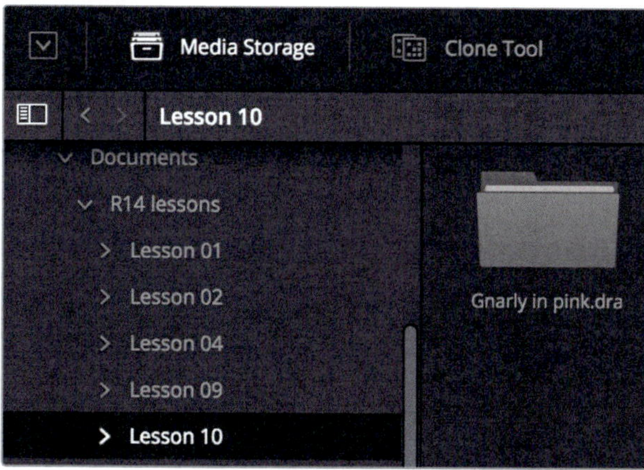

The first step in fixing this relink problem is to manually add the clips from the **MediaFiles** folder to the Media Pool.

3 Right-click the **MediaFiles** folder, and in the menu, choose "Add Folder and Subfolders into Media Pool (Create Bins)".

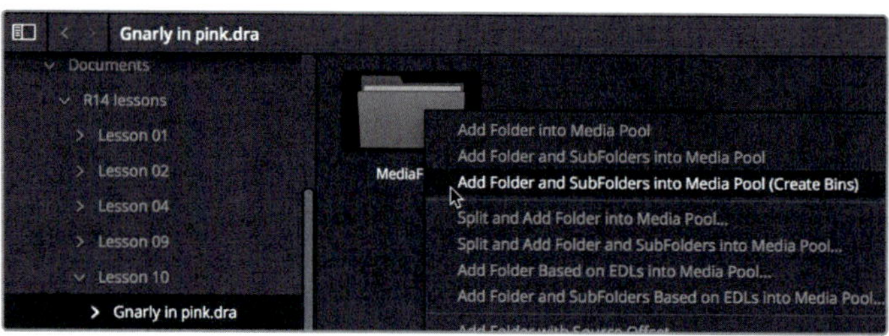

4 Click the list view button in the upper-right corner of the Media Pool.

5 Widen the File Name column by dragging the divider bar to the right until you can see the full name and file extension for all the listed clips, and then scroll through the list.

All of these clips are MOV files. Remember the Import Log showing all of the clips that DaVinci Resolve couldn't find? They all ended in DNG because the XML file was pointing to DNG files. This can happen when you edit with smaller, low-resolution files, often called **proxies**; but now you want to relink to the camera originals for color grading.

> **NOTE** In this lesson we are providing low resolution .mov files to shrink the media download size. However, the process of linking to DNG, high resolution ProRes or DNx files is the same.

6 Click the Edit page button to switch to the Edit page.

The entire timeline is populated using the clips from the bin.

7 Press Shift-Z to see the entire timeline in the window.

Now that you've imported the missing media and it relinked to the timeline, you have one last step in the conform process: confirming that your imported timeline is frame-for-frame accurate to the timeline that created the XML file you imported.

Attaching the offline reference movie

Whenever you conform a timeline from an XML or AAF format, you should always try to receive three items from the person who is creating the XML or AAF file to properly conform a timeline:

- An XML, EDL, or AAF for the timeline import
- Source media that links to the imported timeline
- A single, rendered movie of the full timeline (with graphics and titles) called the **offline reference movie**

Although not always provided, the reference movie can be your single most critical tool to double-check the integrity of the imported timeline. In DaVinci Resolve, any reference movie assigned as the offline reference movie has a very special and unique behavior. Let's find out how it works and
its usefulness in the conform process. You'll begin by adding it to the Media Pool.

1 Click the Media page button.

2 Create a new bin and call it **Reference Clip**.

3 In the media storage browser, navigate to R14 lessons > Lesson 13.

4 Right-click the Gnarly_in_pink_reference clip, and choose "Add as Offline Reference Clip".

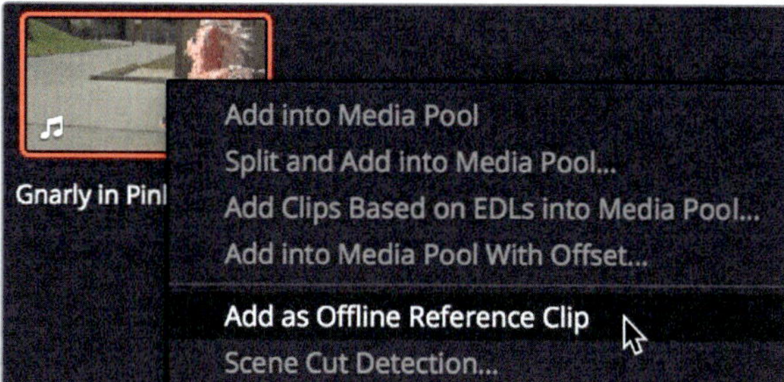

In the Media Pool, you'll see that this clip has a different kind of icon associated with it. Offline reference clips are assigned a checkerboard icon.

Offline reference clips have a unique function in DaVinci Resolve. Their only purpose is to **chase**, or follow along, with the conform timeline you imported, thereby allowing you to refer to the approved, rendered movie in sync with the conform timeline. These offline reference clips can't be edited into timelines like normal media.

The concept is easier to understand in action than to explain.

5 Click the Edit page button to return to the Edit page.

6 Choose DaVinci Resolve > Preferences.

7 In the Preferences window, click the User tab and select the Editing category.

8 Enable the checkbox for "Create Smart Bin for Timelines".

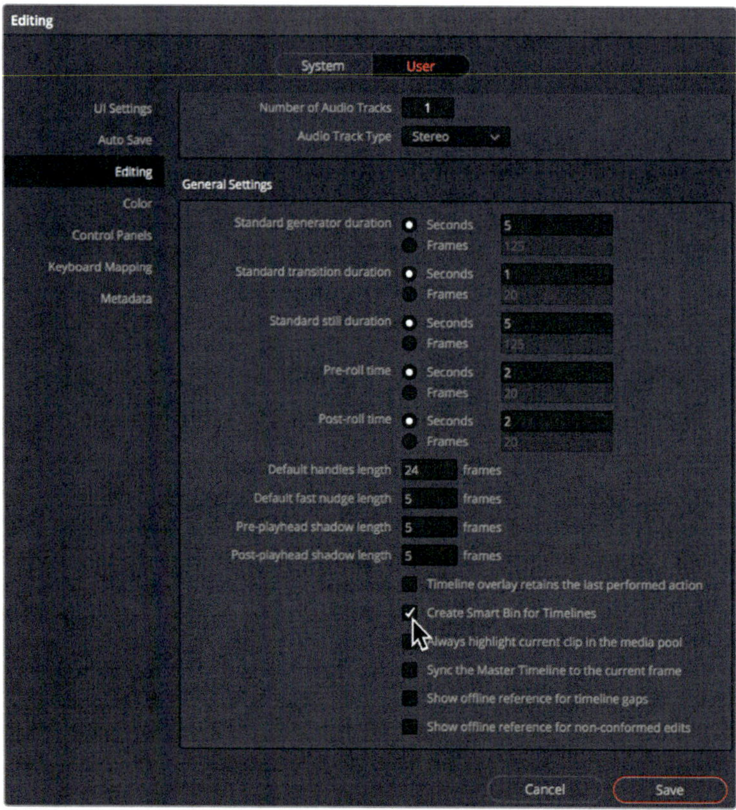

9 Click Save to close the Preferences window.

Creating this smart bin will give you fast and easy access to the imported timeline.

10 On the Edit page, in the Media Pool, choose the Timelines smart bin. Right-click the "Gnarly In Pink from xml" timeline.

11 Choose Timelines > Link offline reference clip > Gnarly_In_Pink_reference.mov.

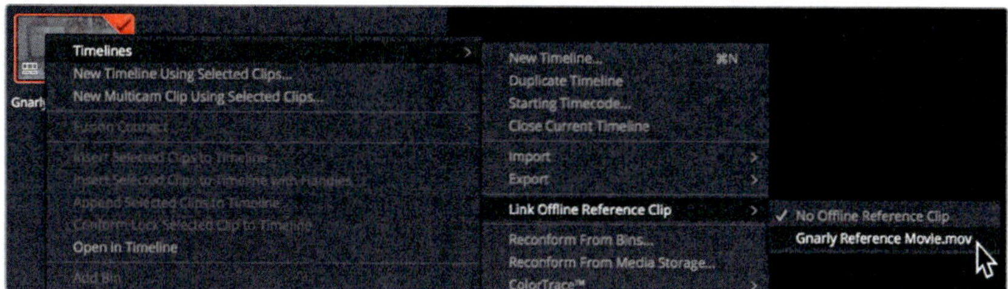

It's not obvious yet, but you've just linked the reference movie to the timeline. You can now select the offline reference movie in the source viewer and watch it play next to the timeline viewer.

12 In the lower-left corner of the source viewer, click the pop-up menu, and choose Offline.

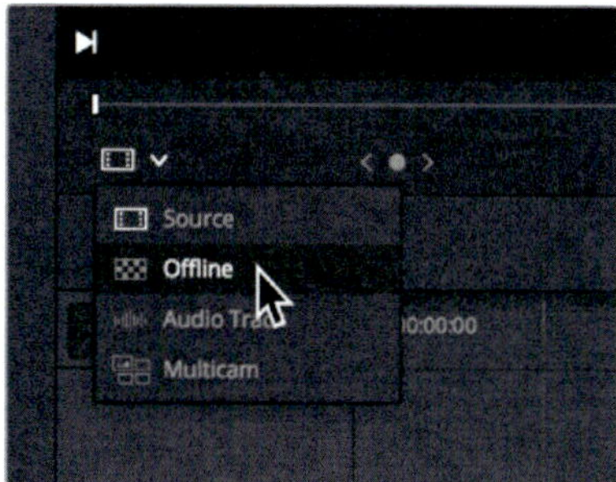

13 Move the playhead to the beginning of the timeline, and press the Down Arrow to go to the next clip. Notice that the source and timeline viewers match.

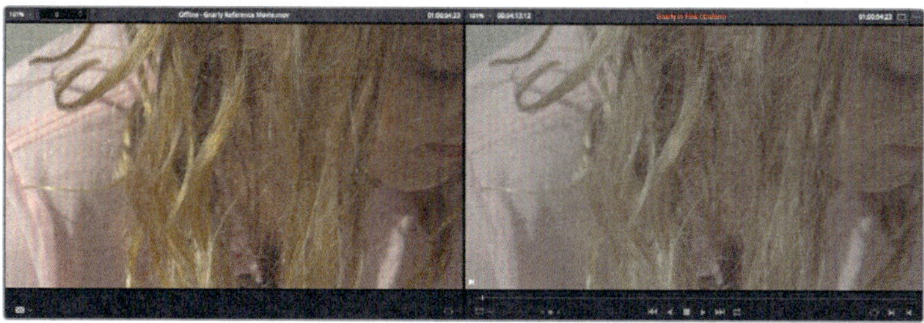

TIP The conform process can be tedious, but it's a critical step. This kind of detail work is required whenever you import a timeline from another non-linear editor. And the work you do here is a great example of why DaVinci Resolve's integration between editing and color correcting is a huge advantage over other editing software.

14 To confirm that the reference clip is in sync through the entire timeline, move the playhead to the middle of the timeline. Press the Down Arrow to go to the next edit point. Press the Left Arrow to go one frame before the edit point.

The images in the source and record viewers should stay in perfect sync.

You should go through the timeline to check its beginning, middle, and end to ensure that the offline reference clip is 100 percent in sync with the imported timeline. Your next step is to check the framing of the imported timeline against the offline reference clip.

Conforming against the offline reference clip

Now that the timeline is imported and the offline reference precisely matches the cuts of the imported timeline, it's time to check the accuracy of the framing.

1 Move the playhead to the last video clip in the timeline.

2 In the timeline viewer, right-click, and in the menu, choose Horizontal Wipe.

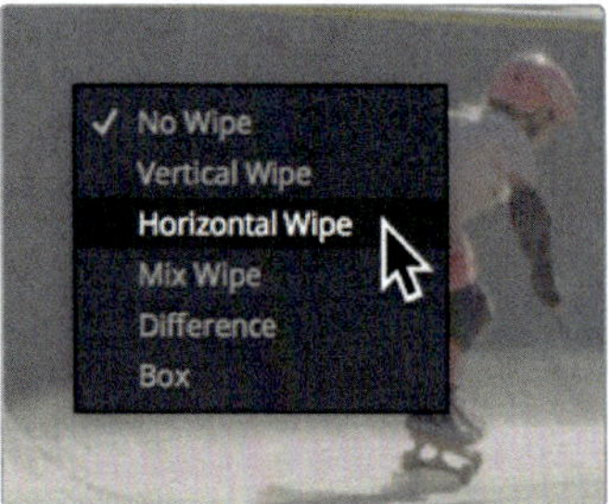

The timeline viewer is now divided between the conform timeline to the left and the reference movie to the right.

3 In the viewer, drag to the left and right to change the position of the wipe.

As you can see, the offline reference matches exactly.

4 Right-click the viewer again, and in the menu, choose Mix Wipe.

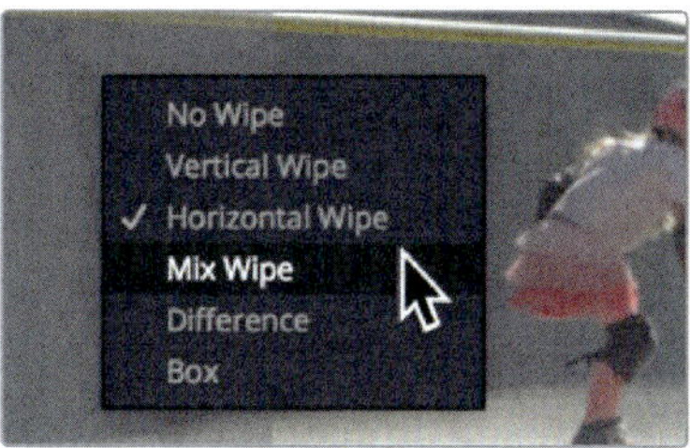

This setting does a 50 percent dissolve between the offline reference and the timeline. It's a terrific setting for finding mismatches in zooms or problems with speed effects.

This shot appears to match perfectly without any of the ghosting artifacts that you would see if the frames were offset from a resize.

5 Right-click the timeline viewer, and in the menu, choose No Wipe.

You have completed the conform. Your timeline is ready for you to begin color grading. Because color grading can be time consuming, you should take advantage of any help DaVinci Resolve can give you. So, before you perform any manual color grading, let's make all your clips a bit more unified in appearance.

Using DaVinci Resolve color management

You're almost ready to start color correcting. But on a project like this—in which the footage is provided to you as flat, low-contrast images—you can change DaVinci Resolve's color science to give your images a better starting point for color correction. You are about to enable the very powerful DaVinci Resolve color management workflow.

1 Drag the playhead to the first video clip in the timeline.

This displays a good image to see the results of the next few steps.

2 Choose File > Project Settings, and select the Color Management category.

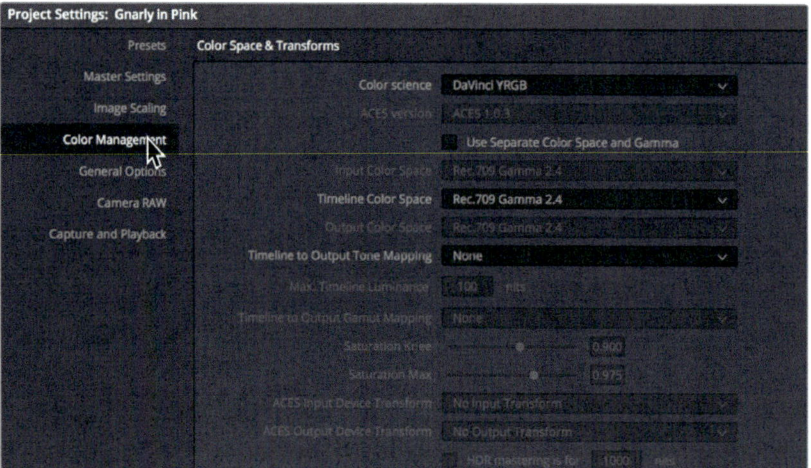

3 In the Color Science pop-up menu, choose DaVinci YRGB Color Managed.

You won't see any changes to the image, but a few new pop-up menus are activated just below the Color Science menu.

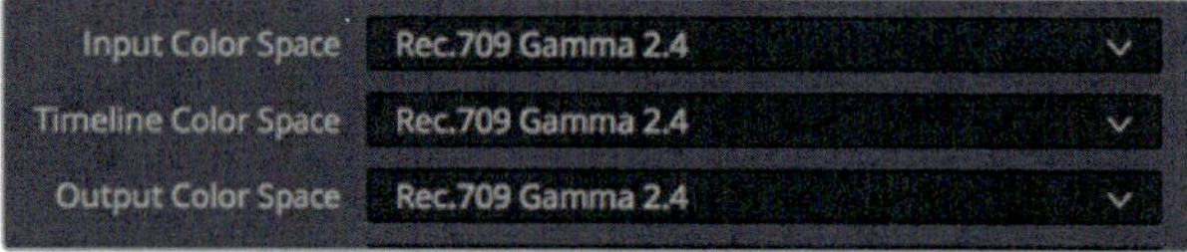

The three color management menus to be aware of are Input Color Space, Timeline Color Space, and Output Color Space. You will leave the Timeline Color Space and Output Color Space at their default values, Rec.709 Gamma 2.4. This is the standard HD color space and you can assume that your project is to be delivered in HD for television.

However, you want to change the Input Color Space value to match the device that recorded the imported clips.

4 Set the Input Color Space to Blackmagic Design 4.6K film.

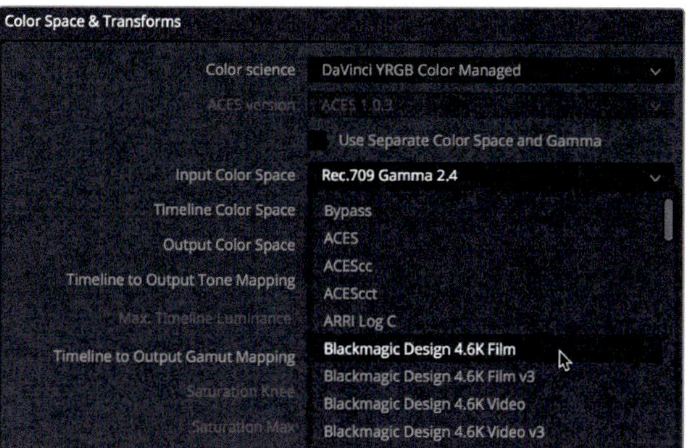

5 Click Save to close the settings, but keep an eye on the timeline viewer.

You may have to wait a second or move the playhead, but you should notice that the blacks become darker and the highlights become brighter. Sometimes the effects of DaVinci Resolve color management are subtle, sometimes they are very obvious. The results depend on the camera and how the image was originally exposed.

TIP If you're editing in DaVinci Resolve, you can make these DaVinci Resolve color management changes at any time in the process. If your footage was shot flat, the preceding technique is a quick and easy way to get extra contrast without color correcting.

NOTE Also, be sure to scroll down the list of Input Color Spaces. You'll see almost every major camera represented in this list. Typically, you'll choose an Input Color Space that matches the record settings of the camera-original footage.

In the source viewer, notice how over-saturated the reference movie now appears.

That's because DaVinci Resolve thinks this clip was also captured using the Blackmagic Design 4.6K film color space. It wasn't. Because not every clip in your project will be recorded by the same device, you can override the color management setting on a clip-by-clip or bin-by-bin basis.

6 In the reference movie bin, right-click the **Gnarly_In_Pink_reference** clip, and hover your mouse over the Input Color Space menu.

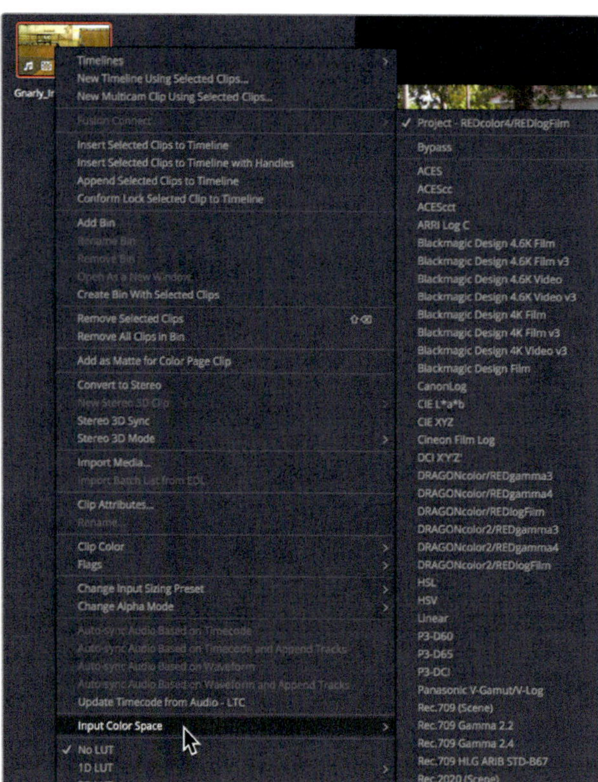

Doing so displays the same list of color spaces as in the project settings panel. Here you can override the project settings on a clip-by-clip basis.

7 In the Input Color Space menu, choose Rec 709 Gamma 2.4.

   The Source viewer will once again switch to source from showing the offline clip so you need change it.

The reference movie no longer looks blown out and contrasty. It appears normal again. Now your color science is optimized for your final HD delivery format.

Imported timelines can be very tedious to set up. The problems you solved here are typical and frequent. Yet, DaVinci Resolve's color-correction toolset is so powerful and feature rich that artists around the world endure the preliminary tedium because it pays off with terrific images that reinforce the story.

By now, you should also understand why the timeline integration between the Edit page , the Fairlight page and Color page is a game-changing innovation. After 20 years of XML, EDL, and AAF workflow headaches, DaVinci Resolve eases that pain with its integrated workflow.

Lesson 14

Delivering a Final Program

Everybody's workflow is different, and the Deliver page is designed to give you flexible options for creating output from low-resolution editing proxies to the highest quality master files. You can batch render multiple files to multiple resolutions, compression formats, and destinations, all from the same project. With so many different options available on the Deliver page, you might think it would be difficult to learn. But it isn't.

In this lesson, you'll use the Deliver page to create two separate outputs: one for high quality web streaming, and another custom output to create files for visual effects work.

Time

This lesson takes approximately 20 minutes to complete.

Goals

Creating a web streaming file

Almost every project ends up on a web-streaming service in some form, even if it's just as promotional material. To output the most common file-based output formats, DaVinci Resolve 14 includes easy setups, presets that automatically configure all the parameters for an output type that you select.

1. In DaVinci Resolve, open the Age of Airplanes project that you worked on in previous lessons.
2. In the Rough Cuts bin, double-click the Fairlight timeline to open it into the timeline window.

   The currently displayed timeline is the one that you will prepare for export on the Deliver page.
3. At the bottom of the DaVinci Resolve window, click the deliver button to open the Deliver page.

   The Deliver page is divided into five areas.

The toolbar buttons switch between file rendering and tape output.

The viewer shows images from the selected timeline.

The Render Queue contains all of the jobs waiting to render.

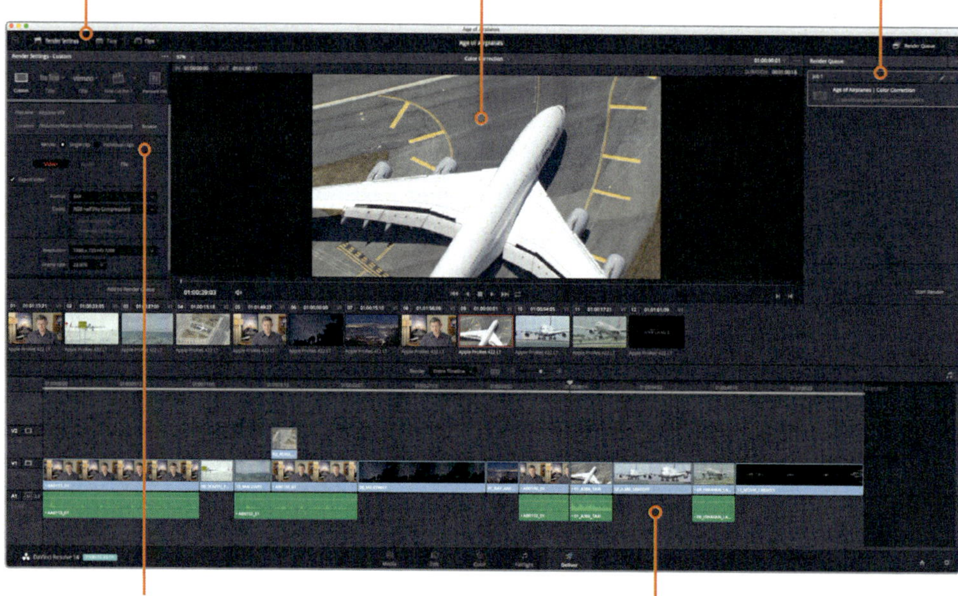

The render settings configure the output format.

The timeline displays a graphical representation of the timeline you are outputting.

You'll use the toolbar to select tape-based output or to adjust Render Settings for file-based output. When Render Settings are selected, as they are by default, you can choose a render preset to configure for your output.

4 At the top of the Render Settings, click the Vimeo preset.

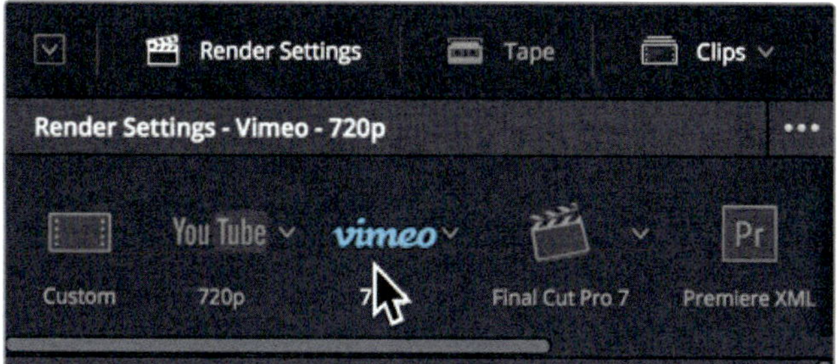

When you choose one of the presets, DaVinci Resolve configures the Render Settings to output a file that conforms to the destination's requirements. If you wanted to stop at this point and output the file, you could do so. The only parameter you would need to set would be the destination of the output file.

5 Just below the row of presets, click the Browse button, and choose a location for the output file. Click the OK button on the dialog.

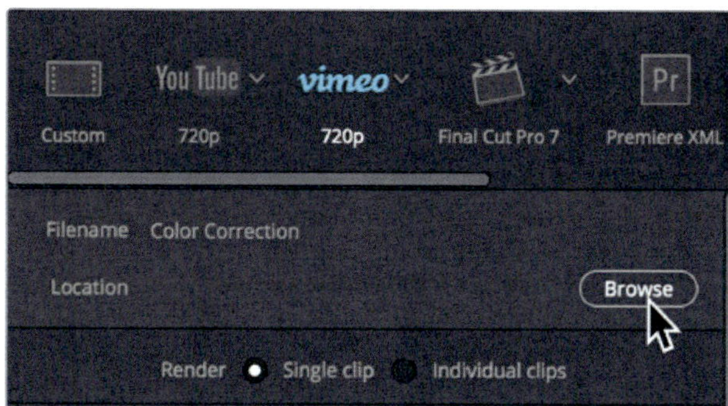

Although that is all you must do to output a file, you also have many options that you can apply to presets. Let's examine some options that can be useful when creating files for streaming.

6 Below the Browse button, click the Video tab to view the video-related options.

Although you can customize the video compression and format parameters in the Video tab, when you choose one of the easy presets many of the basic settings are optimized for you.

When the Vimeo preset was selected, QuickTime H.264 was chosen as the format and codec because those parameters are commonly used for web-streaming videos. The preset also configures the resolution and frame rate menus to the current timeline settings.

The bit rate, or data rate, settings offer various recommendations depending on the target streaming service. For example, the default 10000 kbs (10 Mbps) rate is the maximum for a 720p video on Vimeo, but it could be too high for use with other services. Let's lower the bit rate to ensure acceptable performance on some of the other web-streaming services.

7 Scroll down, and in the Quality Kb/s areas, select "Restrict to", and in the value field, enter **7500** Kb/s.

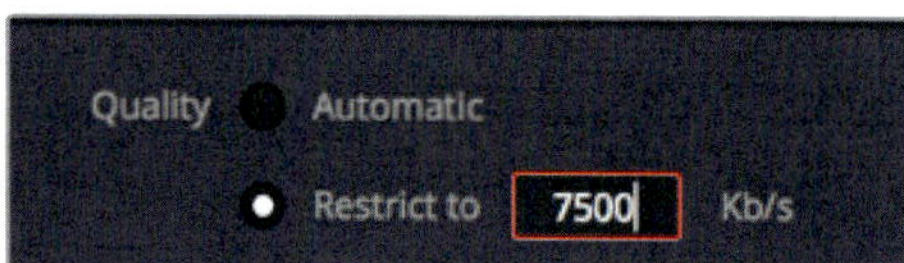

You also have the option to perform a single-pass encoding or a multi-pass encoding. In almost every case, you will want to choose multi-pass encoding. Multi-pass encoding uses the first pass to gather information about the images, then uses subsequent passes to efficiently encode the frames, choosing the optimal number of bits to use based on motion within the frames. That sounds as if it would always be the best choice, but the main reason you choose single-pass is to save time. Multi-pass encoding can take two or three times as long to process as single-pass encoding.

Because this is a short movie, choose multi-pass.

8 In the Passes menu, choose Multiple.

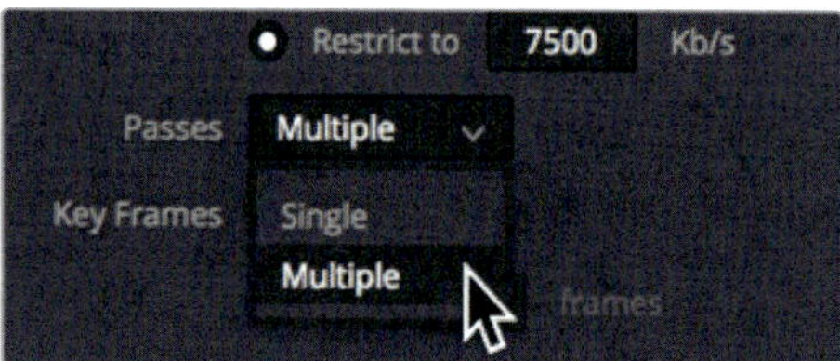

Encoding profiles represent another group of settings that you can change but the default Auto profile is usually a good starting place. If you are aware of the specific device your content will be played on, then changing the Encoding Profile might be a good choice. If you are going directly to iPhone or Android, those devices prefer the Base (or Baseline) profile setting which is a lower-power consumption profile. If you are going to upload to Vimeo, then the High profile is your choice because it was originally created as a digital format for Blu-ray discs. Choosing High will give you the best quality, but creates a file of larger size that also requires more computing power to decode.

9 In the Encoding Profile menu, choose High.

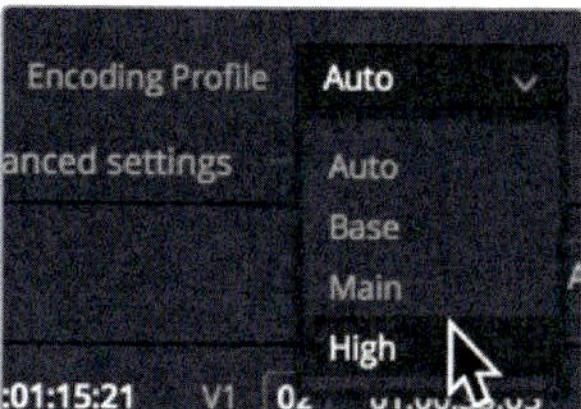

When you have all the video settings configured, you can switch to the Audio tab.

10 Below the Browse button, click the Audio tab to view the audio-related options.

Most of the time, audio in a QuickTime H.264 file uses AAC compression because AAC is the preferred choice for streaming audio on most sites. However, you have a few significant options that you can set.

The Bit Rate Strategy setting allows you to choose between Constant and Variable Bit Rates. A Constant Bit Rate will give you a fixed target bit rate for the entire audio file. This is generally preferred by streaming media sites and podcasting guidelines because of the bit-rate predictability. Variable Bit Rate encoding squeezes the same quality audio into smaller file sizes but will vary widely in its bit rate. A third option might just give you the best of both worlds. The Variable Bit Rate Constrained option gives you the efficiency of Variable Bit Rate limited to a target maximum bit rate (in Kbps) similar to a constant bit rate.

11 In the Bit Rate Strategy menu, choose Variable Bit Rate Constrained.

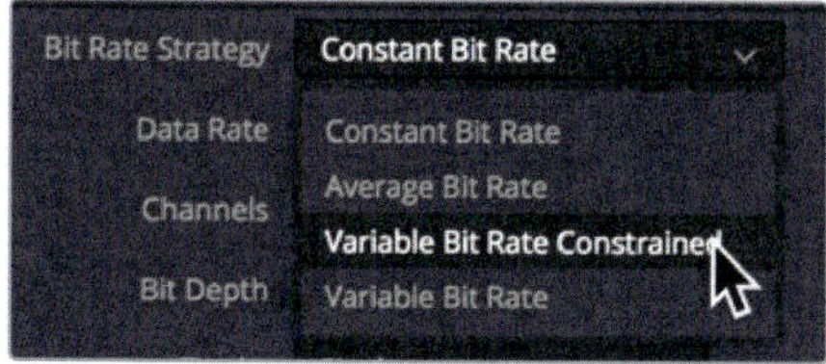

Now that audio and video encoding is configured, you can name the file that will be output.

12 Below the Browse button, click the File tab to view the file naming options.

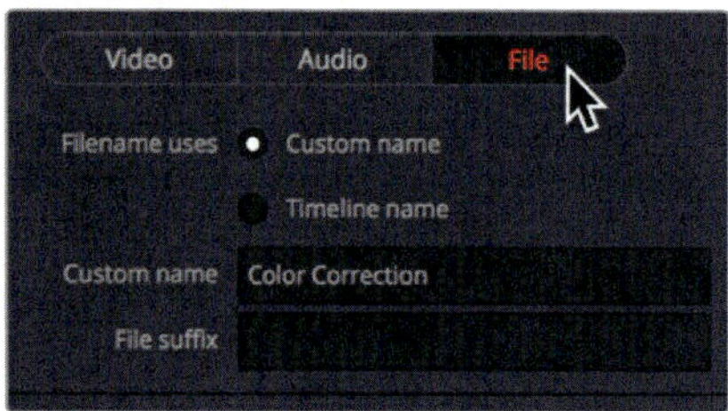

13 In the "Custom name" field, type **Training Movie for Streaming**.

That is the easiest tab you will work with in the Deliver page. Before you output this file, let's add one more addition to the view.

Adding a watermark

You can add a text-based watermark, similar to a title, that overlays the video. Most often this watermark is used to superimpose a "Do Not Distribute" text or to display a running timecode. The watermark is created and formatted using the Data burn-In controls on the Color page.

1 At the top of the Render Settings, click the Video tab.

2 Scroll down and click the "Advanced settings" disclosure arrow.

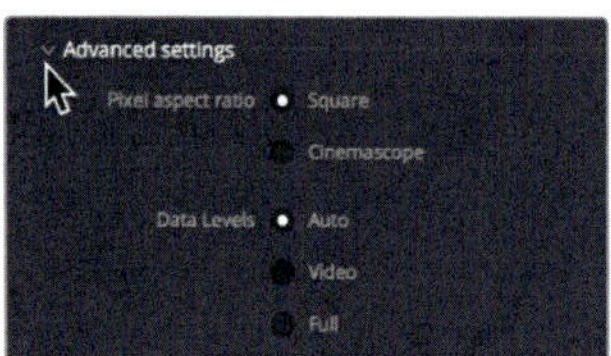

3 Set the "Data burn-in" pop-up menu to "Same as project".

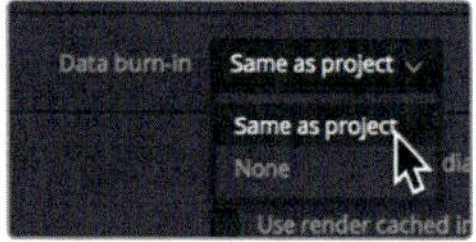

The Burn-in controls can be accessed from any page.

4 Choose Workspace > Data Burn-In to open the Data Burn window.

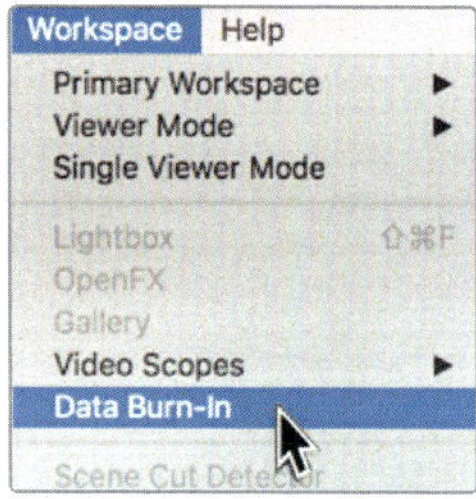

The left side of the Data Burn window includes a list of possible data types to overlay on the video. Selecting the checkbox next to any option enables it.

5 In the list of burn-in data, select the Record Timecode checkbox to overlay the record timecode on the output movie file.

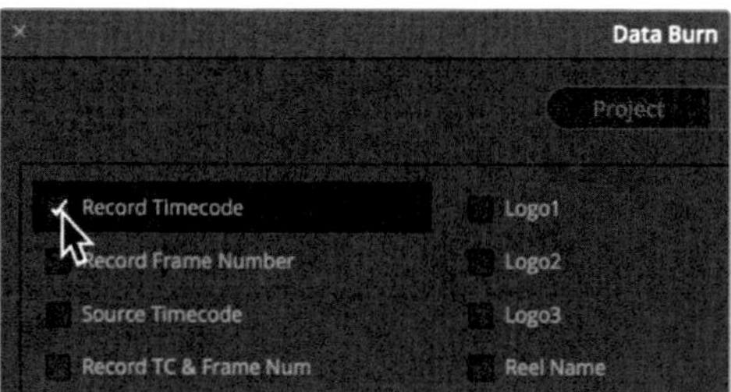

6 Select the Custom Text1 checkbox to add a text watermark.

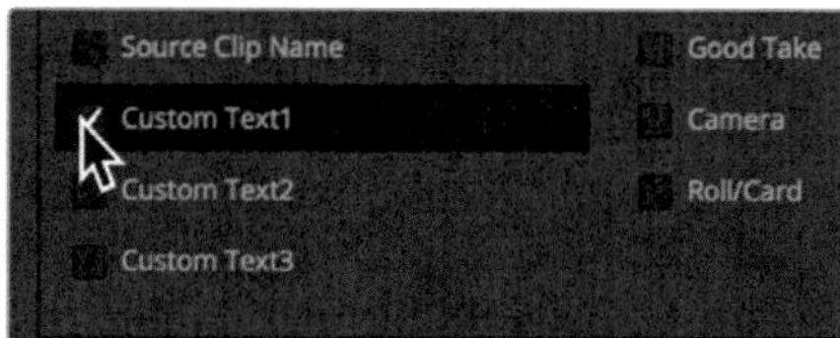

7 In the Custom Output Text field, enter **DO NOT DISTRIBUTE**.

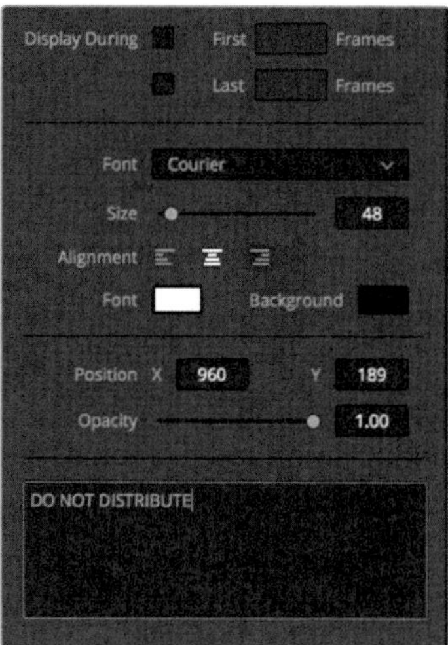

8 Drag the Size slider to the right to increase the text size to 55.

The viewer displays all the changes to the Data burn-In settings as you make them. You are now ready to output this vimeo file.

9 Close the Burn window to return to the Deliver page.

10 At the bottom of the Render Settings, click the “Add to Render Queue” button.

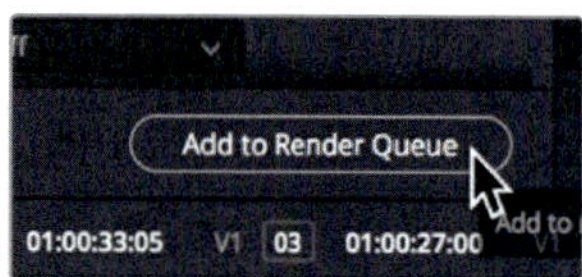

The job is tranferred to the Render Queue to the right.

The Render Queue acts as a temporary holding area for jobs that you want to output from DaVinci Resolve. You can add as many jobs to the Render Queue as you need to output. Although you could render this file immediately, let's wait and prepare another job for the queue.

Creating a custom preset

In some situations, you will need to render out only a portion of your timeline for others to work on. When you have a few clips that need to receive motion graphics or be integrated into visual effects, you will need to render just a section of the timeline and, usually, in a very specific format.

Let's render out a section of the timeline in a format that often is required by visual effects technicians. They need high-quality renders; but because DaVinci Resolve does not have a Visual Effects preset, you'll want to create one of your own, and save it as a preset for later use.

1 At the top of the Render Settings, click Custom.

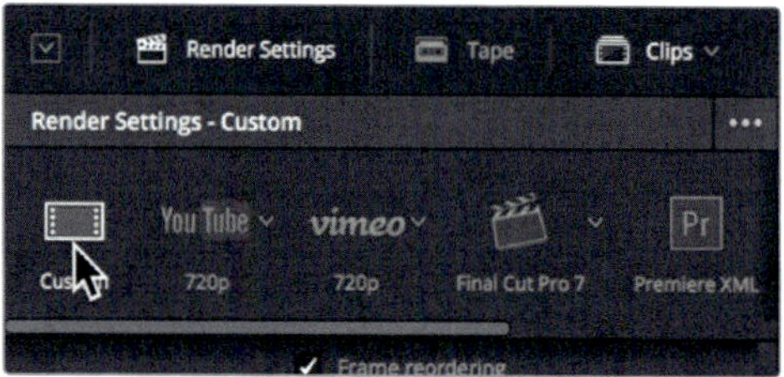

Clicking Custom makes all of the options in the Render Settings available for selection, and begins the process of creating a new preset. First, you want to decide if you want every clip in the timeline to render out as its own individual file or as a single file.

2 At the top of the settings, click the "Single clip" button to output a single movie file.

3 Click the Video tab, and in the Video Format menu, choose EXR.

EXR is a single-frame format like TIFF that is often requested by visual effects studios.

Because these files are needed to produce visual effects, you'll want to disable the watermark.

4 Scroll down to the "Data burn-in" menu, and choose None.

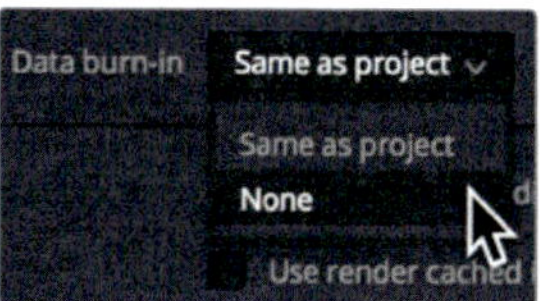

Although the burn-in is disabled, the overlay is still shown in the viewer.

Because EXR files do not have audio tracks, you can skip the Audio tab and go to the File tab

5 Click the File tab to name the files.

6 Enter the Custom name, **Airplane VFX**.

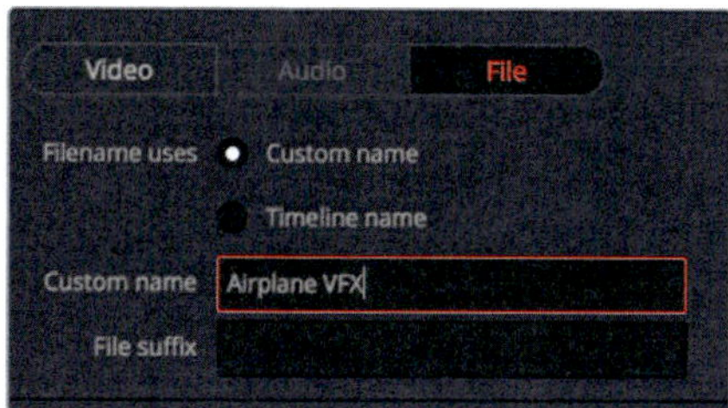

Because each frame will become an EXR file, you need to number the frames so that the visual effect artists will know the order in which the frames should appear.

7 Set "Use # digits in the filename" to 4.

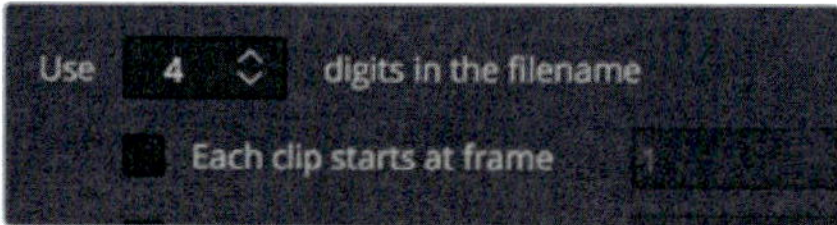

This value will add four numeric digits to the end of each file. None of our shots are over 1000 frames long so four digits should be plenty.

8 Select the "Each clip starts at frame" checkbox, and in the numeric field, enter **001**.

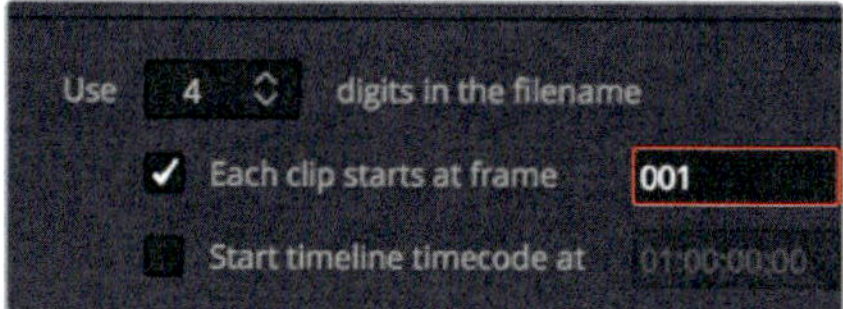

Next, because you want to send only one shot from your timeline to the VFX staff, you need to mark the range you want to render.

9 In the Deliver page timeline, move the playhead over the 08_SOUTH_POLE_DC3 clip.

10 Choose Mark > Mark Clip, or press X, to add in and out points around the South Pole clip.

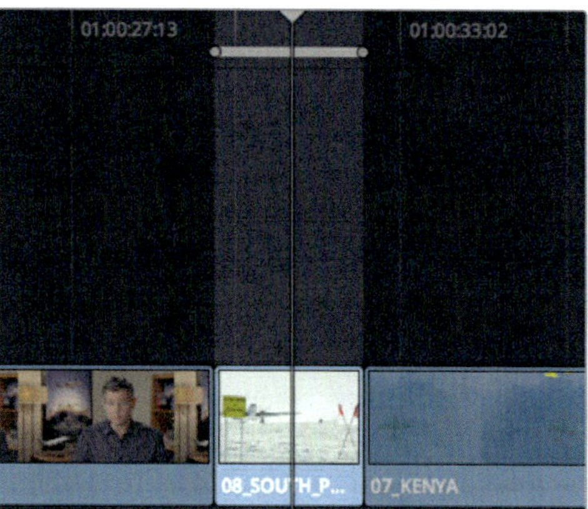

The marks in the timeline identify the region that you will render out. You can now save these settings as your own customized VFX preset.

11 At the top of the Render Settings, choose Options > Save As New Preset.

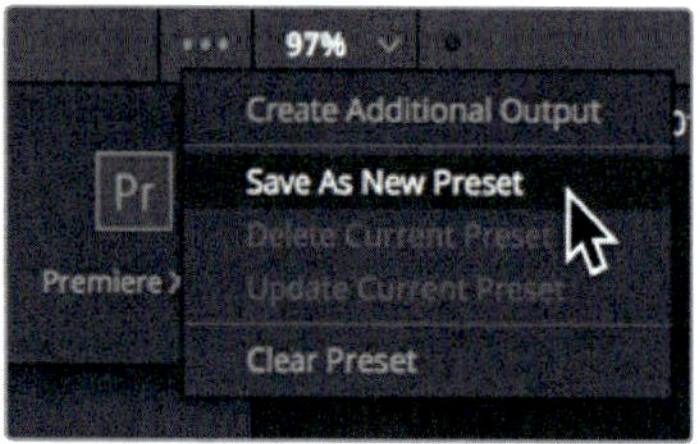

12 In the "Enter preset name" field in the dialog, type **VFX Output**, and click OK.

The new preset's name appears at the top of the Render Settings, indicating that it is the currently active setting. It also is listed in the row of presets.

With all of the settings configured, you can add this job to the Render Queue.

13 At the bottom of the Render Settings, click the Add to Render Queue button to add your second job.

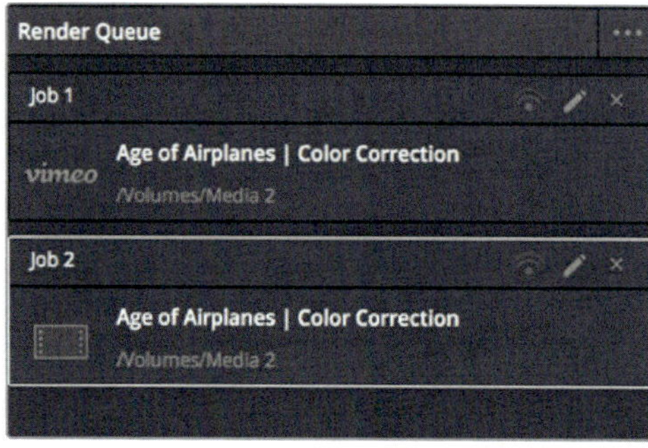

You now have two jobs in your Render Queue. If you save and quit DaVinci Resolve now and return later, those jobs will still be in the Render Queue. Because you may not want to devote disk space to these lesson results, you won't be expected to render them for this exercise.

14 When you do want to render one of your own projects, in the Render Queue, select the items that you want to render. Then click the Start Render button at the bottom of the Render Queue to begin the process. One by one, the jobs will render out and be marked as completed.

15 What you will find useful right now is knowing how to clear the Render Queue. In the upper-right corner of the Render Queue, click the Options pop-up menu.

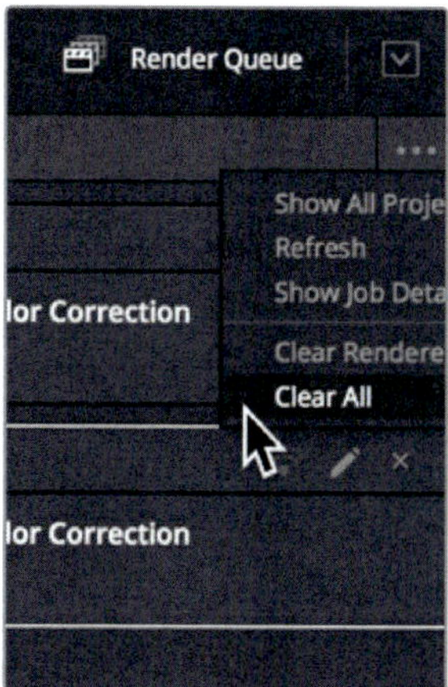

16 In the menu, choose Clear All to remove all the current jobs from the queue.

The Deliver page has many more options for outputting various file formats, and for supporting a variety of workflows; but you now understand the most commonly used workflows and how to set up your own custom presets.

Lesson 15

Managing Media and Databases

Unlike almost all other editing, audio mixing, or color grading software, DaVinci Resolve 14 uses an industry-standard database to store all of your projects, bins, clips and timelines-basically any item that is not an audio or video media file is kept in the database. As a result, DaVinci Resolve works a little differently from other apps when you want to move projects from one system to another or create backups of your projects. It is not any harder to do those things, and in some cases it can be a lot easier. But it is different.

In this lesson, you will explore ways to back up and move projects and media, as well as learn how to manage the DaVinci Resolve database.

Time

This lesson takes approximately 20 minutes to complete.

Goals

Consolidating media

NOTE The following exercise uses a second hard drive for consolidating media. It is not recommended that you perform these steps on this project. However, you are urged to read through the lesson to gain an understanding of the process.

Before we explore database management, let's start with a topic that is a bit more down to earth: media management.

Almost every project will need some kind of media management. It might be just to copy media from one drive to another, or to convert files from one format to another. The media manager is the tool that performs these types of tasks in DaVinci Resolve. You can use it to copy, move, and even consolidate media.

When your hard drive cannot accommodate a complete copy of the original media, consolidating can help remove media that is no longer needed in your project. Using the media manager, consolidating gives you the option to copy only those pieces of media that are actually in use. For long-form projects with lots of media, consolidating is a great way to free up disk space and make backing up a project a quicker task.

Because the media manager can be used only within a project, let's open the multicamera project that is in the project manager.

1 With Age of Airplanes still open, click the Media page button; and in the Media Pool, click the Rough Cuts bin. Select all the timelines.

2 Choose File > Media Management.

When timelines are selected, the Media Management window opens with the Timelines button selected.

3 Click the Copy button.

Choosing Copy first creates a duplicate set of media files on the portable hard drive, but leaves the originals in place.

4 Click the Browse button, and in the File Destination dialog, navigate to a location where you want to copy the media. Click OK.

> **TIP** You can click the New Folder button to create a folder for the consolidated files that you create. If you don't do so, you might end up with too many files on the top level of a hard drive and no way to know which files to keep and which to delete.

The option to consolidate media requires that you select one or more timelines to determine which files to keep. You have two choices for consolidating your media: you can choose to copy the clips that you have used in the selected timelines in their entirety, or to copy just the portions of those clips that you have used in the selected timelines (including handles).

5 Choose "Copy and trim used media keeping".

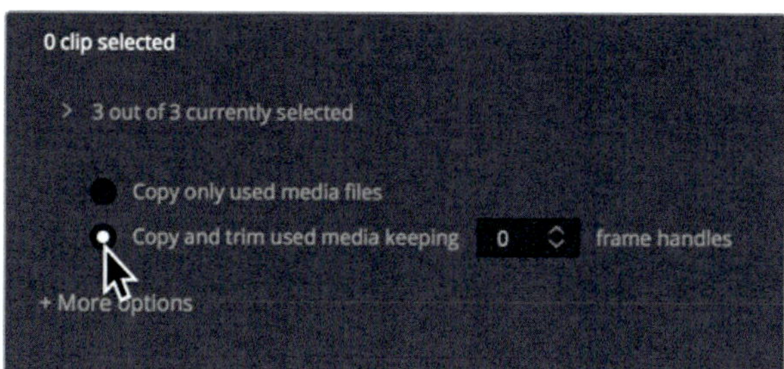

Choosing this option will trim the master clip files down to only those clip portions used in the selected timeline(s). A field is available in which you can enter the number of frame handles you would like added to each side of a clip in case you will later need to trim or create transitions.

> **TIP** Compression formats that use interframe, temporal compression, such as H.264, cannot be used with the "trim unused media" option in the Media Management window.

6 Enter **12** as the number of frame handles to add. Because your project runs at 23.979 frames per second, entering this value this will enable a half-second of extra media on both ends of the clip for trimming and transitions.

7 Click "More options" to view additional settings.

8 Select "Relink to new files" to use the new, copied media in the timeline instead of the existing media.

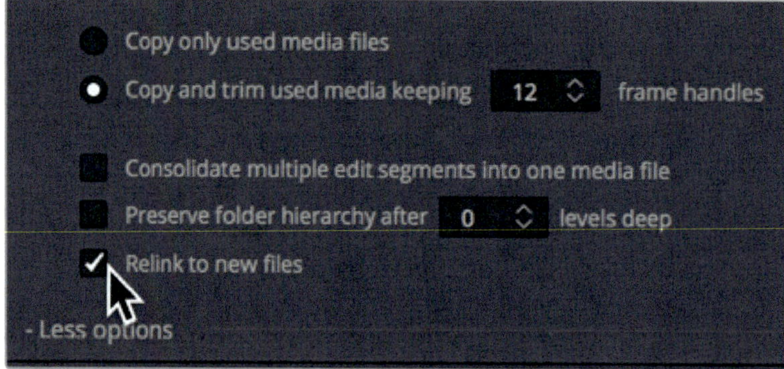

If you were finalizing these steps, you would now click Start to begin consolidating the media, but in this case do not do so.

After you started the consolidation process, the files used in the timeline would be copied to the destination drive, and then trimmed so that only the clip portions in use were saved to disk—along with a half-second handle on each side of a clip.

The clips in your bin would now link to these new, shorter media files, but all of the original clips would still remain in their original locations. It would remain for you to delete them when you were confident that you would no longer need them.

Copying projects and media to a new hard drive

Although consolidating media can save you the most storage space, it addresses only the media. Your timelines, clips, and metadata are not included. The easiest way to copy, back up, or move a single project and all of its content from one computer to another is to use DaVinci Resolve's archive and restore features. Archiving a project collects all your files (even if they are on different drives) and places them in the destination folder of your choice along with the project file. To archive a project, you must do so using the project manager window.

NOTE The following exercises use a second hard drive for copying and transcoding media. If you do not have a second hard drive or do not want to take up valuable disk space by copying and/or moving the training media from this book, you can still read through the lesson to gain an understanding of the process.

1 Open the project manager, right-click the Age of Airplanes project thumbnail, and choose Archive.

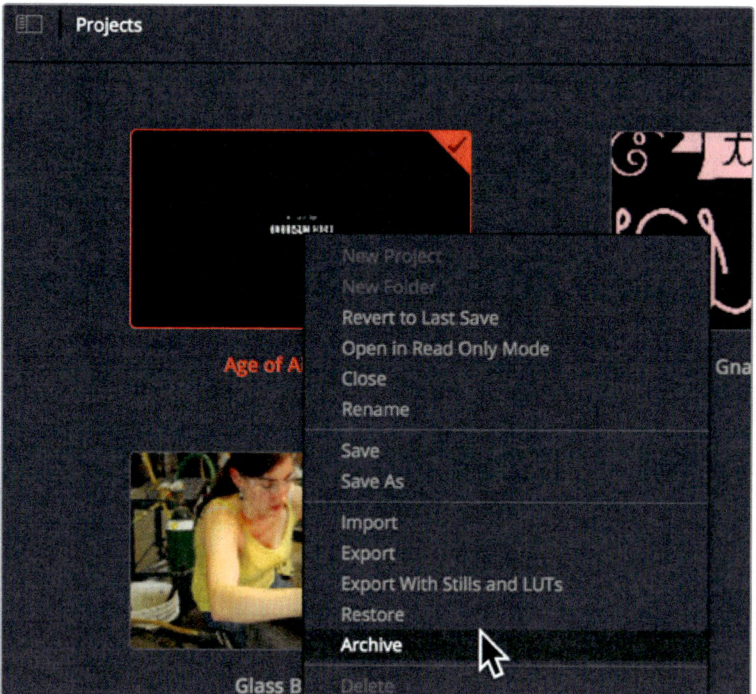

2 In the Archive dialog, navigate to a drive where you want to save the project, and click Save.

All of the media and the project file containing your bins and timelines are copied into a folder with the extension .dra. This folder contains everything you need to open your project on another computer.

Be assured that the original media remains in their original locations, and that the two sets of media are completely identical. It is up to you to decide if you want to delete the original media from those original locations or leave them to work on later.

When you copy the archived folder to another computer, you must restore it in DaVinci Resolve to begin working with it.

NOTE Because you already have the project from the archive on your computer, the following steps are included for educational purposes and are not meant to be followed.

3 To restore an archived project, you would open the project manager.

4 Then, you would right-click anywhere in the project manager, and choose Restore.

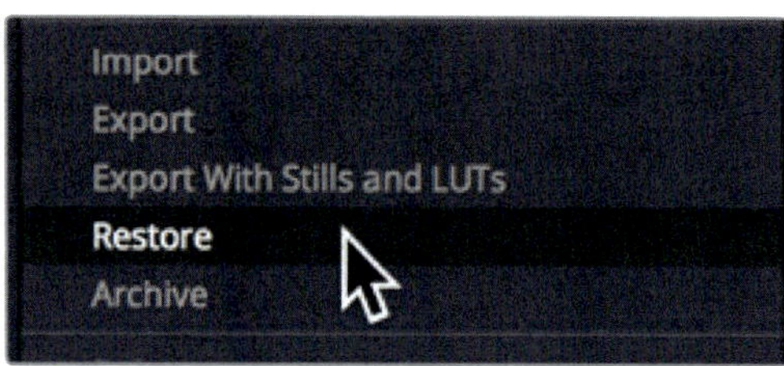

5 In the dialog, you would navigate to the ".dra" folder of the archive you wanted to restore, and click Open.

The archived project would open into the project manager and you could begin working on it.

Working with the DaVinci Resolve database

If you want to move or backup all of your projects and timelines, then you must do so within the database. DaVinci Resolve must be open because projects and all their associated bins, clips, and timelines are not independent files that you can find (and freely manipulate) on your hard drive. And while that may suggest a process that is more complicated than you may be accustomed to, it brings a lot of benefits. A database adds a level of organization to your projects without you doing a thing. You don't spend time keeping track of projects and bins because they already exist in the database for every project.

Also, because projects are kept in a database, they can be stored on a server that can connect to multiple DaVinci Resolve workstations with the flexibility to access any project on any computer or work collaboratively on any project.

NOTE The following exercise for switching databases is required if you wish to perform the last lesson for multicamera editing.

Creating and switching databases

When you first open DaVinci Resolve, it automatically creates a disk database, so it is largely transparent to the user that a database is in use. In Mac, the database is stored in the Library folder; whereas in Windows, it is in the Program Data folder. Even though you know where the database is stored, the first rule of databases is to not rename or modify any of the folders or files in a database and certainly not to delete them!

In this exercise, you'll learn the right way to switch between databases, create your own database, and back up a database without ever altering it at the operating system level. You'll start by switching from the database you've used throughout this book to a database that has the projects for the remaining multicamera lesson in this book.

1. In the upper-left corner of the project manager, click the databases button to open the list of available databases.

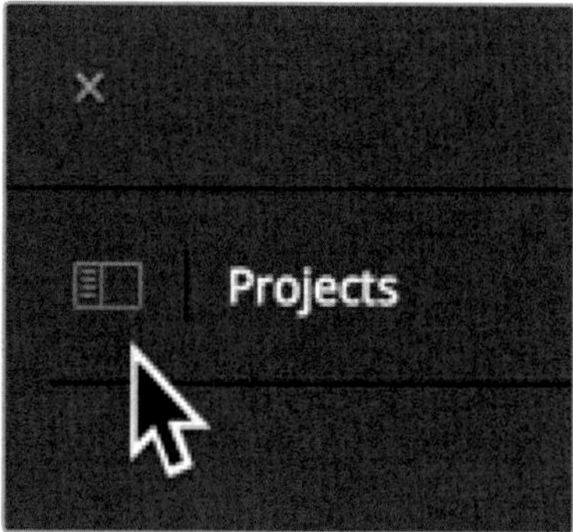

Notice the Local database listed in the sidebar. It is the current default database that DaVinci Resolve created when you first started.

2. Click the New Database button to create a new database.

The New Database dialog gives you two options that are displayed as tabs at the top of the window. You can either Connect to an existing database that you have not previously connected to, or Create a new database. Creating a new database can be useful for organization; for instance, you might create a new database at the start of every year or at the start of each school semester. Switching to a new database is useful in your current situation because when you downloaded the content for this book, you also downloaded a database that was created for the upcoming lessons. So, before you continue, let's connect to that downloaded database.

3 At the top of the New Database dialog, click the Connect tab to switch to a new database.

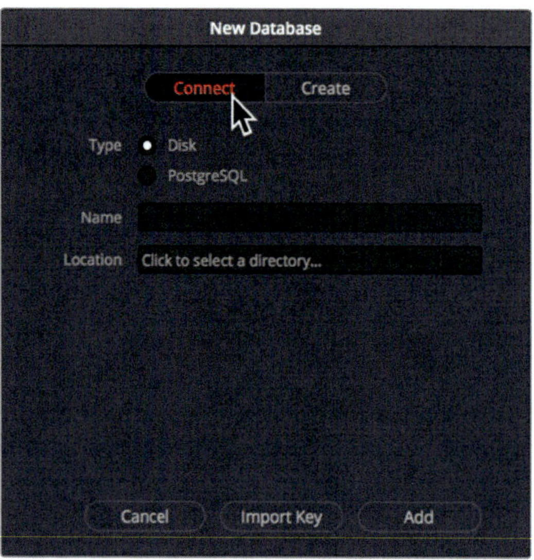

4 In the Name field, enter **resolve101**, and click in the Location field to find the database.

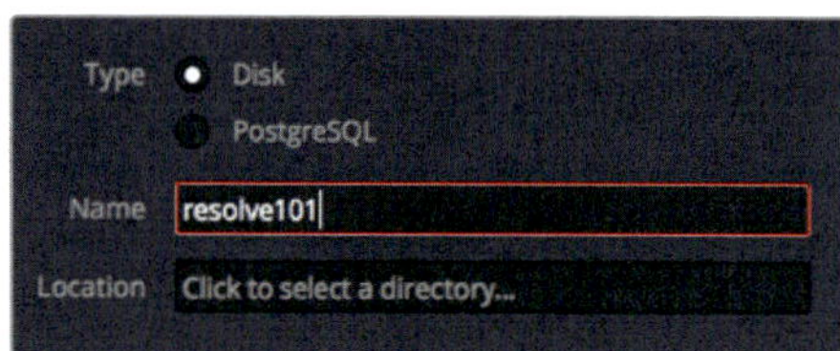

5 Navigate to Documents > R14 lessons > Lesson 15, select the bmd101 folder, and click Open.

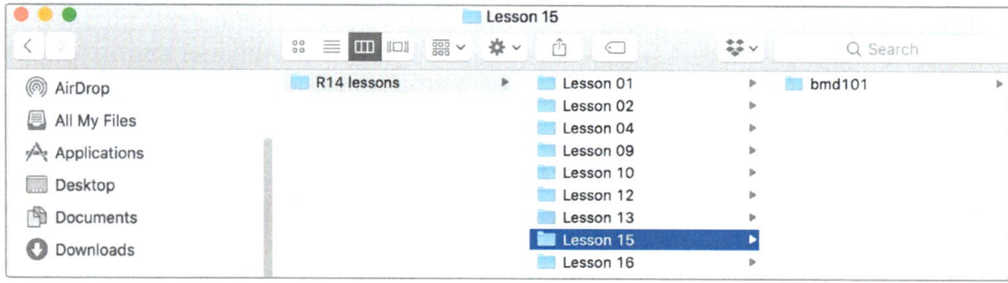

The database is added into the database dialog where you can now add it to DaVinci Resolve.

6 Click Add in the database dialog.

The resolve101 database now appears in your list of disk databases. Clicking any of the databases listed will switch to that database. The newly added resolve101 database is already selected and the projects in that database are displayed on the right of the project manager.

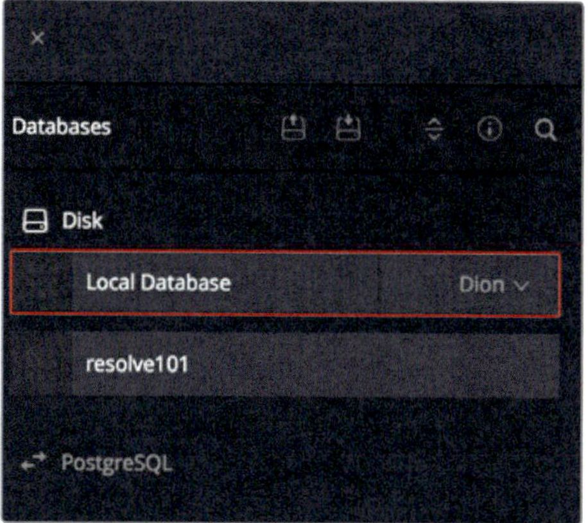

TIP Databases with fewer and smaller projects will save and operate faster than databases with a greater number of large projects.

Backing up databases

Although databases do not contain media, they do contain all of the metadata for every project. What's more, they also contain all of the timelines for every project you have worked on. So, you'll want to regularly create a backup of that database to prevent your work from being accidently erased or deleted due to a hard drive failure or other mishap. Conveniently, you can back up your database within the database manager window.

NOTE The following exercise uses a second hard drive or cloud-based storage to store your database backup. If you do not have access to a second hard drive or cloud storage, you can still read through the exercise to gain an understanding of the process.

1 In the list of databases, make sure the **resolve101** database is selected.

2 At the top of the window, click the backup button.

Generally, databases range in size from a couple of hundred megabytes to a couple of gigabytes; but unlike media, they can easily be saved to a cloud backup storage system or a small local hard drive.

3. Navigate to the hard drive or cloud-based storage where you want to back up your database, and click Save.
4. Once the save is completed, click the confirmation dialog, and click the databases button to close the database window.

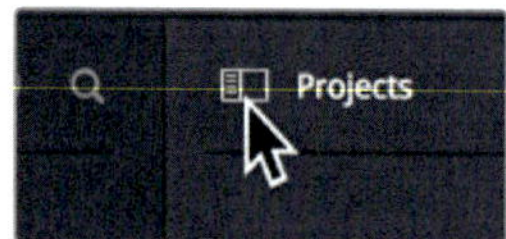

After saving a backup, a .diskdb file is created. This file contains your entire database which you can simply copy to another drive if, for instance, you want to move your projects from one computer to another. On the new computer, you can click the Restore button in DaVinci Resolve to open and use the database.

With all that house keeping out of the way, you can begin the last lesson in this book. You've reached multicam editing!

Lesson 16

Editing Multicamera Productions

When you work on specific types of productions—such as interviews or performances that were shot using multiple cameras running simultaneously—you can take advantage of the multicam mode in DaVinci Resolve 14. Multicam mode editing allows you to group and synchronize two or more source clips that show different angles on the same performance. You can then display them simultaneously in the viewer and cut your timeline in real time. It's an incredibly fast way to edit concerts, reality TV, interviews, or any production in which several cameras recorded the same event.

Time

This lesson takes approximately 30 minutes to complete.

Goals

Note

To use the project in this lesson, you must have completed Lesson 15, "Managing Media and Databases."

Syncing clips

As with every new project, you have to be familiar with your footage before you start working with it. When performing a multicam edit, the assumption is that multiple clips depict different angles of the same event, all shot at the same time. Let's look at your source material to see if you have content suitable for a multicam edit.

1 Open DaVinci Resolve 14, if necessary.

2 Double-click the Multicamera Editing project.

   Let's first reset your user interface and return to the Edit page.

3 Choose Workspace > Reset UI Layout, then click the Edit page button to return to the Edit page.

   This project has two bins: one for the source media and the Master bin. Let's first look at your source media.

4 Select the Multicamera Source Clips bin, and double-click **01 Master Camera** to load that video clip into the viewer.

5 Play a few seconds of the clip to see and hear the band begin to play.

6 Do the same for the remaining three video clips in the Multicamera Source bin.

   These clips show four different but simultaneously recorded angles of the band playing. It's a classic case in which multicamera editing can get the job done faster. The bin also contains a master audio file. While the sound recorded by each camera will be good for syncing the camera angles together, only the master audio file sounds good enough to use in your final project.

7 Double-click the **Master Final Mix** audio clip and play it to hear the difference in sound quality compared to the other clips.

   The best way to use this **Master Final Mix** clip as the final audio track in your timeline is to replace one clip's audio track with this better-sounding master.

DaVinci Resolve includes three ways to sync and replace camera audio with a better-sounding audio file. Two are automatic and one is manual. Because the camera operator recorded rough camera audio, you can use one of the automatic methods to sync and replace the camera audio with the master audio.

8 Select the 01 Master Camera video clip, and Cmd-click (Mac) or Ctrl-click (Windows) the Master Final Mix clip.

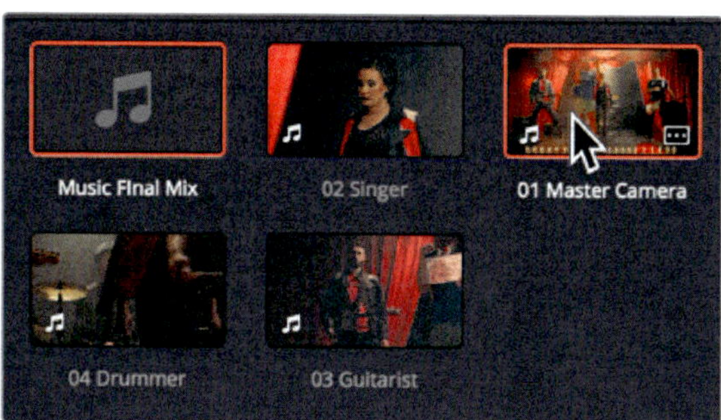

It makes sense to replace the 01_Master Camera audio with the better Master Final Mix file because the 01_ Master Camera will probably become the most-used angle.

9 Right-click the 01_Master Camera video clip to open the contextual menu.

In the middle of the contextual menus you'll find four options to auto-sync audio and video clips. Two options can be used when you have matching timecode on both clips. Those won't work in this case because the timecode is different on each clip. The other two options align the audio waveforms on both clips to sync them. This option will work here because you have audio recorded with the video clip. Now you can choose between appending the audio track to the video clip, which retains the audio tracks, or just replacing the camera audio with the master audio track. For this project, let's replace the camera audio entirely.

10 In the contextual menu, choose "Auto-sync Audio Based on Waveform".

It will take a few seconds to sync the two clips; but when it is done, the 01_Master Camera will play back with the Master Final Mix audio.

Double-click the 01_Master Camera to load it into the source viewer, and press Spacebar to play the clip with its improved audio.

When you begin to edit your program, you'll now be able to use the audio from the 01_Master Camera throughout the entire song.

Creating a multicam clip

To prepare the clips for multicam editing, you must sync them so that the band is playing the same part of the song at the same time on every clip.

1 In the Multicam Source bin, select all four video clips but not the Music Final Mix.

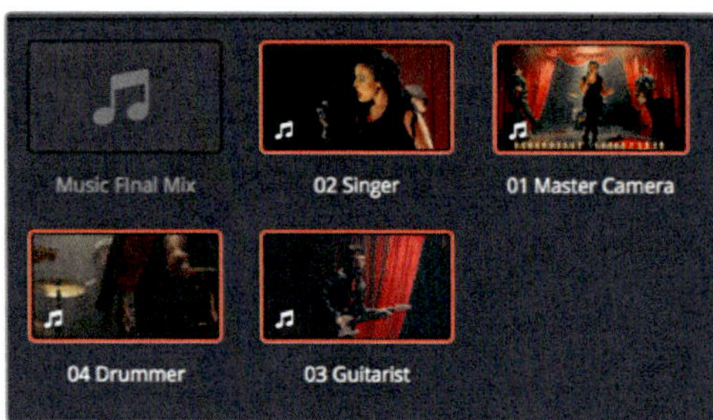

2 Right-click any of the selected clips, and in the contextual menu, choose New Multicam Clip Using Selected Clips.

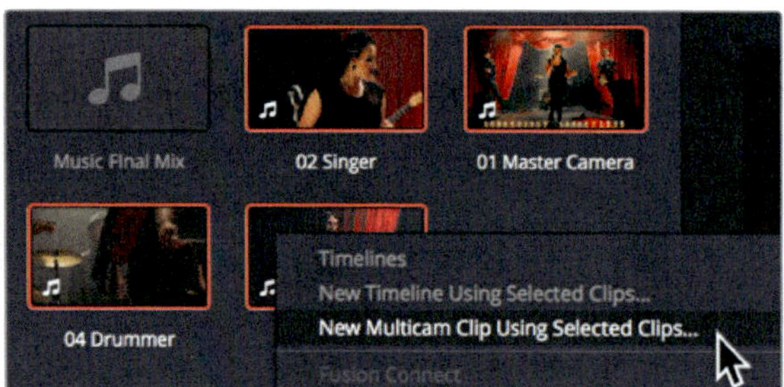

In the New Multicam Clip dialog that appears, you can determine which clip attributes you will use to synchronize them. You select sync options in the Angle Sync menu.

3 Click the Angle Sync menu to view the sync options.

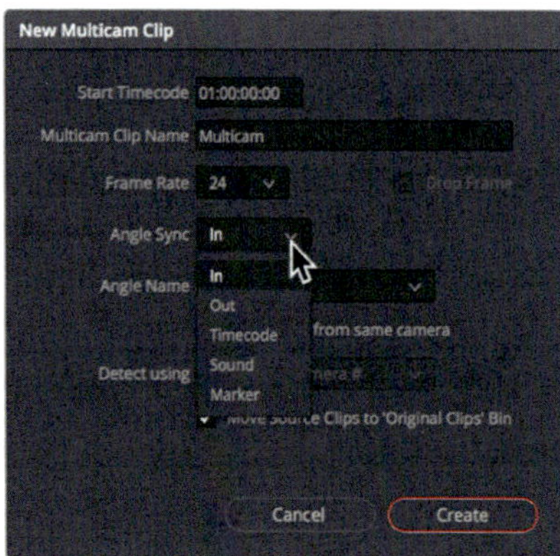

The Angle Sync menu includes four methods for aligning clips. You can choose to manually align them by marking an in point or an out point on all the clips that identifies the same moment in time. This method is fairly easy when a clapper is present in all of the angles. If all of the clips have matching timecode, then Timecode might be the most effective sync method. The last method, Sound, is the same technique you used previously to sync the audio, and video alignment is based on each clip's audio waveforms. Because in points have already been set for each clip, you will sync the clips based on those in points.

4 In the Angle Sync menu, choose In to sync the clips based on the In point.

5 In the Angle Name pop-up menu, choose Clip Name to display the clip names for each camera angle.

6 Deselect the "Move Source Clips to 'Original Clips' Bin" checkbox.

If you were to select that checkbox, DaVinci Resolve would create a bin called Original Clips and move the source clips into it, leaving the multicam clip alone in its bin and easier to find. This clip organization can be helpful when you have a large number of source clips to sync.

7 Click the Create button to begin syncing the clips and making a multicam clip.

In the Multicamera Source bin, a new multicam clip appears that contains all of the selected clips, fully synced.

Viewing multiple angles

You are now ready to edit with your new multicam clip. You can edit a multicam clip into the timeline just as you edit any clip.

1 Drag the Multicam thumbnail into the empty timeline.

A new timeline is created with the **Multicam** clip.

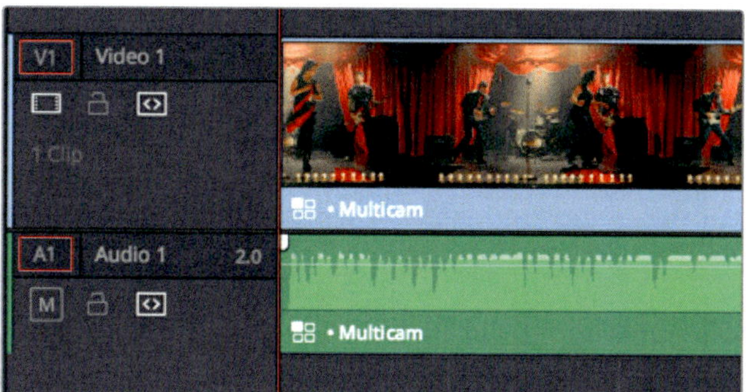

2 Drag the playhead to the start of the timeline.

When you edit a multicam clip into the timeline, it appears as a single clip with only one visible camera angle. To view all the angles in a multicam clip, you must load the multicam source clip into the source viewer.

From the Multicamera Source bin, drag the **Multicam** clip into the source viewer. All four camera angles now fill the source viewer.

The source viewer can display up to nine angles at once, although your multicam clip can contain more clips/angles.

3 Drag the playhead under the source viewer to the end of the source viewer's jog bar and then back to the beginning.

As you move the source viewer's playhead, all the camera angles in the source viewer update to match the current source viewer playhead location.

As usual, the timeline playhead doesn't move when you move the source viewer playhead. However, to perform multicam editing, you need to link the source viewer to the timeline playhead, so that when you move in the timeline, the source viewer updates and allows you to see all the camera angles available for that point in time.

4 To enter multicam mode, in the source viewer mode pop-up menu, choose Multicam.

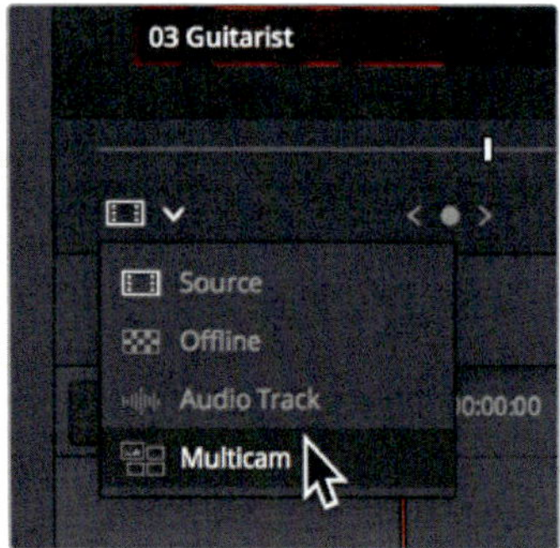

The source viewer now displays controls for multicam editing. Instead of transport controls, you have access to audio/video selection buttons to determine whether you are going to cut audio only, video only, or both. Because the music will remain the same for the entire song (remember the master audio clip that you have waiting in the wings), you will just cut video.

5 Under the source viewer, click the video button.

Any edits you make while in multicam mode will now be performed only on the video tracks.

You are now ready to begin editing and enjoying the benefits of using a multicam clip.

Editing in multicam mode

In multicam mode, the source viewer subtly changes to allow each angle displayed in the viewer to act as a switching control. During real-time playback, clicking any angle within the viewer will place a cut in the timeline at the current playhead position that switches to the selected angle.

1 In the timeline, position the playhead about 20 seconds into the clip, just before the singer turns to look toward the drummer.

2 In the source viewer, click the 04 Drummer clip.

A cut is added to the timeline video track, and in the source viewer, a blue outline appears around the drummer clip to indicate that it is the current camera angle.

3 In the timeline, position the playhead at about 24 seconds, just after the singer begins to take off her jacket, and click 01 Master Camera, a shot that shows the entire band.

A cut is made in the timeline, and 01 Master Camera becomes the current camera angle.

4 To see the results of these edits, position the timeline playhead at the start of the project, and press the Spacebar. Stop playback when you have seen the last cut that you made.

The cuts look good, but instead of positioning the playhead and clicking in the source viewer, let's speed things up and perform edits while the timeline plays.

During playback, you can switch angles by pressing the 1, 2, 3, and 4 keys at the top of your keyboard or on its number pad. The numbers correspond to the frames within the source viewer. The upper-left frame is 1, the upper-right frame is 2, lower-left is 3, and lower-right is 4.

5 Position the playhead at the start of the timeline.

6 Press the Spacebar to begin playback, and then tap the desired number to switch to each camera angle in real time.

   The blue outline relocates to indicate the currently active angle.

7 When you have finished editing, press the Spacebar to stop playback.

   As easy as it is to edit in multicam mode, you'll typically want to refine some cuts. The multicam timeline responds to all the same editing, effects, and color-grading techniques that you've learned. So, you can use the rolling trim cursor and adjust any cut as you would refine any other timeline edit.

Replacing shots in multicam mode

Unless you have fantastic hand, eye, and thought coordination, chances are you won't get every edit correct in your first pass. Fortunately, you can always go back to replace an angle or two.

1 In the timeline, position the playhead over the third clip in the timeline of 01 Master Camera.

Let's switch this camera angle to the close-up of the singer.

2 Hold down the Option key (Mac) or the Alt key (Windows) and hover the mouse pointer over the 02 Singer clip in the source viewer.

When the modifier key is pressed, the mouse pointer changes from a blade icon to a replace edit icon.

3 In the source viewer, Option-click (Mac) or Alt-click (Windows) the 02 Singer clip.

The 01 Master Camera clip is replaced with the 02 Singer clip.

4 Drag the playhead to the start of the timeline, and press Spacebar to play the program.

When you are done with your multicam edit, you can switch back to source mode.

5 To exit multicam mode, in the source viewer mode pop-up menu, choose Source.

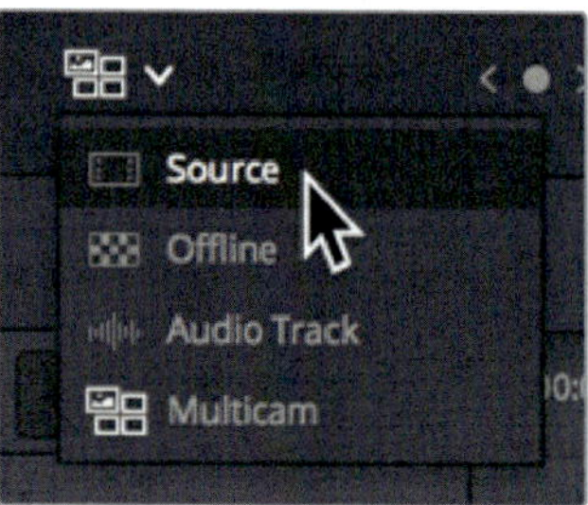

Congratulations!

You have completed **The Definitive Guide to DaVinci Resolve 14** and
are ready to explore more editing, color grading, and audio mixing functionality using the additional certified books in this series. Completing all the lessons in this book have prepared you to become a certified DaVinci Resolve user. You can take the online exam located on the DaVinci Resolve web page to earn your certificate.

We also invite you to become part of the DaVinci Resolve community by joining the web forum on the Blackmagic Design web site. There, you can ask further questions about the creative aspects of editing, color correction and audio mixing.

We hope that you have you have found DaVinci Resolve 14's professional non-linear editing and world-class color correction tools to be intuitive to learn and a perfect fit to become the hub of your entire creative workflow.

Test your skills by taking the online Certification exam **https://goo.gl/Lb5KEx**

Index

J

K

L

M

N

O

P

Q

R

S

T

V

W

About the Author

Paul Saccone has been working in the post production industry for over 20 years and is an editorial workflow expert. He's based in San Francisco, California where he lives with his husband, Ryan, and their dogs, Chloe and Gracie. This is his second book, which means that someone must have read his first one. From 1998 to 2013 Paul worked at Apple and was the Product Manager for Final Cut Pro versions 2 through 7, including Final Cut Studio. He is well known around the world as a speaker, demo artist, and educator for his dynamic and witty presentations. Paul currently works for Blackmagic Design and spends his free time editing with DaVinci Resolve 14.

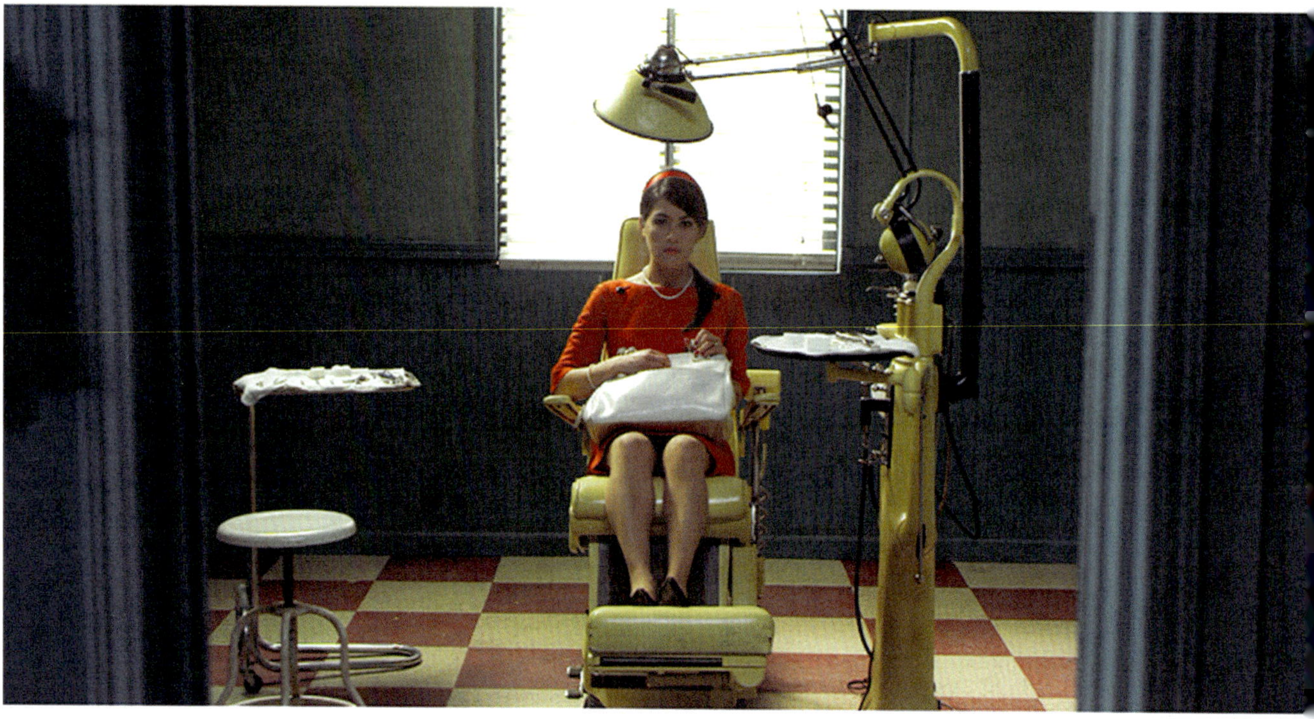

75595088R00206

Made in the USA
San Bernardino, CA
02 May 2018